D1742880

EcoScope
User's Guide

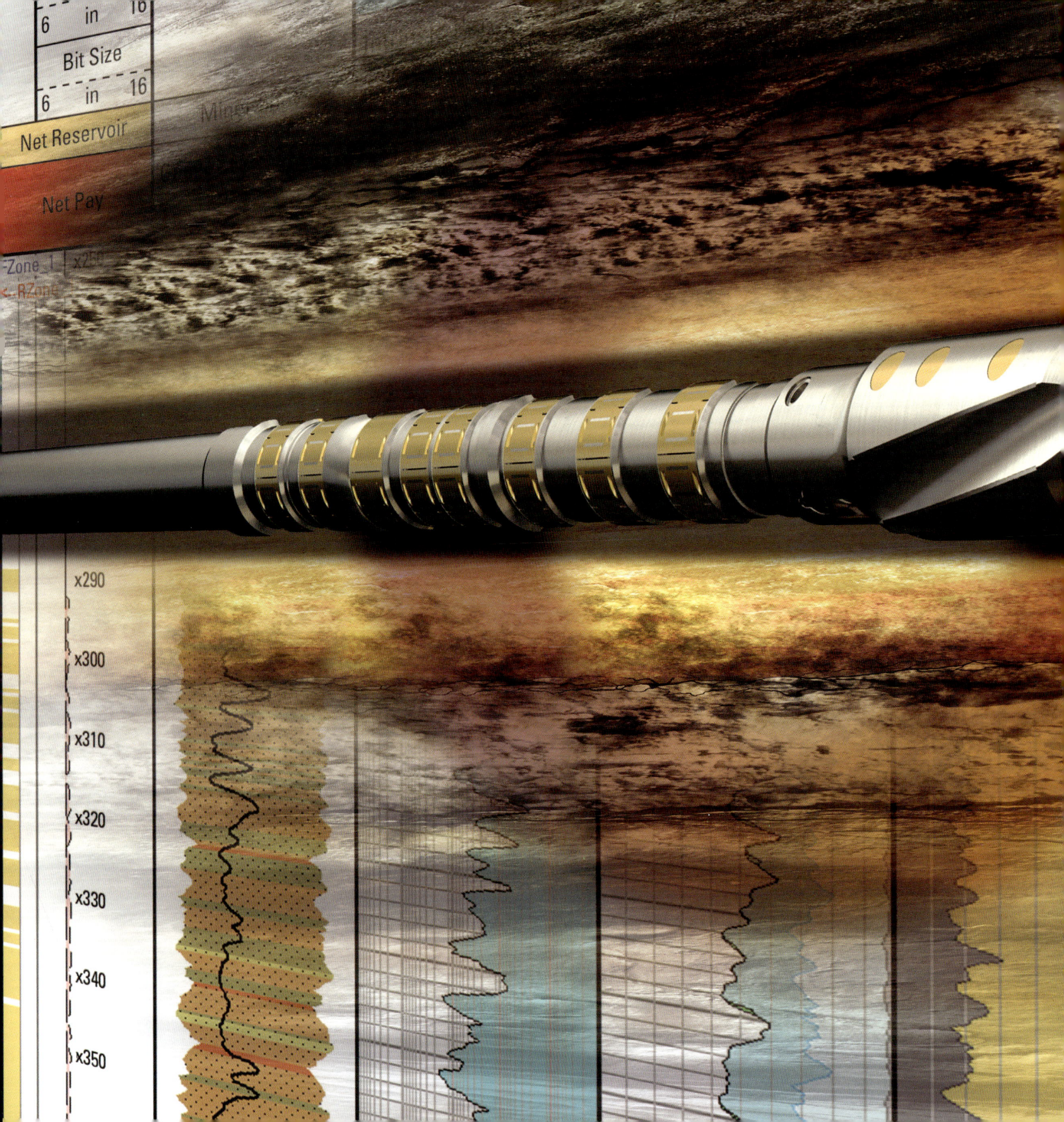

EcoScope
User's Guide

Roger Griffiths

Schlumberger

**Dedicated to the numerous team members
who have contributed to the EcoScope project,
and to the memory of Thierry Leroy,
the "voice of the field"
during the concept and design phases
of the EcoScope development.**

Schlumberger

225 Schlumberger Drive
Sugar Land, Texas 77478
www.slb.com

EcoScope User's Guide
By Roger Griffiths
9.25 x 9.25 x 1.5 inches

ISBN 978-0978853308-2

Copyright © 2010 Schlumberger. All rights reserved.

No part of this book may be reproduced, stored in a retrieval system, or transcribed in any form or by any means, electronic or mechanical, including photocopying and recording, without the prior written permission of the publisher. While the information presented herein is believed to be accurate, it is provided "as is" without express or implied warranty. Specifications are current at the time of printing.

10-DR-0248

An asterisk (*) is used throughout this document to denote a mark of Schlumberger. Other company, product, and service names are the properties of their respective owners.

†Japan Oil, Gas and Metals National Corporation (JOGMEC), formerly Japan National Oil Corporation (JNOC), and Schlumberger collaborated on a research project to develop LWD technology that reduces the need for traditional chemical sources. Designed around the pulsed neutron generator (PNG), EcoScope service uses technology that resulted from this collaboration. The PNG and the comprehensive suite of measurements in a single collar are key components of the EcoScope service that deliver game-changing LWD technology.

Contents

EcoScope System Introduction

EcoScope System The tool has been designed to deliver a comprehensive suite of measurements positioned as close to the bit as possible, thereby delivering data for drilling, formation evaluation, and well placement optimization as soon as possible after the formation has been drilled.

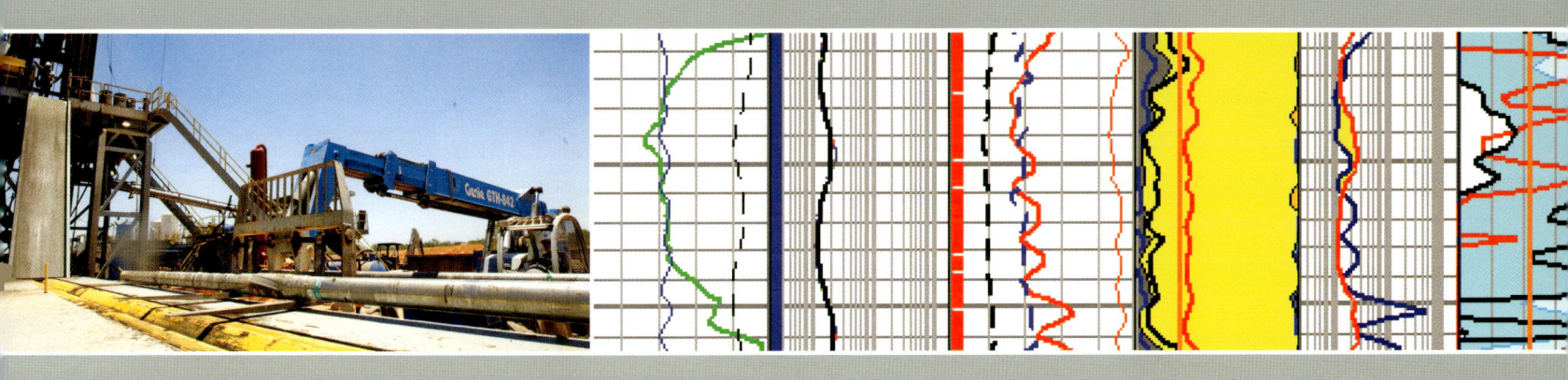

1. EcoScope System Introduction

The EcoScope* multifunction logging-while-drilling (LWD) service[†] offers a comprehensive suite of measurements in a single 25.2-ft-long collar, as shown in **Figure 1-1**. All of the measurements are acquired within 16 ft of the bottom of the tool.

Figure 1-1. The EcoScope system for drilling optimization, formation evaluation, and well placement in a single 25.2-ft collar.

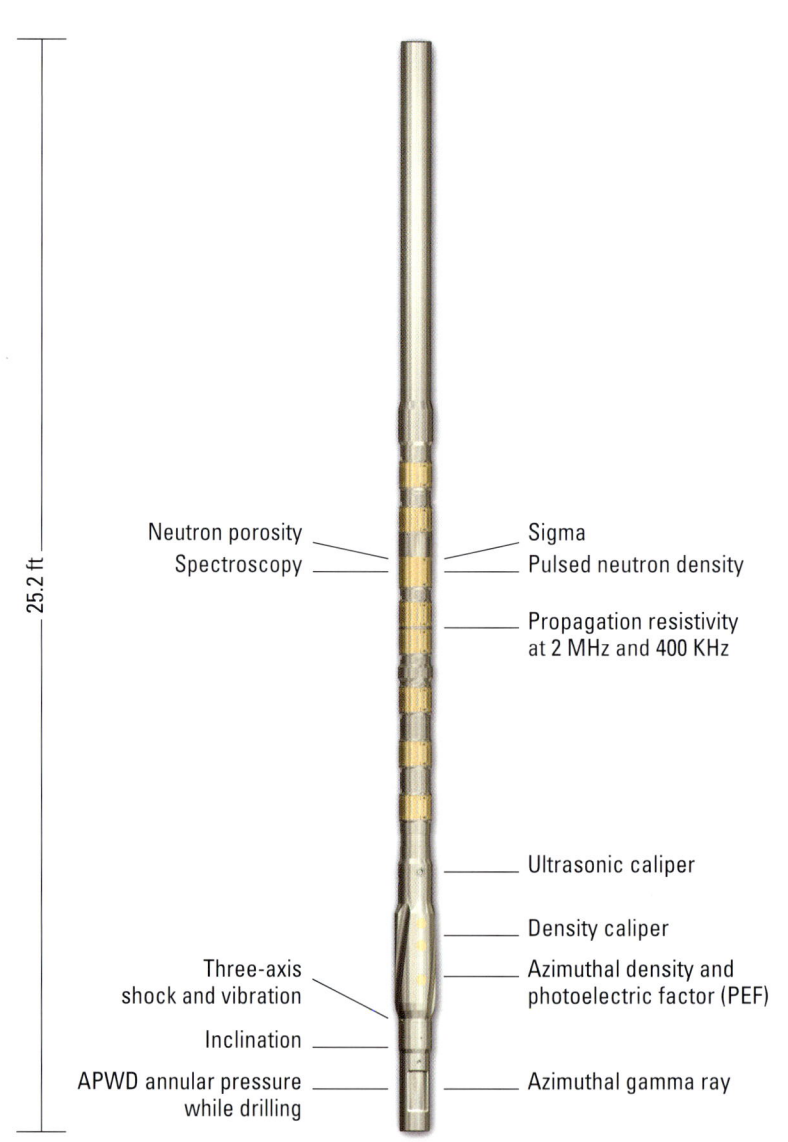

25.2 ft

Neutron porosity

Spectroscopy

Sigma

Pulsed neutron density

Propagation resistivity at 2 MHz and 400 KHz

Ultrasonic caliper

Density caliper

Three-axis shock and vibration

Azimuthal density and photoelectric factor (PEF)

Inclination

APWD annular pressure while drilling

Azimuthal gamma ray

The tool has been designed to deliver a comprehensive suite of measurements (shown in **Table 1-1**) positioned as close to the bit as possible, thereby delivering data for drilling, formation evaluation, and well placement optimization as soon as possible after the formation has been drilled. Data is recorded to the tool memory (recorded mode) or transmitted to surface in real time while drilling. When the EcoScope system is used in conjunction with the TeleScope measurement-while-drilling (MWD) system, real-time data transmission rates have been improved through the implementation of data compression techniques.

Table 1-1. The EcoScope measurement portfolio.

Measurement	Channel Name	Recorded-Mode Data	Real-Time Data	Recorded-Mode Image	Real-Time Image
Azimuthal gamma ray	GRMA	Yes	Yes	Yes	Yes
Propagation resistivity	PxxH AxxH	Yes	Yes	No	No
Azimuthal bulk density (and correction)	RHOB (DRHO)	Yes	Yes	Yes	Yes
Azimuthal PEF/volumetric PEF (U)	PEF/U	Yes	Yes	Yes	Yes
Thermal neutron porosity	TNPH	Yes	Yes	No	No
Hydrogen index (HI)	BPHI	Yes	Yes	No	No
Neutron-gamma density (NGD)	RHON	Yes	Yes	No	No
Elemental spectroscopy	DWxx Wxxx	Yes	Yes	No	No
Sigma	SIFA	Yes	Yes	No	No
Azimuthal density caliper	DCAV	Yes	Yes	Yes	No
Azimuthal ultrasonic caliper	UCAV	Yes	Yes	Yes	No
Near-bit inclination	DEVI	Yes	Yes	NA[†]	NA
Annular pressure	DHAP	Yes	Yes	NA	NA
Annular temperature	DHAT	Yes	Yes	NA	NA
Three-axis shock and vibration	SHKxx VIBxx	Yes	Yes	NA	NA

[†] Not applicable.

Enabling tight integration of these measurements, a pulsed neutron generator (PNG) positioned under the resistivity array removes the need for a chemical neutron source. The electronically controlled pulsed neutron source delivers more neutrons of higher energy than the chemical americium beryllium (AmBe) source that it replaces. Improved neutron source control enables sigma, spectroscopy, and sourceless density measurements, all of which are new to the LWD domain. The PNG also represents a significant reduction in radiation risk, both for personnel and the environment (the half-life of the Am in the AmBe source is 432 years). The PNG introduces significant operational efficiencies by eliminating neutron source storage, transportation, loading, and unloading. Further operational efficiency is gained through rapid side-loading of the cesium-137 (^{137}Cs) source for the azimuthal density measurement.

The driller-friendly spiraled stabilizer design delivers optimal formation density measurements using a cesium-137 chemical gamma source (half-life of 32 years). The density measurement principle is the same as conventional LWD density tools, but the source is side-loaded into a position closer to the formation, which gives improved measurement statistics and precision. Increased detector spacing delivers a slightly deeper density measurement with reduced standoff sensitivity. Note that while the cesium-137 source is not fishable and will have to be abandoned inhole in cases of stuck pipe or bottomhole assembly (BHA), a detailed safety assessment has shown that this gives lower overall system risk than fishing the relatively low half-life source.

1.1. EcoScope measurement overview

The EcoScope tool uses an integrated array of sensors to deliver its suite of measurements. **Figure 1-2** shows a schematic configuration of the major sensors in the tool alongside a matrix outlining the measurements derived from the major sensors (dark green markers). As the array of measurements characterizes many of the features of the formation and environment around the tool, most of the EcoScope measurements are environmentally corrected using measurements from other tool sensors (light green markers).

1. Azimuthal natural gamma ray
2. Propagation resistivity
3. Azimuthal bulk density
4. Azimuthal PEF
5. Hydrogen index
6. Thermal neutron porosity
7. Neutron gamma density
8. Elemental spectroscopy
9. Thermal neutron capture cross section (sigma)
10. Azimuthal density caliper
11. Azimuthal ultrasonic caliper
12. Near-bit inclination
13. Annular pressure
14. Annular temperature
15. Three-axis shock and vibration

■ Used for fundamental measurement   ☐ May be used for environmental correction

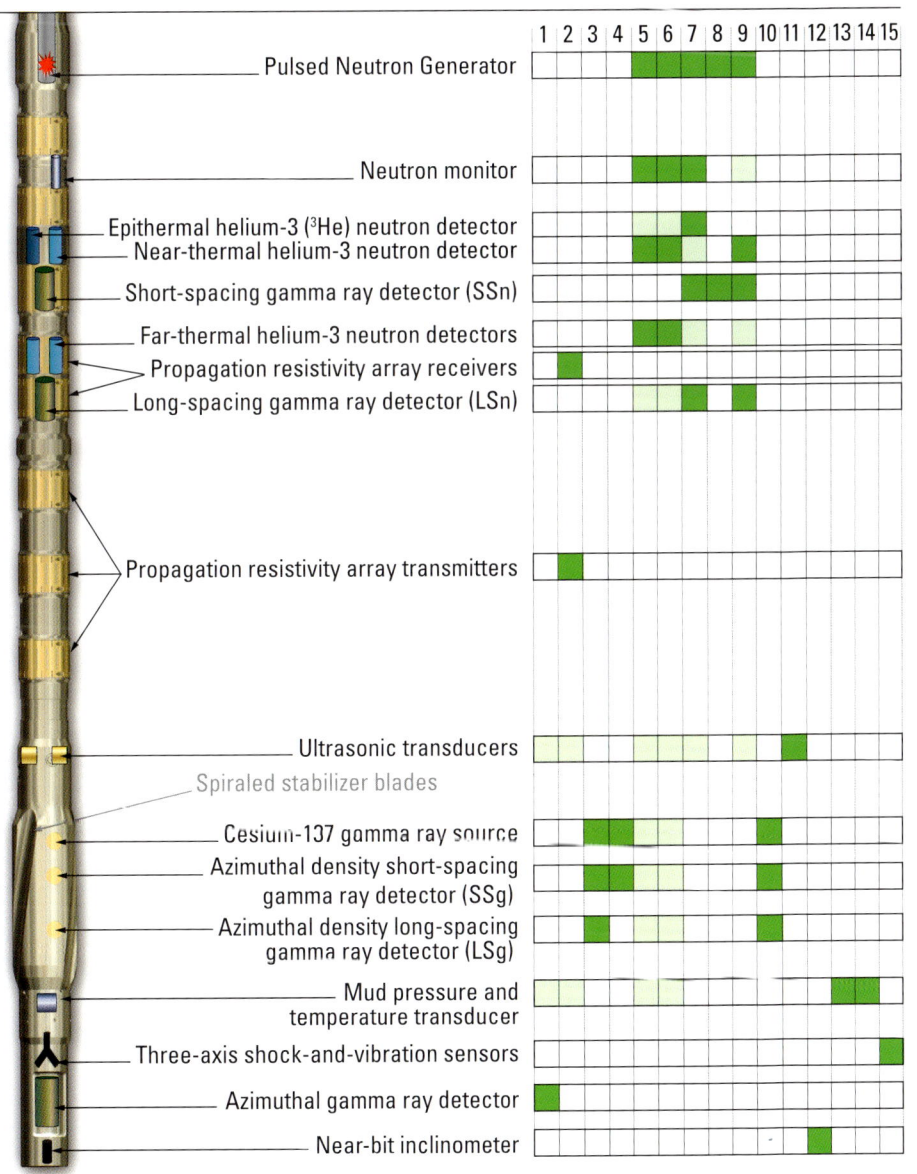

Integrated array of sensors (measurements 1–15):

Sensor	1	2	3	4	5	6	7	8	9	10	11	12	13	14	15
Pulsed Neutron Generator					█	█	█	█	█						
Neutron monitor					█	█	█	░							
Epithermal helium-3 (³He) neutron detector				░	█	█									
Near-thermal helium-3 neutron detector						█	█	░							
Short-spacing gamma ray detector (SSn)							█	█	█						
Far-thermal helium-3 neutron detectors					█	█									
Propagation resistivity array receivers		█													
Long-spacing gamma ray detector (LSn)						░	█	░							
Propagation resistivity array transmitters		█													
Ultrasonic transducers	░		░	░	░						█				
Cesium-137 gamma ray source			█	░						█					
Azimuthal density short-spacing gamma ray detector (SSg)			█	░						█					
Azimuthal density long-spacing gamma ray detector (LSg)			█	░	░										
Mud pressure and temperature transducer	░	░	░	░	░								█	█	
Three-axis shock-and-vibration sensors															█
Azimuthal gamma ray detector	█														
Near-bit inclinometer												█			

Spiraled stabilizer blades

Figure 1-2. Integrated array of sensors, which makes multiple measurements. The close proximity of the measurements means that they can be used to environmentally correct each other.

The following measurements are available from the EcoScope system. For further details on these measurements, please refer to Chapter 4.

- **Azimuthal natural gamma ray** is a total gamma ray measurement acquired using a large, shielded detector with high azimuthal sensitivity (making it capable of delivering a gamma ray image).

- **Propagation resistivity** includes borehole-compensated phase shift and attenuation resistivities at two frequencies (2 MHz and 400 kHz) and five spacings (40, 34, 28, 22, and 16 in) with measurements updated every 2 seconds. ARCWizard processing provides automated interpretation of the resistivity responses.

- **Azimuthal bulk density** uses a side-loaded chemical cesium-137 gamma ray source positioned in the collar closer to the formation wall than in conventional LWD tools. This proximity results in better statistics, improved measurement depth of investigation (DOI), and reduced borehole sensitivity. It also permits a density measurement update every 10 seconds. The outgoing and returning gamma rays can be focused to create an azimuthal measurement of the formation. This density measurement is referred to as azimuthal, or gamma-gamma bulk, density to distinguish it from sourceless, or neutron-gamma, density.

- **Azimuthal PEF** is induced by the low-energy gamma rays used for the azimuthal density measurement. It is a shallow, focused measurement acquired every 10 seconds and is generally used for formation lithology identification and imaging.

- **Hydrogen index** is enabled by enhanced neutron energy, count rate, and a suite of detectors. It is less sensitive than the corresponding thermal neutron porosity to environmental conditions such as borehole effects, salty formation water, or shale, thereby simplifying interpretation.

- **Thermal neutron porosity** is a thermal neutron response emulating the thermal neutron porosity measurement from the previous-generation azimuthal density neutron tool.

- **Neutron gamma density** is derived utilizing the PNG and a suite of detectors to determine formation density from gamma rays induced by the interaction of high-energy neutrons with the formation. This is referred to as the sourceless, or neutron-gamma, density. The NGD can be used to replace the traditional gamma-gamma density measurement, allowing conventional formation evaluation without the use of chemical nuclear sources. However, the NGD does not deliver the photoelectric measurement associated with the azimuthal density measurement, nor can it be focused azimuthally to form an image. The NGD measurement is also deeper and less dependent on good contact with the formation than the azimuthal measurement.

- **Elemental yields** from prompt neutron capture spectroscopy are derived from the short-spaced gamma ray detector. Elemental yields are determined from spectral analysis of the gamma rays emitted following neutron capture in the formation. The spectroscopy computations are performed downhole, allowing real-time identification of the elemental composition of the formation, formation matrix nuclear response properties, and lithology. Volumetric lithology quantification and matrix properties are important for accurate porosity and saturation determination.

- **Thermal neutron capture cross section**, or sigma, is determined by the time decay constant of the neutron capture gamma ray spectrum. Sigma is primarily sensitive to the presence of chlorine (Cl). The measurement has a variety of applications, including distinguishing the fluids (water, oil, or gas) in the formation and identification of low-resistivity pay (LRP) zones. In a saline water environment, it can be used to calculate the formation oil saturation independent of resistivity measurements.

- **Azimuthal density caliper** is derived from the magnitude of the standoff compensation implicit in the azimuthal density measurement used to determine the size of the borehole around the density stabilizer. The EcoScope density caliper incorporates information from the shallow PEF response for a more robust caliper measurement under a wide range of conditions.

- **Azimuthal ultrasonic caliper** is determined by two 180°-opposed ultrasonic transducers located just above the stabilizer to measure the borehole diameter, even while sliding. A 16-sector (8-diameter) image of the borehole shape is acquired while rotating.

- **Near-bit inclination** is derived by a single-axis accelerometer located near the base of the tool.

- **Annular pressure** is measured by a pressure transducer in the tool, allowing effective mud weights to be monitored for borehole stability applications.

- **Annular temperature** is measured at the same point as the annular pressure. This provides insight into mud condition, the temperature experienced by the LWD tool, and the operating temperature required for environmental correction of some of the formation measurements.

- **Three-axis shock-and-vibration** measurements allow recognition of unfavorable drilling states and permit diagnosis of the causes so that remedial action to minimize consequences can be taken. Axial, lateral, and torsional shock amplitude, shock frequency, and vibrational energy to which the collar is subjected are measured.

- **Azimuthal images** represent sectors of the borehole rather than the average around the borehole and are acquired by using BHA rotation to scan a single set of sensors around the inner surface of the borehole. The measurements are binned into azimuthal sectors, and the azimuthal aperture, or width, of the sectors depends on the degree to which the measurement can be focused. Neutron measurements are difficult to focus and hence, are generally presented as an azimuthal average. Raw gamma ray measurement data is acquired and images displayed in sectors, the quantitative measurements are generally provided in quadrants or as an azimuthal average. Quadrants are defined in the EcoScope system as the bottom, left, up, and right (often called B-L-U-R) quadrants. For the density and PEF measurements, each of these quadrants can be further subdivided into 4 sectors, giving a total of 16 sectors around the borehole in a density or photoelectric image. **Figure 1-3** shows examples of quadrants and sector measurements. Note that the density and photoelectric images are subdivided into 16 sectors, while the density and photoelectric measurements are delivered as quadrants. This is because these statistical measurements require counts from four sectors to have the statistical precision required for use as quantitative measurements.

Figure 1-3. The azimuthal resolution of each measurement, which depends on how tightly the measurement can be focused. Quantitative measurements generally require averaging across several sectors to improve measurement signal-to-noise, so while the raw data is acquired and images displayed in sectors, the quantitative measurements are generally provided in quadrants or as an azimuthal average.

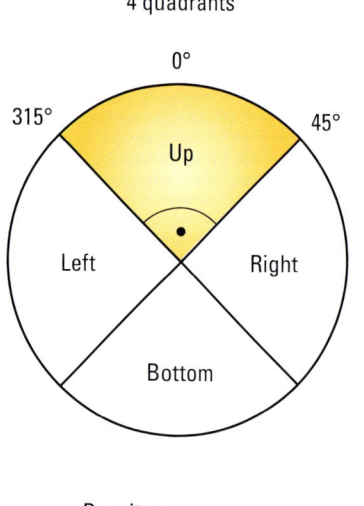

4 quadrants

Density measurement
Photoelectric measurement
Gamma ray measurement

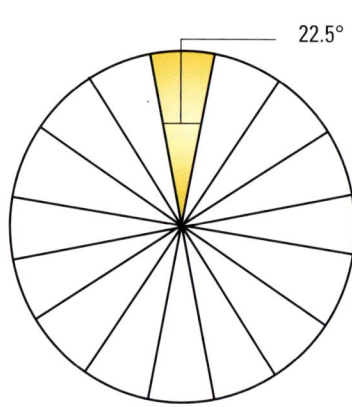

16 sectors

Density image
Photoelectric image
Gamma ray image
Ultrasonic image

Two-dimensional (2D) LWD images are generally displayed as if the borehole has been split along the top and unfolded so that the center of the image corresponds to the bottom of the borehole while the outside edges of the image correspond to the top of the borehole. Color is used to encode information about the magnitude of the formation parameter used in the creation of the image. For example, the density information in each of the 16 sectors is converted into an image by mapping the density value to a color, where darker colors generally indicate lower density, and lighter colors indicate higher density, as shown in **Figure 1-4**.

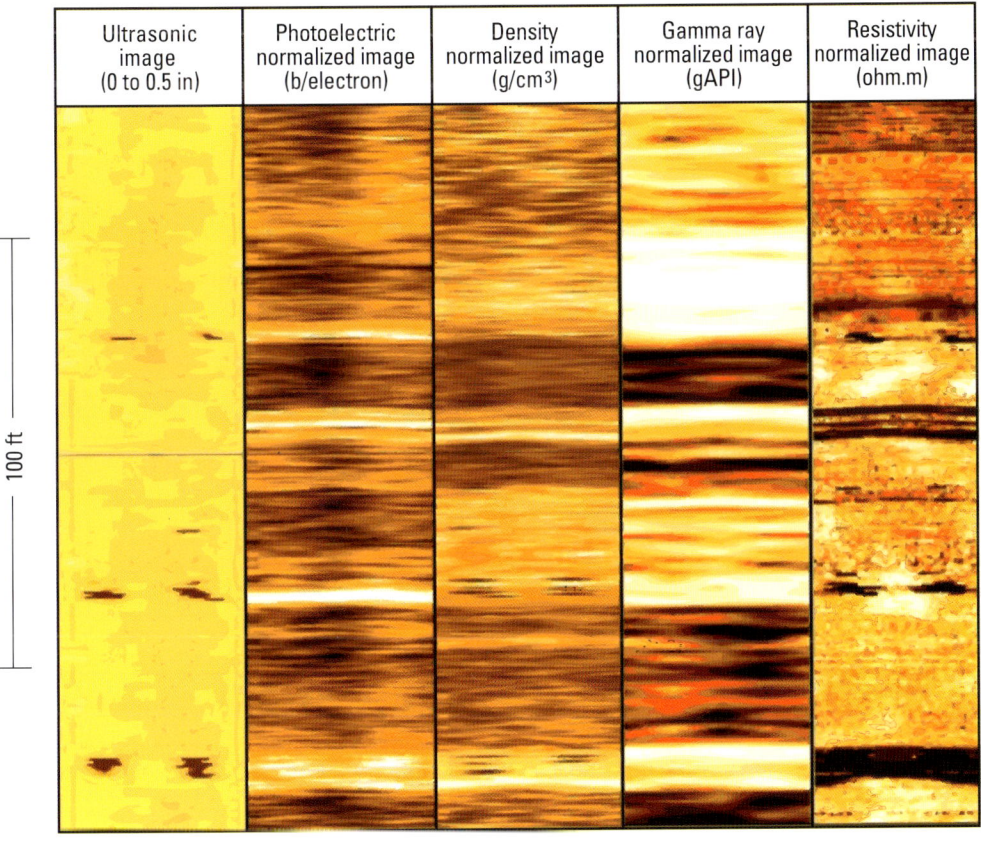

Figure 1-4. The suite of imaging measurements from the EcoScope service, responding to different formation properties and displaying different formation features.

Image data is useful for identifying features crossing the borehole and can be used to quantify the dip of these features, as shown in **Figure 1-5**.

Figure 1-5. Imaging while drilling up- and down-section. Images A through E depict 2D azimuthal LWD data. Conceptually, the borehole is split along the top and unfolded such that the middle of the image corresponds to the bottom of the borehole. In images F through J, inclination is adjusted from vertical through horizontal to drilling up at 100°. The bedding plane is parallel to the wellbore at an inclination angle of approximately 75°. Images F, G, and H reflect drilling down-section (the layer crossing the borehole creates a "sad face" on the image). Image I shows the parallel "railroad tracks" characteristic of drilling along at the same angle as the bedding plane. Image J shows the "happy face" that occurs when the borehole drills up through a layer. In each case, the amplitude of the sinusoid is characteristic of the incidence angle between the borehole and the layer.

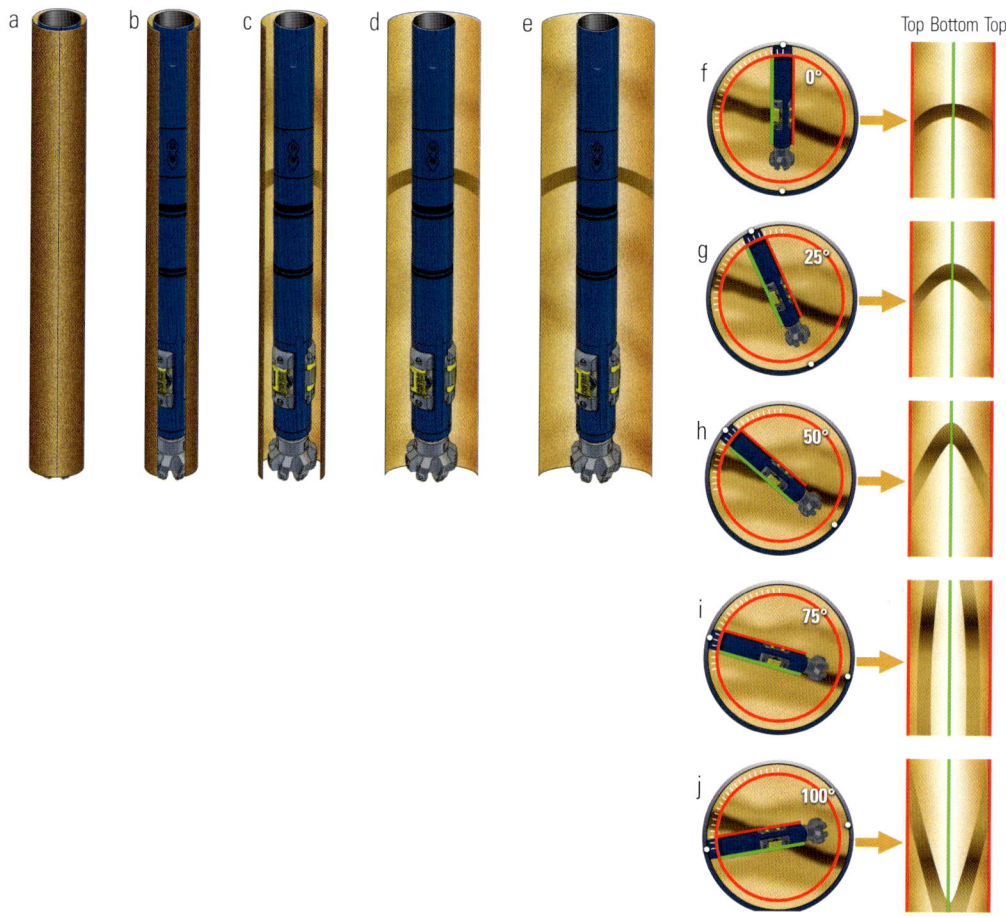

If a borehole drills down into a layer, the bottom of the borehole will see the layer first, then the sides, and finally the top. On the image, the layer will first appear in the middle and then create a sinusoidal shape that ends at the edges of the image corresponding to where the last of the layer is seen at the top of the borehole. The amplitude of this sinusoid is related to the incidence angle between the wellbore and the layer.

Even without calculating this incidence angle, the features resembling happy and sad faces on an image due to drilling up and down, respectively, through layers can assist in well placement decision making, as they indicate where the wellbore is positioned relative to the layering. For example, if "happy face" features are seen on the image while drilling in a reservoir, drilling up through the sequence is indicated, requiring a drop in well inclination to come parallel to the layering so as to remain in the reservoir.

As shown in **Figure 1-6**, the incidence angle between the borehole and formation layering can be deduced using trigonometry by solving for the properties of the triangle created by their intersection.

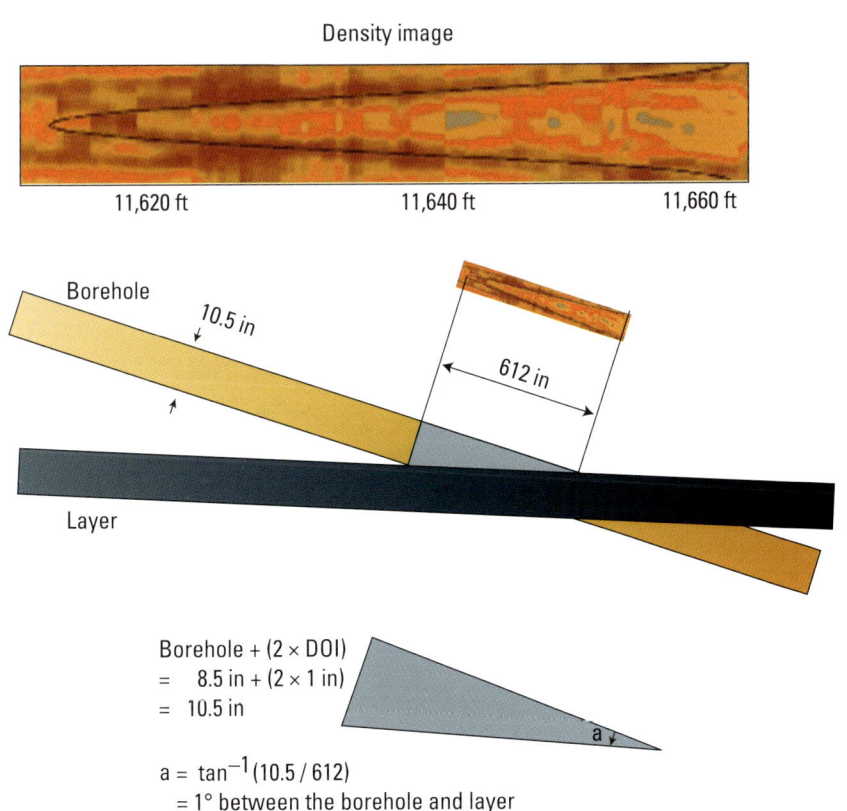

Figure 1-6. The incidence angle between the borehole and formation layering, deduced by solving the properties of a triangle.

Solving for the incidence angle is represented trigonometrically as

$$\tan(\theta) = o \,/\, a \qquad \text{(Equation 1-1),}$$

where

tan = tangent

o   = opposite side from the angle

a   = adjacent side to the angle.

For the adjacent side, the amplitude of the sinusoid on the image is measured along the borehole. The measured depth (MD) amplitude of the sinusoid must be converted to the same units as the borehole diameter.

For the opposite side, images do not scan the surface of the borehole. They represent the formation property at the DOI of the measurement from which they are derived. Hence, the DOI of the imaging measurement must be added on each side of the borehole diameter. In the case of the density measurement, this is approximately 1 in, so for an 8.5-in borehole, the diameter at which the image is acquired is 10.5 in. For a laterolog resistivity image, the electrical penetration depth is approximately 1.5 in, so 3 in should be added to the diameter of the borehole when calculating a formation dip from an LWD resistivity image. Figure 1-6 shows a schematic of a wellbore crossing a layer with the corresponding image and incidence angle calculation.

Note that this is the projection of the incidence angle between the borehole and formation in a vertical plane along the length of the borehole.

1.2. EcoScope applications overview

For a more detailed description of EcoScope applications, please see Chapter 2.

1.2.1. Drilling optimization

The EcoScope system has been designed to facilitate drilling operations through increased reliability and efficiency. Rig floor operations are made faster by

- fewer BHA components to handle because of the integrated tool design

- fast side-loading of only one source (gamma) rather than two (gamma and neutron); this reduces source-loading to approximately 5 minutes from 30, and when run in sourceless mode, there is no source-loading

- fast data dump using a high-speed data transfer protocol

- faster tool programming through simplified engineer interface and significantly improved memory capacity.

Drilling operations are made faster by

- fewer measurement-related drilling rate-of-penetration (ROP) limitations, as the nuclear measurement statistics (the constraining factor) are improved; also, measurements are all acquired as fast as the physics of the respective measurements permits, ensuring quantitative measurement data at ROPs up to 900 ft/h

- less rathole drilling for measurements, as they are much closer to the bit (for example, neutron and density are 44 and 49 ft, respectively, closer to the bit than in the conventional triple-combo BHA)

- all measurements within 16 ft of the bottom of the tool, allowing earlier real-time well placement and formation evaluation decisions; distances from the bottom of the tool to some of the measurement points are

  - 5 ft for inclination, annular pressure, and gamma ray

  - 9 ft for density/PEF/volumetric equivalent of PEF

  - 11 ft for ultrasonic caliper

  - 16 ft for resistivity, neutron, sigma, spectroscopy, and sourceless density measurements.

1.2.2. Well placement

Real-time well placement decisions are made easier by the collocation of measurements close to the bit. Compared with conventional LWD tools, this delivers critical information earlier after a formation has been drilled, allowing better trajectory control. The unique suite of measurements translates into a greater choice for steering. Formation sigma and spectroscopy can be used in addition to the conventional measurements. Furthermore, azimuthal images of the formation density, PEF, and gamma ray are provided. To track the position of the tool, a single-axis inclinometer provides continuous inclination at the base of the tool.

1.2.3. Formation evaluation

With innovative and unique measurements, including direct lithology identification and a strict quality control (QC) process, the EcoScope system provides unparalleled formation evaluation while drilling. For interpreting the integrated measurements, the EcoView answer product extracts real-time reservoir information through quantitative mineralogy, porosity, and water saturation from the measured data. It also provides analysis of drilling-related measurements, such as annular pressure and temperature, BHA shock and vibration, hole caliper, and breakout orientation, which permits optimization of the drilling process.

1.3. EcoScope log quality control

Real-time log log quality control (LQC) is particularly critical for LWD when decisions are made based on the limited number of measurements transmitted to surface while drilling. The EcoScope tool introduces an improved approach to LQC. Several of the logs have colored stripes called flags. These flags are green when the tool is operating normally and change to either yellow or red when a predetermined threshold or tolerance is exceeded. There are flags for every major hardware subassembly and for all processed log outputs.

Green indicates that the complete subsystem is functioning correctly and providing measurements that are within tolerance. Yellow means that the measurement may be affected by environmental or tool operating conditions. Red indicates that the measurement may be out of range or under bad operating conditions.

The EcoScope system's strict QC process is shown in **Figure 1-7**. Starting at the tool hardware level right up to the most integrated answer products, QC processes have been implemented, allowing validation of the tool answers at every stage. QC flags on the final products allow instant verification of the quality of the data and assist in the quick identification of the source of questionable data in a logical and hierarchical structure.

QC flags are generally presented in the right track of the log presentation, as shown in **Figure 1-8**.

Figure 1-7. The EcoScope log quality control hierarchy provides hardware, processing and answer product quality indicators for both real-time and recorded mode data.

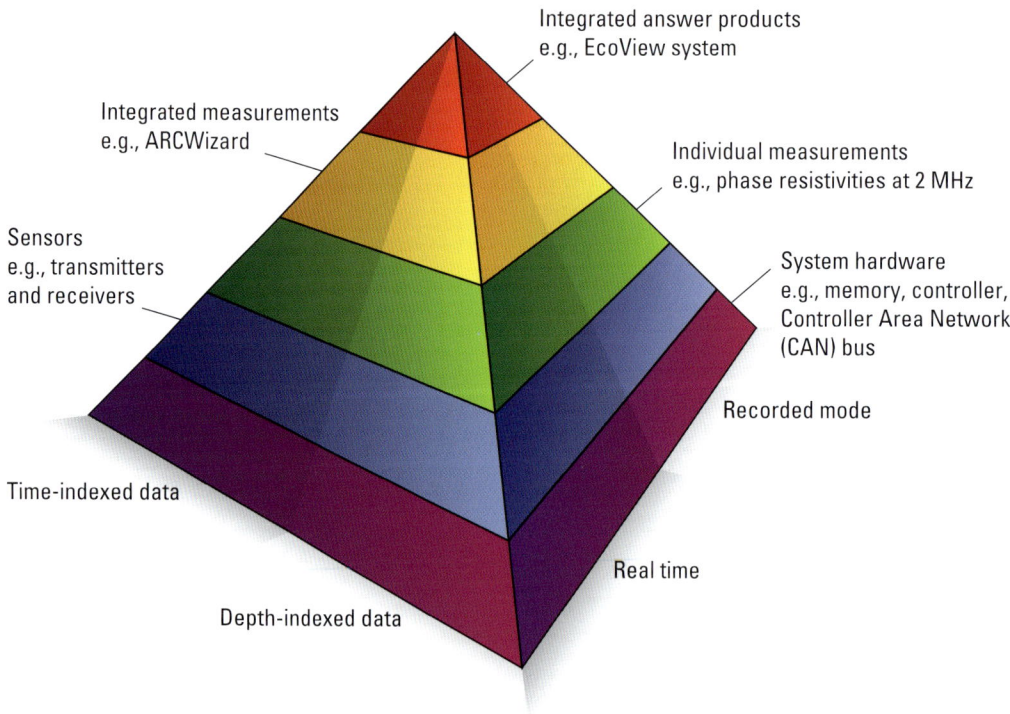

Integrated answer products
e.g., EcoView system

Integrated measurements
e.g., ARCWizard

Individual measurements
e.g., phase resistivities at 2 MHz

Sensors
e.g., transmitters
and receivers

System hardware
e.g., memory, controller,
Controller Area Network
(CAN) bus

Recorded mode

Time-indexed data

Real time

Depth-indexed data

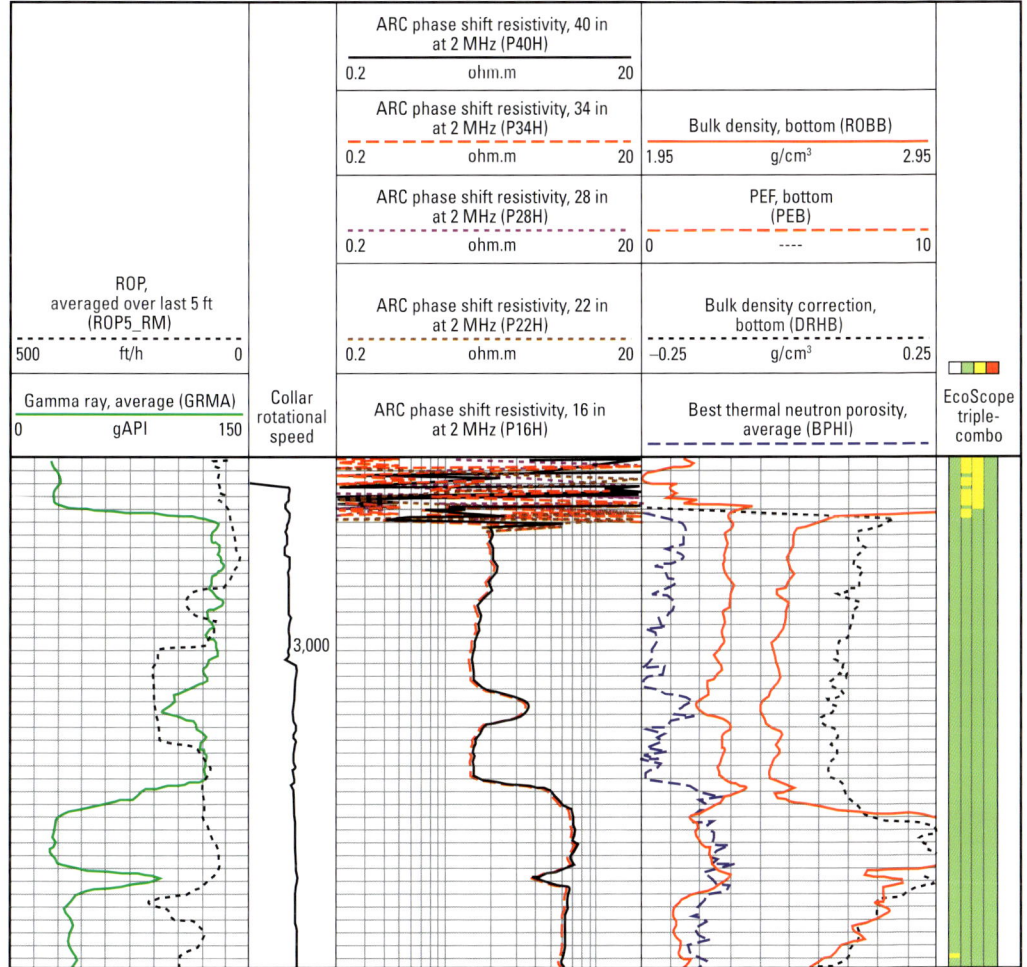

ROP, averaged over last 5 ft (ROP5_RM)		Collar rotational speed	ARC phase shift resistivity, 40 in at 2 MHz (P40H)				EcoScope triple-combo
500	ft/h	0		0.2	ohm.m	20	
			ARC phase shift resistivity, 34 in at 2 MHz (P34H)			Bulk density, bottom (ROBB)	
			0.2	ohm.m	20	1.95 g/cm³ 2.95	
			ARC phase shift resistivity, 28 in at 2 MHz (P28H)			PEF, bottom (PEB)	
			0.2	ohm.m	20	0 ---- 10	
			ARC phase shift resistivity, 22 in at 2 MHz (P22H)			Bulk density correction, bottom (DRHB)	
			0.2	ohm.m	20	−0.25 g/cm³ 0.25	
Gamma ray, average (GRMA)			ARC phase shift resistivity, 16 in at 2 MHz (P16H)			Best thermal neutron porosity, average (BPHI)	
0	gAPI	150					

3,000

Figure 1-8. EcoScope log, showing the QC flags in the right track. Yellow flags in the upper section indicate that the resistivity, density and PEF neutron measurements are affected by casing at this depth.

1.3.1. Quality control flags

The QC flags on the logs use a color system of green, yellow, and red to validate the measured data. The meanings of the various flag colors are outlined in **Table 1-2**.

Table 1-2. EcoScope QC flag color description.

Color	Hardware	Answer Product or Measurement Status	Remarks/User Actions
Green	Normal, operating.	Normal, within tolerance.	No problem detected. Data and results are useable.
Yellow	Warning. Some limits or tolerances may be close or slightly exceeded.	Warning. Result may be affected by environmental or tool operating conditions.	User should investigate the cause of this indication (using lower-level displays) to verify the impact on the desired result. Results may be adversely affected.
Red	Failure. Limits exceeded, unstable.	Answer or corrections out of range. Wrong operating conditions.	Hardware failure or wrong settings will likely result in bad (unrecoverable) data. A reliable answer cannot be computed.
Clear	Not present.	Not computed, not relevant.	Absence of a measurement or computation.

1.3.2. Calibration

Calibration of the EcoScope measurements is performed to ensure valid, quantitative data from the tool. Resistivity and nuclear calibrations are performed in the workshop before transportation of the tool to the wellsite. The propagation resistivity is calibrated in a nonconductive environment, such as hanging in air. Nuclear calibrations generally consist of measuring the background counts to determine the counts received from the environment, a positive counts check when the source of a known strength is applied, and a response validation check.

The shop calibration should take no more than 3 months before the data acquisition is completed. Ideally, it should be done before every job. Nuclear calibration should be performed

- if it has been longer than 3 months since the last calibration

- if the stabilizer has been changed

- after a job if significant wear is noted on the stabilizer

- when a new source is to be used in the tool.

Figure 1-9 shows a typical EcoScope shop calibration found on the logs. Dates, times, and equipment serial numbers are provided, along with the measurement description and their units. The value for each calibration measurement is given in the area named "Shop" and is colored green if it falls within the specified limits presented in the "Min./Nominal/Max." column.

Detailed Calibration Record

DV6MT: EcoScope Integrated LWD Tool—6.75 Calibration Resistivity Calibration

Primary Set Components	Description	Tool Element	Serial Number
	EcoScope 6.75—Electronics chassis	DVME	811
	EcoScope drill collar—6.75	DVMC	726

Calibration Dates	Shop Calibration
Date and time/date validity Calibration source	09 May 2008 08:36:54 pm— Valid time frame file

Calibration Type	Resistivity: Air

Description	Minimal/Nominal/Maximal	Shop	Unit
ATT1F2AIR Attenuation T1 at 2 MHz	7.0/9.0/11.0	8.8	dB
ATT2F2AIR Attenuation T2 at 2 MHz	4.0/6.0/8.0	5.6	dB
ATT3F2AIR Attenuation T3 at 2 MHz	3.5/5.5/7.5	5.4	dB
ATT4F2AIR Attenuation T4 at 2 MHz	2.5/4.5/6.5	4.0	dB
ATT5F2AIR Attenuation T5 at 2 MHz	2.0/4.0/6.0	4.0	dB
PST1F2AIR Phase shift T1 at 2 MHz	-4.0/0/4.0	1.4	°
PST2F2AIR Phase shift T2 at 2 MHz	-4.0/0/4.0	-1.5	°
PST3F2AIR Phase shift T3 at 2 MHz	-4.0/0/4.0	1.3	°
PST4F2AIR Phase shift T4 at 2 MHz	-4.0/0/4.0	-1.5	°
PST5F2AIR Phase shift T5 at 2 MHz	-4.0/0/4.0	1.4	°
ATT1F4AIR Attenuation T1 at 400 KHz	7.0/9.0/11.0	8.8	dB
ATT2F4AIR Attenuation T2 at 400 KHz	4.0/6.0/8.0	5.7	dB
ATT3F4AIR Attenuation T3 at 400 KHz	3.5/5.5/7.5	5.4	dB
ATT4F4AIR Attenuation T4 at 400 KHz	2.5/4.5/6.5	4.1	dB
ATT5F4AIR Attenuation T5 at 400 KHz	2.0/4.0/6.0	3.9	dB
PST1F4AIR Phase shift T1 at 400 KHz	-4.0/0/4.0	0.7	°
PST2F4AIR Phase shift T2 at 400 KHz	-4.0/0/4.0	-0.7	°
PST3F4AIR Phase shift T3 at 400 KHz	-4.0/0/4.0	0.7	°
PST4F4AIR Phase shift T3 at 400 KHz	-4.0/0/4.0	-0.7	°
PST5F4AIR Phase shift T5 at 400 KHz	-4.0/0/4.0	0.7	°

DV6MT: EcoScope Integrated LWD Tool—6.75 Calibration Gamma Density Long-Spacing Window 3 Calibration Run 1

Primary Set Components	Description	Tool Element	Serial Number
	EcoScope 6.75—Electronics chassis	DVME	811
	EcoScope 6.75—Gamma-gamma density source	GGLS	672
	EcoScope drill collar—6.75 in	DVMC	726
	EcoScope 6.75—Stabilizer	DVMS	

Calibration Dates	Shop Calibration
Date and time/date validity Calibration source	10 May 2008 04:14:10 am— Valid time frame file

Calibration Type	Density: Long-Spacing Window 3

Description	Minimal/Nominal/Maximal	Shop	Unit
LSW3_BG Long-spacing window 3—Background	50.0/70.0/90.0	64.7	1/s
LSW3_BG Long-spacing window 3—Aluminum (Al)	350.0/575.0/800.0	568.0	1/s
LSW3_BG Long-spacing window 3—Magnesium (Mg)	2,200.0/3,350.0/4,500.0	3,417.2	1/s
RHOL_H_0 Long-spacing water density	1.026/1.043/1.059	1.034	g/m³

DV6MT: EcoScope Integrated LWD Tool—6.75 Calibration Gamma Density Short-Spacing Window 1 Calibration Run 1

Primary Set Components	Description	Tool Element	Serial Number
	EcoScope 6.75—Electronics chassis	DVME	811
	EcoScope 6.75—Gamma-gamma density source	GGLS	672
	EcoScope drill collar—6.75 in	DVMC	726
	EcoScope 6.75—Stabilizer	DVMS	

Calibration Dates	Shop Calibration
Date and time/date validity Calibration source	10 May 2008 04:14:10 am— Valid time frame file

Calibration Type:	Density: Short-Spacing Window 1

Description	Minimal/Nominal/Maximal	Shop	Unit
SSW1_BG Short-spacing window 1—Background	50.0/75.0/100.0	73.4	1/s
SSW1_AL Short-spacing window 3—Aluminum (Al)	2,300.0/3,550.0/4,800.0	3,227.0	1/s
SSW1_MG Short-spacing window 1—Magnesium (Mg)	4,560.0/6,830.0/9,100.0	6,495.3	1/s

DV6MT: EcoScope Integrated LWD Tool—6.75 Calibration Gamma Density Short-Spacing Window 3 Calibration Run 1

Primary Set Components	Description	Tool Element	Serial Number
	EcoScope 6.75—Electronics chassis	DVME	811
	EcoScope 6.75—Gamma-gamma density source	GGLS	672
	EcoScope drill collar—6.75 in	DVMC	726
	EcoScope 6.75—Stabilizer	DVMS	

Calibration Dates	Shop Calibration
Date and time/date validity Calibration source	10 May 2008 04:14:10 am— Valid time frame file

Calibration Type:	Density: Short-Spacing Window 3

Description	Minimal/Nominal/Maximal	Shop	Unit
SSW3_BG Short-spacing window 3—Background	270.0/370.0/470.0	356.0	1/s
SSW3_AL Short-spacing window 3—Aluminum	7,600.0/11,550.0/15,500.0	11,402.4	1/s
SSW3_MG Short-spacing window 3—Magnesium	11,100.0/16,700.0/22,300.0	17,059.2	1/s
RHOS_H_O Short-spacing water density	1.221/1.256/1.291	1.228	g/cm³

DV6MT: EcoScope Integrated LWD Tool—6.75 Calibration Neutron Calibration Run

Primary Set Components	Description	Tool Element	Serial Number
	EcoScope 6.75—Electronics chassis	DVME	811
	EcoScope 6.75—Gamma-gamma density source	GGLS	672
	EcoScope drill collar—6.75 in	DVMC	726
	EcoScope 6.75—Stabilizer	DVMS	

Calibration Dates	Shop Calibration
Date and time/date validity Calibration source	10 May 2008 02:55:49 am— Valid time frame file

Calibration Type:	Neutron: Water Tank

Description	Minimal/Nominal/Maximal	Shop	Unit
FAR1_GAIN Far 1 gain	0.7/1.0/1.3	1.0	
FAR1_OFFSET Far 1 offset	-3.0/0/3.0	0.6	1/s
FAR2_GAIN Far 2 gain	0.7/1.0/1.3	1.0	
FAR2_OFFSET Far 2 offset	-3.0/0/3.0	0.9	1/s
THNR_GAIN Thermal near gain	0.7/1.0/1.3	1.1	
THNR_OFFSET Thermal near offset	-500.0/0/500.0	-23.2	1/s

Figure 1-9. EcoScope calibration listings that ensure valid, quantitative data from the tool.

EcoScope Applications *Used for drilling optimization, formation evaluation, and well placement, the service's innovative measurements improve efficiency throughout the drilling process.*

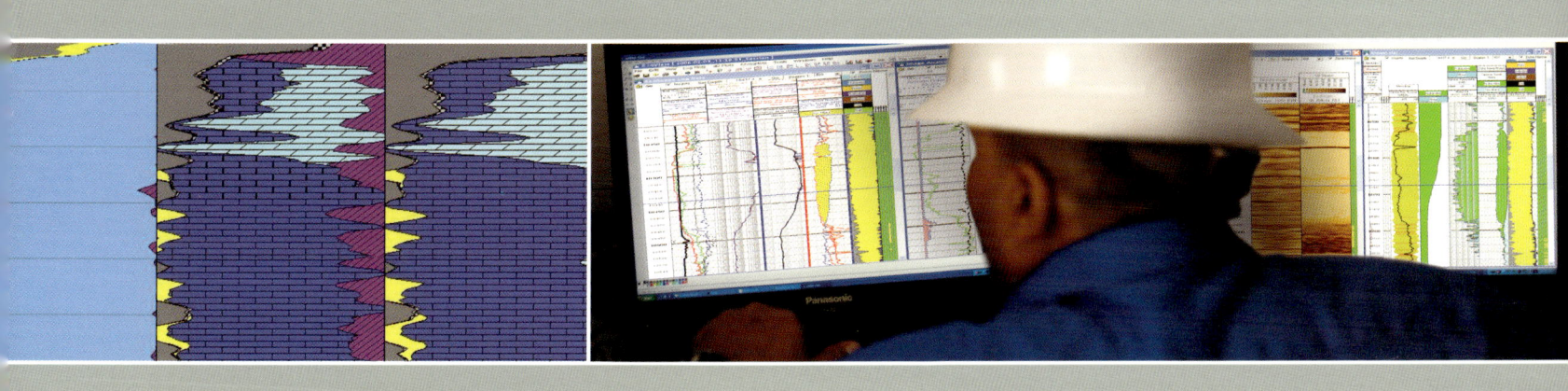

2.1. Drilling optimization

The EcoScope multifunction logging-while-drilling (LWD) service is focused on enhancing drilling efficiency, both as a bottomhole assembly (BHA) component and through delivery of the downhole information required to optimize the drilling process.

Features that enhance drilling optimization

- Integration to deliver a comprehensive suite of measurements in a single collar—this feature means fewer BHA components to handle, making BHA makeup and breakout operations faster

- Sleeve stabilizer design that allows the tool to be configured with various stabilizer diameters to fit with the BHA drilling tendency (build, hold, or drop) requirements (note that the density measurement must be recalibrated if the stabilizer is changed)

- Elimination of transportation, storage, loading, and unloading for the chemical neutron source—the cesium (Cs) density source is side-loaded, significantly reducing the rig-floor time required to load the source and prepare the tool

- Faster tool programming and data retrieval processes as compared with previous-generation LWD tools—the high-speed communication protocol, simplified programming procedures, and increased memory capacity of the tool allow for this feature

- Fewer measurement-related drilling rate-of-penetration (ROP) considerations, as the limiting factor of nuclear measurement statistics have been significantly improved—the pulsed neutron generator (PNG) delivers more counts for the neutron measurements than the americium beryllium (AmBe) chemical neutron source used in previous-generation tools, and the density source positioned closer to the formation (as a consequence of the side-loading) delivers more counts for the density measurement, both of which ensure quantitative measurement data at drilling rates up to 900 ft/h

- Measurement acquisiton as fast as the physics of the respective measurements permit, a feature possible because of the enhanced memory capacity of the tool

- Less rathole drilling for measurements, as the measurements are significantly closer to the bit—for instance, the neutron and density measurements are 44 and 49 ft, respectively, closer to the bit than in the conventional triple-combo BHA

- All measurements within 16 ft of the bottom of the tool, allowing more efficient and effective drilling by ensuring that the information required to maintain the wellbore in the target layer is available for earlier real-time well placement and formation evaluation decisions

- Availability of drilling-focused measurements to ensure that the behavior of the BHA, mud system, and borehole can be monitored so that appropriate action can be taken to avoid unwanted drilling events

Measurements for drilling optimization

- Triaxial shock and vibration
- APWD annular pressure while drilling
- Near-bit inclination
- Calipers—density and ultrasonic

- Real-time lithology from spectroscopy
- Formation dip from images for drilling tendency evaluation—density, photoelectric, ultrasonic, and gamma ray

2.2. Formation evaluation

The EcoScope service delivers unparalleled formation evaluation, both through enhanced conventional measurements and delivery of innovative measurements to the LWD domain.

Features that enhance formation evaluation

- Improved azimuthal density measurement statistics, depth of investigation (DOI), and reduced borehole and standoff sensitivity

- Improved neutron measurement statistics and DOI

- Introduction of a neutron-derived hydrogen index (HI) measurement to the LWD domain

- Improved gamma ray measurement statistics

- Introduction of a strongly azimuthal gamma ray measurement, permitting azimuthal formation evaluation and imaging

- Reduced borehole effect on the propagation resistivity measurement

- Introduction of formation thermal neutron capture cross section, or sigma, a measurement with many uses, including resistivity-independent saturation evaluation in saline water environments and simultaneous water salinity and saturation evaluation

- Introduction of elemental spectroscopy while drilling—downhole processing extracts the elemental information from an induced gamma ray spectrum, permitting direct, real-time lithology identification

- Introduction of a sourceless density measurement

- Collocated measurements that result in responses from similar volumes of formation and are acquired at the same time under the same environmental conditions, permitting meaningful comparison

- Improved quality control (QC) processes, including a green-yellow-red flag system to simplify identification of data requiring further scrutiny

- Integrated answers

Measurements for formation evaluation

- Azimuthal natural gamma ray

- Propagation resistivity

- Azimuthal bulk density

- Azimuthal photoelectric factor (PEF)

- Thermal neutron porosity

- HI

- Neutron-gamma density (NGD)

- Elemental spectroscopy

- Formation thermal neutron capture cross section (sigma)

- Calipers—density and ultrasonic

- Near-bit inclination

- Formation dip from images—density, photoelectric, ultrasonic, and gamma ray

- Annular pressure

- Annular temperature

- Three-axis shock and vibration

2.3. Well placement

The EcoScope service supplies appropriate real-time information for timely well placement decisions, thereby enhancing reservoir exposure and production.

Features that enhance well placement operations

- Measurements close to the bit for early detection of changes in the formation

- Collocation of measurements to allow comparison of various responses, permitting earlier decisions to be made; for example, an increase in resistivity may occur because of increasing oil saturation, decreasing porosity, or proximity to another layer, and having the density and neutron measurements close to the resistivity enables certain possibilities to be eliminated, resulting in greater certainty in the well placement decision process

- Near-bit inclination for assistance in determining the location of the wellbore and optimizing trajectory control

- A variety of measurements available for steering—the conventional suite of resistivity, density, neutron, and gamma ray is enhanced by the availability of real-time lithology, sigma, azimuthal gamma ray, and sourceless density

- Reliable caliper data for hole shape and stability evaluation

- A choice of formation parameters—density, PEF, and gamma ray—that can be imaged close to the bit, transmitted in real time, and visualized in three dimensions (3D) to assist in structure visualization

- Downhole data processing and compression, enhancing the information content of the data stream transmitted in real time, which, in conjunction with high telemetry rates, ensures that the required number of channels to make well placement decisions is available with sufficient update frequency, even when drilling at high ROPs

- Improved measurement statistics, resulting in greater confidence in the data when making an assessment for a difficult well placement decision

Measurements for well placement

- Formation dip from images for structural evaluation—density, photoelectric, ultrasonic, and gamma ray

- Azimuthal natural gamma ray

- Propagation resistivity

- Azimuthal bulk density

- Azimuthal PEF

- Thermal neutron porosity

- HI

- NGD

- Elemental spectroscopy—chemostratigraphy (using formation elements to correlate horizons between wells)

- Formation thermal neutron capture cross section (sigma)

- Calipers—density and ultrasonic

- Near-bit inclination

- Three-axis shock and vibration

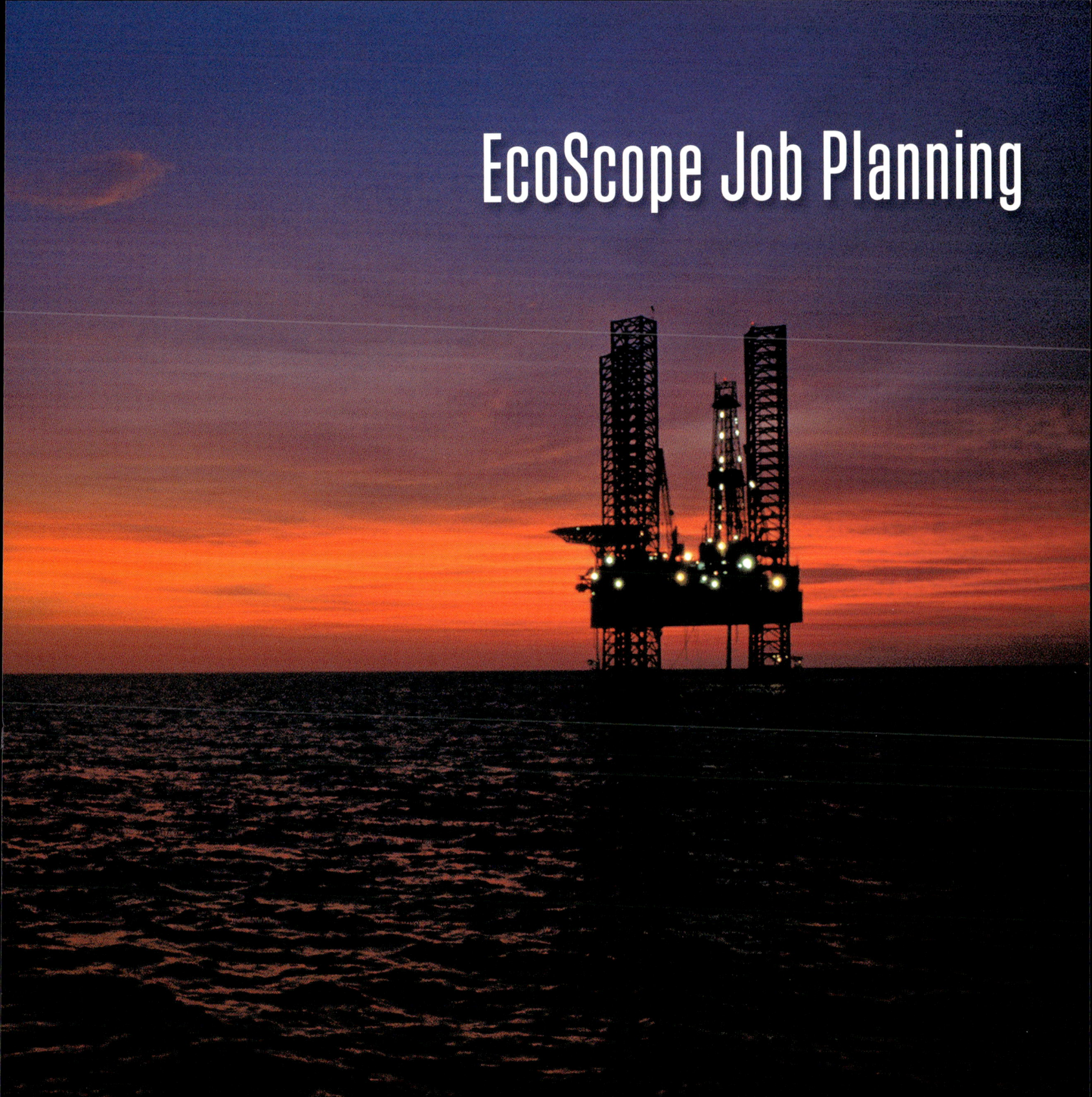

EcoScope Job Planning

EcoScope Job Planning *Adhering to combinability requirements, measurement considerations, and physical specifications allows drilling runs to be properly executed.*

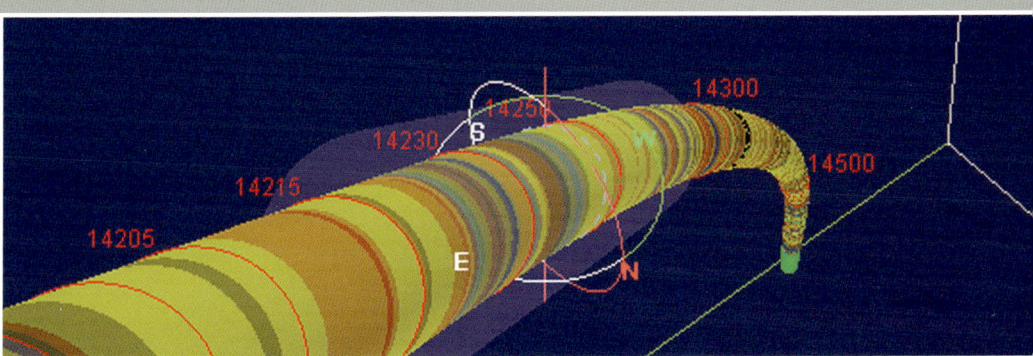

3.1. Service combinability

The EcoScope multifunction logging-while-drilling (LWD) service must be run with the TeleScope high-speed telemetry-while-drilling service to provide pulsed neutron generator (PNG) power control. Typically, the EcoScope tool is placed below the TeleScope tool in the bottomhole assembly (BHA), but it can be run above if required. The EcoScope tool is combinable with all current Schlumberger LWD tools and wired drillpipe data transmission systems.

3.2. Bottomhole assembly configuration considerations

3.2.1. Stabilizer size

The EcoScope system can currently be fitted with three standard stabilizer sizes (7.875, 8.25, and 9.375 in). It is strongly recommended that the size of the stabilizer be agreed upon before tool mobilization to the rig. The stabilizer is not changeable in the field because the density measurement must be recalibrated after any adjustment to the stabilizer.

3.2.2. Steerable assemblies that require sliding

During sliding (nonrotary) with a steerable motor, the density windows of the EcoScope tool should face the low side of the borehole so that the density measurement is not impaired by standoff. An orienting sub can be used to turn the windows with respect to the azimuth of the bent housing. The orienting sub should be placed between the EcoScope tool and the bent sub.

If an orienting sub is not used, it may be necessary to ream through the slide sections in rotary mode to obtain good density measurements.

3.2.3. High-pressure, high-temperature operations

The standard operating temperature and pressure limits of the EcoScope tool are 302 degF and 20,000 psi, respectively. For high-pressure, high-temperature (HPHT) applications, the tool may be modified to extend the operating range.

3.2.4. Secondary gamma ray sensors

Care needs to be taken when running a secondary gamma ray sensor in the BHA, as the PNG can activate the mud, as well as the formation, resulting in an artificially high gamma ray measurement. The secondary gamma ray measurement should be located below the EcoScope tool.

Gamma ray tool above the EcoScope tool

If a gamma ray sensor is run above the EcoScope tool, formation activation by the high-energy neutrons from the EcoScope PNG may cause the gamma ray measurement to read too high. The extent of formation activation depends on

- time at a particular depth—slow rate of penetration (ROP) or time spent circulating at a given depth gives the PNG time to activate the formation; slow ROP gives the activated formation time to decay before the gamma ray detector reaches the activated section, or there is an intermediate ROP, which leads to the largest activation signal

- lithology—the elements in a formation determine how long a formation remains activated; for example, a pure limestone will show very little activation, and sandstone and shale show significant activation.

Gamma ray tool below the EcoScope tool

If a gamma ray sensor is run below the EcoScope tool, the reading may be increased because of activation of oxygen (O) in the mud by the PNG. The magnitude of the increase depends on numerous factors, including mud composition, velocity through the BHA, and the detector configuration in the lower tool. Activation of elements in the formation by the PNG is likely to result in a significant gamma ray reading increase during a ream-up to relog a section.

3.3. Measurement considerations

3.3.1. Invasion

When drilling in permeable zones, drilling mud filtrate may invade the formation around the tool. The diameter and resistivity of this invaded zone cause a separation between the multiple depth-of-investigation (DOI) resistivities. The resistivity separation can be used to deduce the invasion profile and hence, derive the true formation resistivity beyond the invaded zone. ARCWizard processing is available to perform this inversion.

3.3.2. Anisotropy

In layered formations, the propagation resistivity response may vary with well deviation, as the measure current either runs parallel or perpendicular to the layering. In a vertical well through horizontal layering, the measure currents run parallel to the layering and hence, measure the horizontal resistivity. In the same horizontal layering, increasing well angle results in the measure current traversing the layers. A resistivity measurement made perpendicular to the layers is called the vertical resistivity. At varying well angles, the propagation resistivity response is a combination of both the horizontal and vertical resistivities. In an anisotropic formation, vertical resistivity is always greater than horizontal resistivity. ARCWizard processing is available to recognize an anisotropic resistivity signature and derive vertical and horizontal resistivities from the propagation resistivity responses.

3.3.3. Oxygen-activation

The use of a PNG, which emits high-energy neutrons, leads to a small but detectable activation of Oxygen in the mud. The EcoScope gamma ray detector is below the PNG. Mud flowing through the tool is slightly activated as it passes the PNG and then flows down past the gamma ray detector, causing an increase in count rate, which is not associated with natural formation radioactivity. An O-activation correction is provided for the EcoScope gamma ray measurement.

3.3.4. High rate of penetration

The data density of LWD measurements depends on the drilling ROP. Generally, data density is inversely proportional to ROP. The EcoScope measurements are acquired as fast as is reasonable given the physics of the measurements: 12 seconds for spectroscopy, 2 seconds for resistivity, and 4 seconds for most nuclear measurements. The fastest recommended ROP for good nuclear data density is 360 ft/h. Resistivity data can be obtained at one sample every 6 in at an ROP of 900 ft/h.

3.3.5. Spectroscopy

The sensitivity of the spectroscopy measurement to various elements in the formation is a function of the ROP, salinity, formation mineralogy, and porosity. A planner is available to determine the expected spectroscopy precision under various conditions. Further details can be found in Section 4.10.4.

3.3.6. Shocks

The EcoScope system is designed to withstand normal drilling environments. However, exposure to a hostile environment may result in tool failure. The EcoScope tool has sensors that measure the amplitude and frequency of shocks and assign a shock risk level while drilling. Shocks should be monitored and minimized. Level 3 shocks (10 shocks per second over 50 g_n should be avoided because the risk of tool failure is very high.

3.3.7. Measurement-while-drilling signal quality

Measurement-while-drilling (MWD) signal quality is affected by excessive noise and harmonics from the mud pumps. Pulsation dampeners should be adequately charged to ensure that pump harmonics are kept at a minimum. Mud properties also affect the MWD signal. Viscous mud will attenuate the MWD signal, making real-time data transmission difficult.

3.3.8. Dogleg severity

The dogleg severity (DLS) of the well must be considered while planning the job. The EcoScope tool is designed for a maximum DLS of 16°/100 ft while sliding and 8°/100 ft while rotating.

3.3.9. Leakoff test

If a real-time leakoff test (LOT) is to be performed, the approximate duration of the test should be known so that the tool can be programmed with the appropriate LOT configuration prior to running in hole.

3.3.10. Magnetic interference

In hole sections drilled close to casing, magnetic interference may disturb the magnetic orientation of the tool. This situation will normally be avoided during well planning because the magnetic interference requirements for MWD direction and inclination measurements are stricter than those for tool azimuth detection.

3.3.11. Zone of exclusion

All azimuthal measurements require a component of the earth's magnetic field perpendicular to the axis of the tool. If the tool is within 5° parallel to the earth's magnetic field, the tool azimuth cannot be determined. This zone of exclusion (**Figure 3-1**) occurs when both

- the well azimuth is within 5° parallel to the earth's magnetic field

- the well inclination is within 5 to 10° parallel to the earth's magnetic field.

Where possible, the zone of exclusion should be avoided during well planning to ensure that azimuthal data can be acquired.

Figure 3-1. Zone of exclusion around the earth's magnetic vector, in which the tool orientation within the borehole cannot be determined.

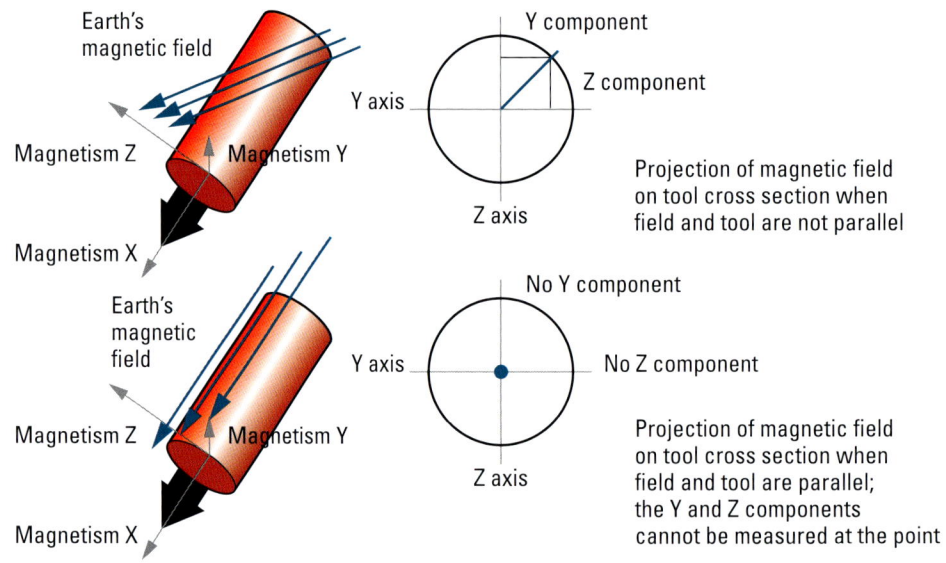

3.4. Physical specifications

The following specifications are correct at the time of publishing. Due to continuous hardware and processing upgrades, we reserve the right to change these specifications at any time. Please contact your local Schlumberger representative for verification of any specifications.

3.4.1. Mechanical specifications

Table 3-1. Mechanical specifications.

		Specification	Comments
Outer diameter (OD)		6.9 in	6.75 in nominal, American Petroleum Institute (API)
Hole size range[†]	Minimum	8.38 in	When run with 7.875-in slick sleeve
	Maximum	Stabilizer OD + 0.5 in	When run with a stabilizer. Common stabilizer sizes: 7.875, 8.25, and 9.375 in
Maximum body OD		7.88 in	On wear upsets
Minimum inner diameter (ID)		2 in	
External upsets		7.88 in	On wear upsets
		7.4 in	On resistivity antennas
		7.75 in	On stabilizer shoulder
Sub fishing neck[‡] (from top)		6.2 ft	
Sub length[§]		25.2 ft	
Length, including crossovers		26.4 ft	Including lower crossover
Weight in air		2,700 lbm	Including lower crossover
Top thread connection		5.5 full hole (FH)	Box, API
Bottom thread connection		5.5 FH	Box, API
Joint yield torque		46,000 ft.lbf	
Nominal makeup torque		25,000 ft.lbf	
Bending strength ratios[††]			
TeleScope collar pin X upper crossover box		2.17	
5.5 FH crossover pin X EcoScope box		2.03	
5.5 FH crossover pin X PD675 flex box		2.17	
5.5 FH crossover pin X CRST upper box		1.9	
Average moment of inertia of collar		87 in^4	6.5-in OD × 2.8-in ID
Equivalent bending stiffness[‡‡]		24.5 ft	24.5 ft of 6.5 × 2.81 in
Magnetic sub		No	

[†] Maximum hole size should not exceed stabilizer size + 0.5 in to guarantee accurate density data.
[‡] Fishing neck is the distance from the face seal of the top connection to the start of the first external upset on the sub.
[§] Sub length is that of the minimum self-contained configuration, EcoScope stand-alone.
[††] Bending strength ratios are specific to particular connections. Values here apply only to the listed connections.
[‡‡] Equivalent bending stiffness is that of the equivalent API collar.

3.4.2. Operating specifications

Table 3-2. Operating specifications.

	Specification	Comments
Maximum tool curvature[†]		
Rotating	8°/100 ft	
Sliding	16°/100 ft	
Maximum weight on bit (WOB)[‡]	80,000,000/L^2 (lbf)	L = distance between stabilizers, ft
Operating rpm	No limit	
Maximum operating torque	16,000 ft.lbf	
Maximum operating tensile load[§]	Refer to **Figure 3-2**	
Maximum jarring load[§]		
Maximum operating pressure—standard[††]	20,000 psi	
Internal to external	5,000 psi	
External to internal	5,000 psi	
Sub pressure drop	Pressure drop (psi) = mud density$^{0.75}$ × flow$^{1.75}$/3,738	Density (lbm/galUS), viscosity (poise), flow (galUS/min)
Temperature—maximum operating	300 degF	
Flow rate—maximum flow limit	800 galUS/min	
Maximum operating flow rate	800 galUS/min	
Minimum operating flow rate	None	
Mud solids—maximum sand content	3%	by volume
Dissolved solid content	No limit	sand < 3%
Maximum dissolved O content of drilling fluid	1 ppm	
Minimum pH of drilling fluid	9	
Lost circulation material (LCM)—maximum size	Medium nut plug	
Maximum concentration	50 lbm/bbl	
Power source	External turbine, Readout port, Internal clock battery	TeleScope MWD
Downhole memory[‡‡]	Yes	1 GB
Telemetry—real-time	Yes	TeleScope MWD
Dump protocol and speed[§§]	14,700,000 bps	
Pulsation dampener requirement	No	Recommended
Filter screen recommendation[†††]	Yes	
Drillpipe filter screen recommendation	Yes	

[†] Maximum tool curvature refers to maximum collar curvature, not to the hole dogleg. The maximum tool curvature is derived from the collar fatigue limit.

[‡] Maximum allowable WOB depends on the unsupported length of the collar, L, which is the distance between the stabilizers.

[§] If exposed to pulling load between operating tensile load and jarring load, tool should be laid down and disassembled for full inspection. If exposed to pulling load beyond jarring load, there is potential damage to chassis and collar.

[††] Pressure rating refers to pressure applied internally to the flow tube and externally to the collar. Any potential source of pressure buildup inside or outside must be taken into consideration to guarantee safe limits.

[‡‡] Standard memory board—the high-temperature version has 0.75-GB capacity.

[§§] Dump speeds up to 7,000,000 bps are possible with a 300-ft toolscope extension cable; dump speeds up to 14,700,000 bps are possible with a 50-ft toolscope extension cable.

[†††] A filter screen is placed in line between the pump and the standpipe manifold.

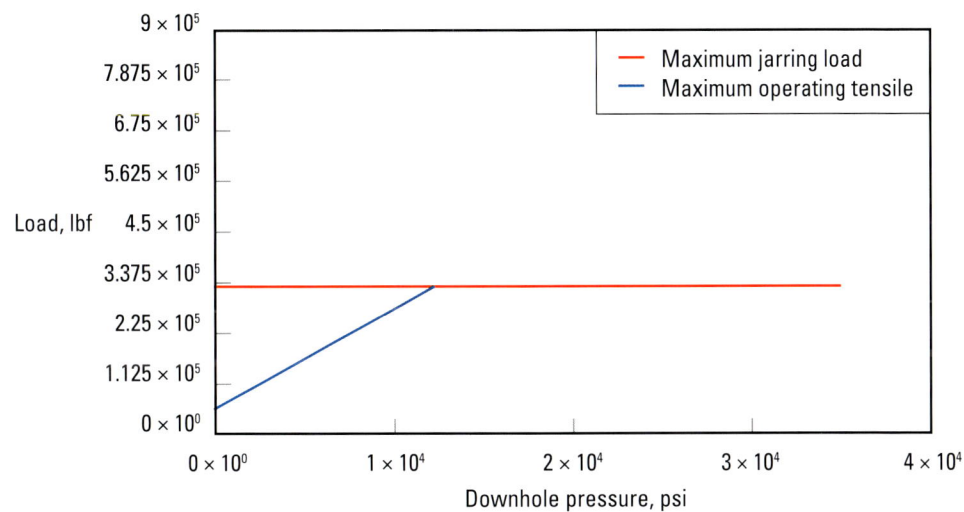

Figure 3-2. Maximum operating tensile and jarring loads—30 minutes at Shock Level 3 (50-g_n threshold) or 200,000 shocks above 50 g_n.

EcoScope Measurements

EcoScope Measurements *provide a wide range of data for drilling optimization, well placement and sophisticated formation evaluation in both real-time and for post-drilling analysis.*

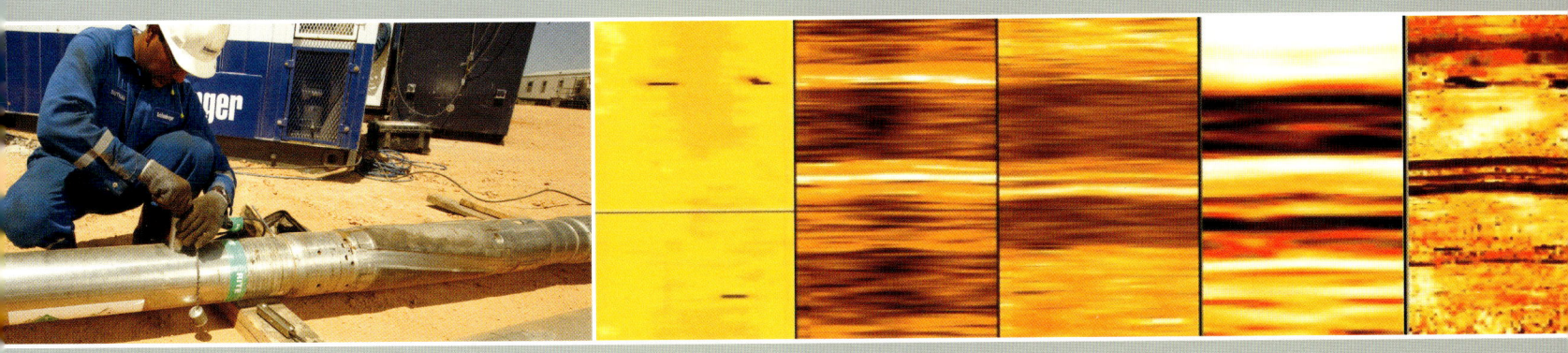

4.1. Downhole data categories

Data from the EcoScope multifunction logging-while-drilling (LWD) service can be categorized according to whether it is

- acquired relative to (indexed by) time or depth

- azimuthally focused or azimuthally symmetric

- transmitted in real time or recovered from the memory of the tool when it returns to surface.

4.1.1. Time- and depth-indexed data

All downhole measurements are initially recorded versus time. For some measurements, such as the annular pressure at a given depth during a leakoff test (LOT), the data remains indexed in time during interpretation. Most of the formation measurements, however, are time-merged with the time-depth data acquired at surface to change the measurement index from time to depth so that it can be presented as a depth log for interpretation.

A number of the drilling-related measurements are presented on both time and depth indices. Shocks, for example, are relevant as time data to show the historical performance of the bottomhole assembly (BHA) and the shocks to which it has been subjected. They are relevant as depth data to see if there is a correlation between the formation type and the shocks.

4.1.2. Azimuthal and nonazimuthal data

Downhole LWD measurements can be subdivided into two main groups, azimuthal and nonazimuthal.

Azimuthal measurements are those for which the physics permit the measurement to be focused on an azimuthal sector of the borehole, resulting in directional sensitivity. An azimuthal, or azimuthally focused, measurement has one or more directions perpendicular to the surface of a logging tool from which it receives most of its signal. Examples are the density and gamma ray measurements.

Nonazimuthal measurements are those for which the physics does not permit azimuthal focusing, so the measurement is acquired simultaneously from the 360° circumference of the borehole. A nonazimuthal, or azimuthally symmetric, measurement is one that measures equally in all directions around the tool. Examples are the propagation resistivity and neutron measurements.

4.1.3. Real-time and recorded-mode data

Data recorded at surface or transmitted to surface while drilling is called real-time data. The transmitted data is generally a small but critical subset of the suite of measurements taken downhole. The downhole measurement data is usually depth-converted, merged with the surface measurements, and presented as a log so that the formation characteristics can be determined while drilling. This real-time data is used for critical operations decisions while drilling, such as for well placement.

On retrieval of the BHA to surface, the memory of the tools is recovered. This memory is called the recorded-mode data and consists of all the measurements. Though there may be small differences associated with limited environmental corrections on the real-time data, the recorded-mode data should closely match the real-time data. The recorded-mode data, with a considerably greater number of data points, will usually have better resolution than the real-time data.

4.2. Azimuthal natural gamma ray

Acquired using a large shielded detector, this total gamma ray measurement has high azimuthal sensitivity, making the measurement capable of delivering a gamma ray image.

4.2.1. Gamma ray theory of measurement

Natural gamma rays are present in geologic formations because of the decay of naturally occurring radioactive isotopes. The three isotopes responsible for natural gamma ray radiation are

- potassium (K)-40

- uranium (U)-238

- thorium (Th)-232.

Gamma rays emitted by the decay of these isotopes and their daughter species impact the structure of a scintillation detector in the tool, resulting in the emission of a photon of light within the detector. The number of light photons released within the detector is proportional to the absorbed gamma ray energy. A photomultiplier tube coupled to the detector converts the photons into an electrical pulse that can be measured electronically.

The natural gamma ray detector in the EcoScope tool counts the total number of gamma rays that impact the detector. It does not discriminate between gamma rays from the three isotopes. The resulting count rate is converted into American Petroleum Institute (API) units to express the gamma ray activity of the formation. This conversion uses a single factor determined experimentally on a master tool during characterization at the standard reference facility for gamma ray well logging.

The gamma ray detector in the EcoScope tool is larger than gamma ray detectors in previous LWD tools and is mounted in such a way that it is preferentially sensitive to gamma rays coming from one azimuthal direction. This sensitivity is achieved by back-shielding the detector to reduce the gamma ray signal from the back of the tool and by reducing the collar wall thickness in front of the detector to enhance the gamma ray count rate from the front of the tool (**Figure 4-1**). As a consequence, the EcoScope gamma ray measurement has excellent azimuthal sensitivity and can be used for imaging.

Figure 4-1. A cross section of the EcoScope tool, showing the natural gamma ray detector with back-shielding and reduced collar thickness in front of the detector to enhance the azimuthal sensitivity of the measurement.

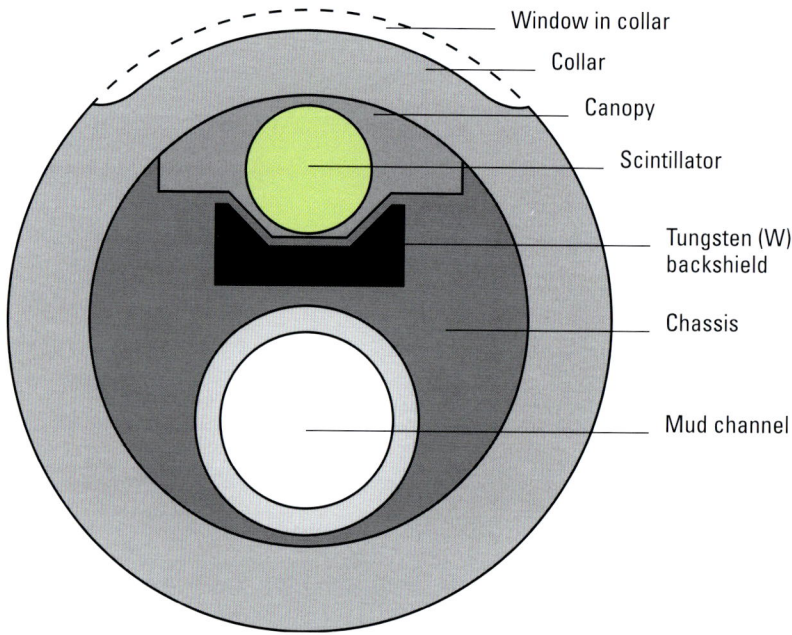

Window in collar
Collar
Canopy
Scintillator
Tungsten (W) backshield
Chassis
Mud channel

4.2.2. Gamma ray environmental corrections

The gamma ray measurement is affected by

- gamma rays emitted by activated nuclei in the mud or formation
- gamma rays emitted by decay of radioactive material (generally potassium-40) contained in the drilling fluid
- density of the drilling fluid
- size of the borehole.

4.2.2.1. Oxygen-activation correction

The emission of high-energy neutrons (with energies greater than 10.2 MeV) by the pulsed neutron generator (PNG) causes activation of oxygen (O) nuclei in the mud. The neutron is absorbed as it knocks out a proton and converts the nucleus into nitrogen (N)-16 which then decays back to Oxygen-16 with a half-life of 7.13 seconds. The Oxygen-16 produced is initially in an excited state, but promptly decays to the ground state by the emission of one or more gamma rays. Most of the gamma rays (approximately 90%) have an energy of 6.13 MeV.

This effect artificially increases the gamma ray measurement. The amount of activation registered by the natural gamma ray detector depends on

- detector size, position, and shielding

- oxygen content of mud (number density of oxygen)

- neutron output of PNG

- flow velocity

- distance from the PNG to the gamma ray detector

- detector distance to bit (upflow)

- borehole size.

As shown in **Figure 4-2**, mud flowing through the tool is activated once as it passes the PNG on the inside of the tool and again as it passes the PNG on the outside of the tool. High mud velocities result in a short exposure time to the activating neutrons, so the effect is small, but it is corrected to ensure a quantitative gamma ray measurement.

Figure 4-2. Activation of oxygen in the mud as it passes the PNG, affecting the gamma measurements above and below the PNG.

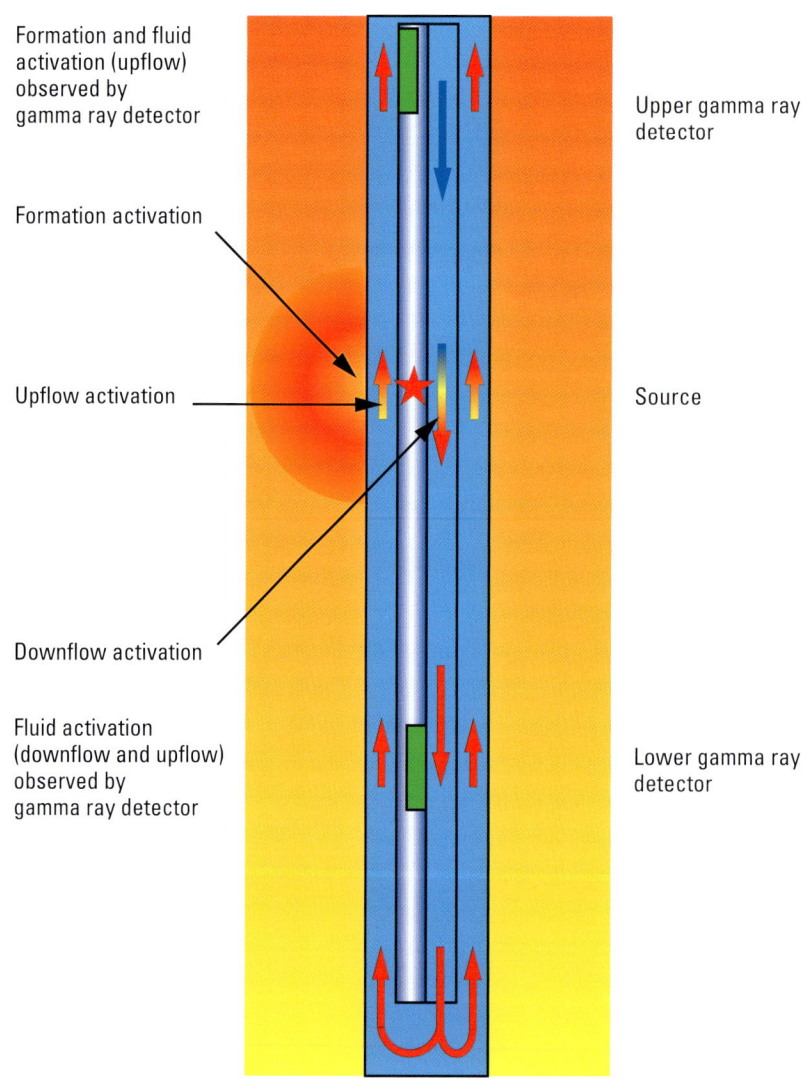

The correction is based on the fact that activated oxygen emits gamma rays with energies between 5 and 6.13 MeV, as shown in **Figure 4-3**. This energy is far higher than the highest natural gamma ray energy of 2.61 MeV, which is emitted by thorium. However, down-scattering of gamma ray energies means that even in the natural gamma ray energy range, some additional gamma rays will be detected.

An energy threshold of approximately 3.3 MeV is applied to the detected gamma rays. Those above the threshold originate from activated oxygen, and those below are there mainly from natural isotopes, with a small proportion down-scattered from the high-energy oxygen-activation gamma rays. Using an association factor to relate the number of down-scattered gamma rays to the count rate above the threshold (which indicates the extent of activation), the count rate below the threshold because of oxygen-activation is computed and subtracted from the total.

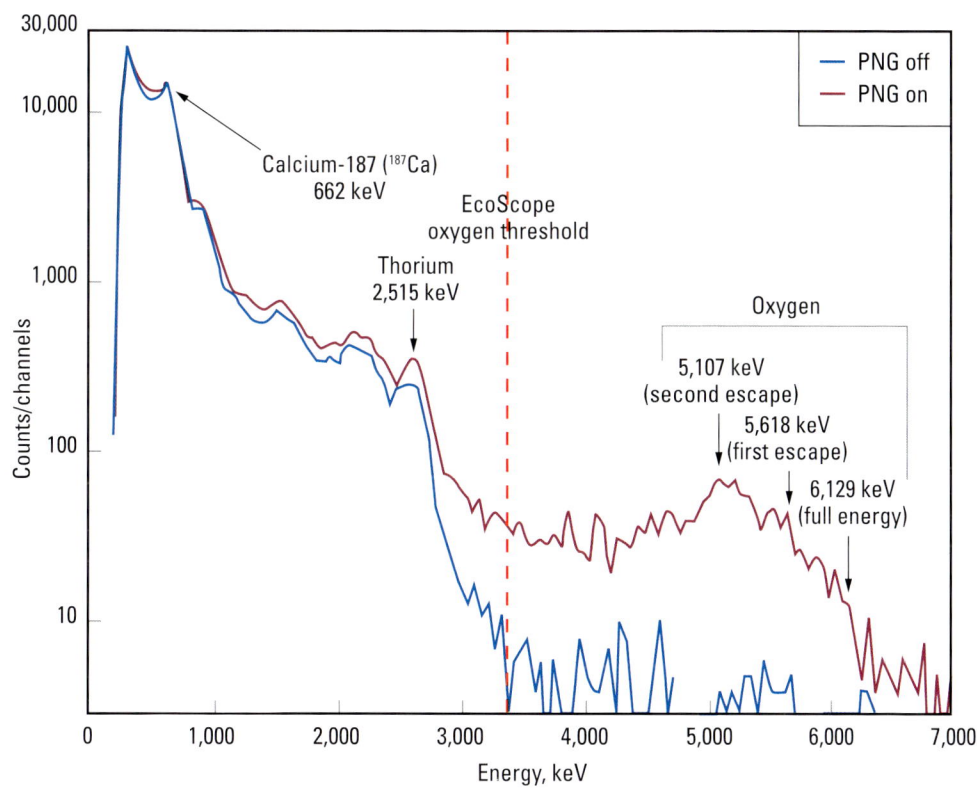

Figure 4-3. The difference between the gamma ray energy spectrum measured with the PNG on (blue) and off (purple), displaying the high-energy-activation gamma rays created by oxygen activation. An energy threshold is applied to distinguish the activation gamma rays from the natural gamma rays, allowing the measurement to be corrected for the oxygen activation effect.

4.2.2.2. Potassium correction

The potassium correction is computed as a function of the mass fraction of potassium in the mud, the mud density, and the borehole size. The predicted additional count rate due to the mud must be subtracted from the average count rate before the borehole correction is applied. This correction factor is exact for a tool centered in the borehole.

4.2.2.3. Borehole size and mud density correction

As gamma rays are stopped by dense material, increasing mud weight will result in loss of gamma ray counts from the formation. Increasing borehole size will also result in a decrease in the number of gamma rays able to traverse the borehole to the detector. The borehole correction is applied after the correction for the mud density and potassium concentration. All these corrections are applied after the oxygen-activation correction. Numerical models have been used to calculate the proportion of gamma rays stopped as a function of borehole size and mud weight. The measured count rate is corrected by dividing by the appropriate attenuation factor (smaller than 1).

4.2.3. Gamma ray applications

The gamma ray log is often used for correlation, as it is the formation measurement least affected by changes in formation porosity and fluid type.

The gamma ray is sometimes used to determine the proportion of clay in a formation because there tends to be a higher concentration of radioactive elements in clay. However, some uranium salts are soluble in water and can migrate through the formation dissolved in water before depositing elsewhere. Zones enriched with uranium by this process have a higher gamma ray reading than their clay content would normally provide and may therefore be mistaken as having higher clay content than is actually the case. This situation can be resolved by using a spectral gamma ray measurement, which distinguishes the energies of the incoming gamma rays and determines whether the gamma rays originated from a uranium, thorium, or potassium atom. The relative proportions of thorium and potassium can then be determined, and the contribution from uranium can be removed, resulting in a more robust clay indicator.

The EcoScope system uses an alternative way to estimate the formation clay content. Elemental spectroscopy determines the relative proportions of a wide range of elements (rather than just the three naturally radioactive isotopes), allowing the clay content to be determined more directly from its elemental composition. Refer to Section 4.10 for further details.

The azimuthal focusing of the EcoScope gamma ray measurement allows the formation gamma ray response to be measured in four quadrants corresponding to the bottom, left, top, and right of the borehole. The gamma ray information is also presented as an azimuthal image, allowing the dips of gamma ray features to be identified as they cross the borehole.

4.2.4. Gamma ray log quality control

The shop calibration must have been performed less than 3 months before the acquisition date and must be in tolerance. Refer to the Schlumberger LWD Quality Control Manual for details.

The EcoScope hardware incorporates checks to verify correct performance of the acquisition detectors and electronics. Additional checks to ensure that the measurements fall within the range expected for downhole formation are outlined in the log quality control (LQC) logic shown in **Table 4-1**.

Table 4-1. Gamma ray LQC flag logic.

QC_GR	Hardware	Measurement
Green	Natural gamma ray high-voltage control loop status is FINE and Natural gamma ray high-voltage control is AUTO and 1,000 V < NGHM < 1,600 V 1,000 V < NGHV < 1,600 V 1,000 V < GRCD < 1,600 V and −60 V < NGHM − GRCD < 10 V −60 V < NGHV − GRCD < 10 V	1 gAPI < GRMA_RT < 248 gAPI (real time) 1 gAPI < GRMA < 1,000 gAPI and rate of penetration (ROP) < 450 ft/h BS ≤ 9⅞ User-selected caliper (BS, Ucal, DCAV, NCAL) < 14 in and QC flag of user-selected caliper yellow or red
Yellow	Natural gamma ray high-voltage control loop status is COARSE or High energy counts too high; NGRTH4A > 15 counts or 800 V < NGHM < 1,000 V or 1,600 V < NGHM < 1,950 V 800 V < NGHV < 1,000 V or 1,600 V < NGHV < 1,950 V 800 V < GRCD < 1,000 V or 1,600 V < GRCD < 1,950 V or NGHM-GRCD out of range (−60,10 V) NGHV-GRCD out of range (−60,10 V)	GRMA_RT < 1 or > 248 gAPI (real time) GRMA < 1 or > 1,000 gAPI ROP > 450 ft/h BS > 9⅞ User-selected caliper (BS, Ucal, DCAV, NCAL) ≥ 14 in or QC flag of user-selected caliper yellow or red
Red	Natural gamma ray high-voltage control loop status is SEARCH or Natural gamma ray high-voltage control is MANUAL or HOLD, meaning the high voltage is not automatically controlled or NGHM out of range (800 to 1,950 V) NGHV out of range (800 to 1,950 V) GRCD out of range (800 to 1,950 V) or DCSW_v (DVME115 status word) or RAM self-test error Bit 0 or Controller Area Network (CAN) bus error (no memory record) Bit 1 or SPI communication issue Bit 3 or NGR test mode Bit 8 or board state error Recorded mode only: Memory usage indicator = full	
White		No new gamma ray tick available for more than 2 ft

4.2.5. Gamma ray parameters and channels

4.2.5.1. Real time

Table 4-2. Gamma ray real-time parameters.

Parameter Name	Description	Unit
GR_O2COR_OPT_RT	Enable gamma ray oxygen-activation correction, real time	Yes/no (default is yes)
OACF_RT	Oxygen-activation correction factor, real time	
K%	Mud potassium percent	weight %
BS_RT	Bit size from current BHA	in
EcoScope tool OD	Tool outer diameter (OD) from current BHA	in
MWIN	Mud weight in	lbm/galUS

Table 4-3. Gamma ray real-time transmitted channels.

Transmitted Channel	Description
GRMA_DH_ECO_RT	Gamma ray, average, real time, computed downhole
GRMB_DH_ECO_RT	Gamma ray, bottom, real time, computed downhole
GRML_DH_ECO_RT	Gamma ray, left, real time, computed downhole
GRMU_DH_ECO_RT	Gamma ray, up, real time, computed downhole
GRMR_DH_ECO_RT	Gamma ray, right, real time, computed downhole
GR_IMG16_STA_TOH_ECO_RT	Gamma ray (16-sector), image oriented top of hole (U, R, B, L, U), real time
QC_GR_DH_ECO_RT	Gamma ray quality indicator, real time, computed downhole
TCOA_DH_ECO_RT	Gamma ray, oxygen activation, real time, computed downhole

Table 4-4. Gamma ray real-time computed channels.

Output Channel	Description
GRMA_ECO_RT	Gamma ray, average, real time
GRMB_ECO_RT	Gamma ray, environmentally corrected, bottom, real time
GRML_ECO_RT	Gamma ray, environmentally corrected, left, real time
GRMU_ECO_RT	Gamma ray, environmentally corrected, up, real time
GRMR_ECO_RT	Gamma ray, environmentally corrected, right, real time
GRIQ_ECO_RT	Gamma ray image from quadrants, real time
GR_IMG16_DYN_TOH_ECO_RT	Gamma ray (16-sector), dynamic image oriented top of hole (U, R, B, L, U), real time
GRMAQ_ECO_RT	Gamma ray, average, computed from quadrants, real time

4.2.5.2. Recorded mode

Table 4-5. Gamma ray recorded-mode parameters.

Parameter Name	Description	Unit
GR_O2COR_OPT	Enabled gamma ray oxygen-activation correction	Yes/no (default is yes)
OACF	Oxygen-activation correction factor (recorded mode)	(Default is 8)
GR_CF	gAPI to counts/s factor	(Default is 1.8)
BS_RM	Bit size (recorded mode)	in
GCSE	Generalized caliper selection	DCAV or UCAV or BS
MW_RM	Mud weight (recorded mode)	g/cm^3
PMUD	Potassium concentration in mud	weight %
STOH	Top of hole sector	Sector 0 to 15

Table 4-6. Gamma ray recorded-mode channels.

Output Channel	Description
Average	
GRMA	Gamma ray, average
RGRA	Raw gamma ray, average
GRMA_CAL	Calibrated gamma ray
GRMA_FILT	Gamma ray, calibrated and filtered, average
QC_GR	Gamma ray quality indicator
TAB_GR	Gamma ray time after bit
TICK_GR	Gamma ray samples
Quadrants	
RGRB	EcoScope raw gamma ray, bottom
GRMB_CAL	Calibrated gamma ray, bottom
GRMB_FILT	Calibrated, filtered gamma ray, bottom
RGRL	EcoScope raw gamma ray, left
GRML_CAL	Calibrated gamma ray, left
GRML_FILT	Calibrated, filtered gamma ray, left
RGRU	EcoScope raw gamma ray, up
GRMU_CAL	Calibrated gamma ray, up
GRMU_FILT	Calibrated, filtered gamma ray, up
RGRR	EcoScope raw gamma ray, right
GRMR_CAL	Calibrated gamma ray, right
GRMR_FILT	Calibrated, filtered gamma ray, right
Images	
RGRI	Raw gamma ray, 16-sector waveform
GRIQ_FILT	Gamma ray image from quadrants, filtered
GR_IMG16_CAL	Gamma ray image from sectors, calibrated
GR_IMG16_FILT	Gamma ray image from sectors, calibrated and filtered

4.2.6. Gamma ray measurement specifications

Table 4-7. Gamma ray measurement specifications at 100 ft/h.

Item	Value	Remarks
Range	0 to 1,000 gAPI	
Axial resolution	12 in	
Accuracy[†‡]	5%	
Precision	1.8 gAPI	At 100 gAPI and 100 ft/h with three-level spatial averaging

[†] Values given for a tool centered in 8.5-in borehole filled with freshwater.

[‡] Tool-to-tool variations are estimated below 5% and are not taken into account.

4.3. Propagation resistivity

The EcoScope propagation resistivity measurements include borehole-compensated phase shift and attenuation resistivities at two frequencies (2 MHz and 400 kHz) and five spacings (40, 34, 28, 22, and 16 in) with the measurements updated every 2 seconds. ARCWizard processing provides automated interpretation of the resistivity responses.

4.3.1. Propagation resistivity theory of measurement

The propagation resistivity measurement is performed using loops of wire wrapped around a collar to transmit an electromagnetic wave into a formation. The difference in the phase (phase shift) and amplitude (attenuation) across a pair of coil receivers is measured, as shown schematically in **Figure 4-4**.

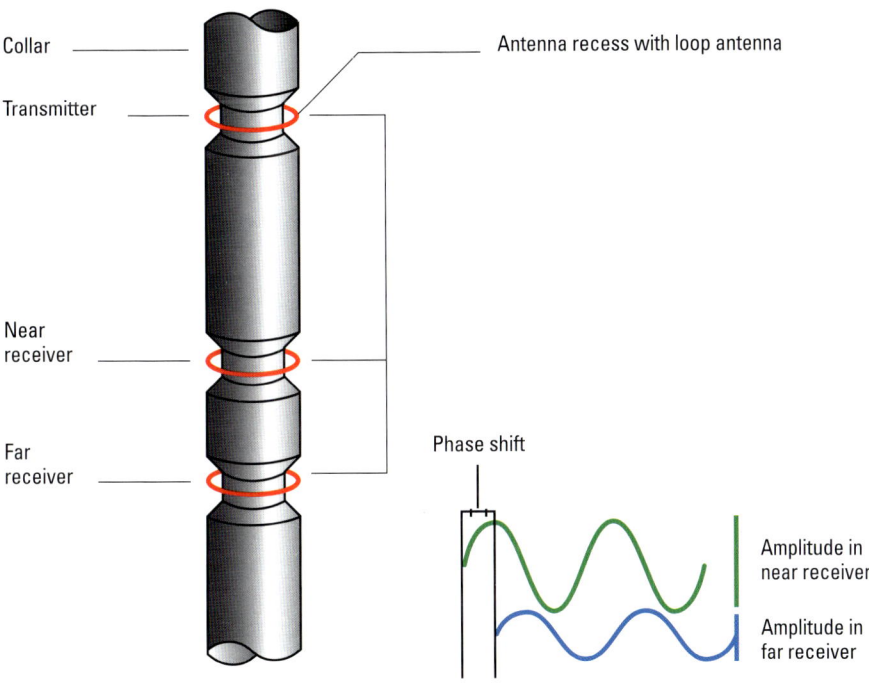

Figure 4-4. Phase shift and attenuation of an electromagnetic wave, propagating from a transmitter through a formation being measured at a pair of receivers.

The impedance of a volume of rock to the passage of electrical current has a component that is independent of the frequency of the current passed through it (the conductance) and a component that depends on the frequency (the capacitance). The response of the propagating electromagnetic wave depends on both the formation conductivity, σ, which controls the conductance, and the dielectric constant, ε, which controls the capacitance.

The formation conductivity is the parameter of primary interest, so the dielectric constant must be determined to estimate formation conductivity. **Figure 4-5** shows a correlation between resistivity and dielectric constant for hundreds of sandstone and carbonate core samples, both fully and partially saturated with water. The equation derived from fitting this data is used in the transform of the raw propagation measurements to conductivity and hence, formation resistivity.

Figure 4-5. The Schlumberger correlation between resistivity (R) and dielectric constant, derived from analysis of hundreds of core samples.

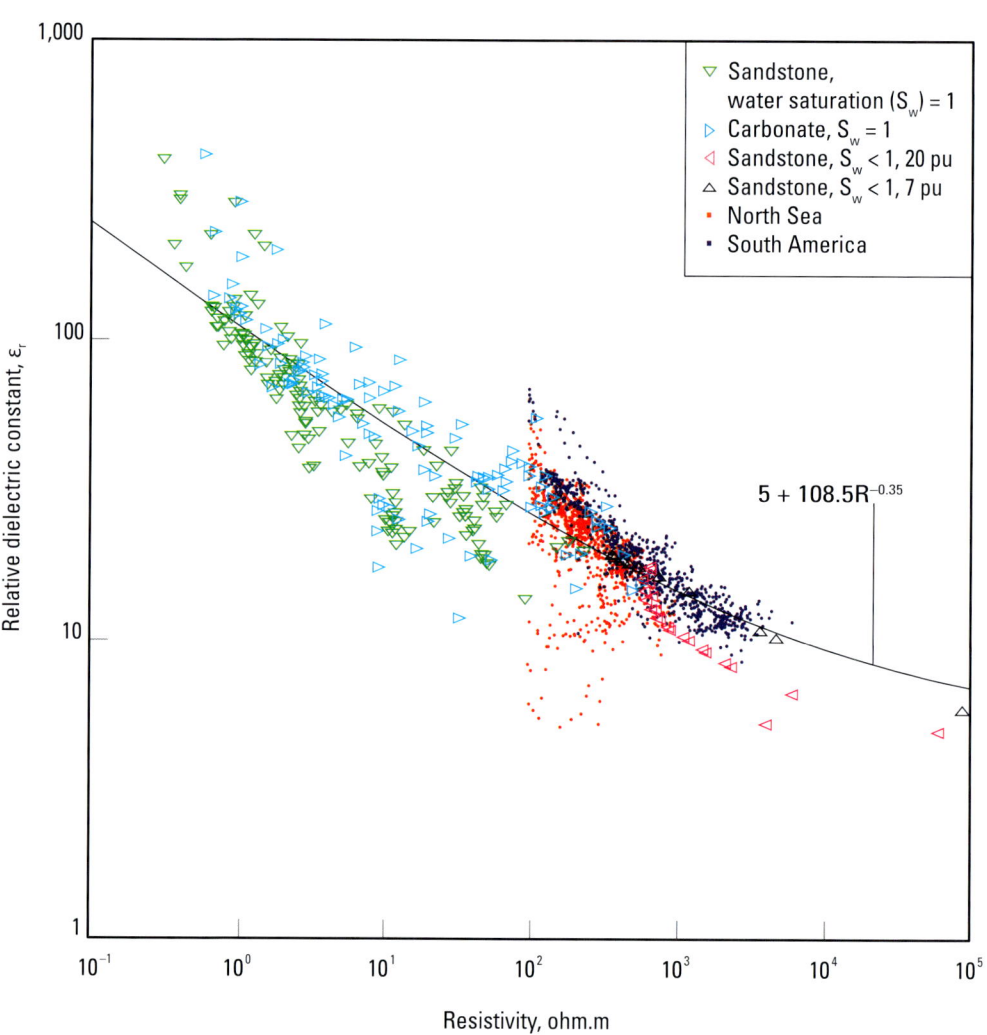

The propagation resistivity measurements are made at frequencies of 2 MHz and 400 kHz. These have been selected as a compromise between minimizing frequency effects (dielectric) on the measured signal (low frequency preferred) while ensuring reliable phase shift and attenuation measurements at high resistivities (high frequency preferred).

4.3.1.1. Borehole compensation

Transmitting from both above and below the receiver pair creates a balanced set of measurements that compensate for effects such as hole rugosity and drifts in receiver electronics. This balancing method is called borehole compensation (BHC). Standard BHC combines data from two transmitters placed symmetrically around the receiver pair to create one compensated measurement (**Figure 4-6**).

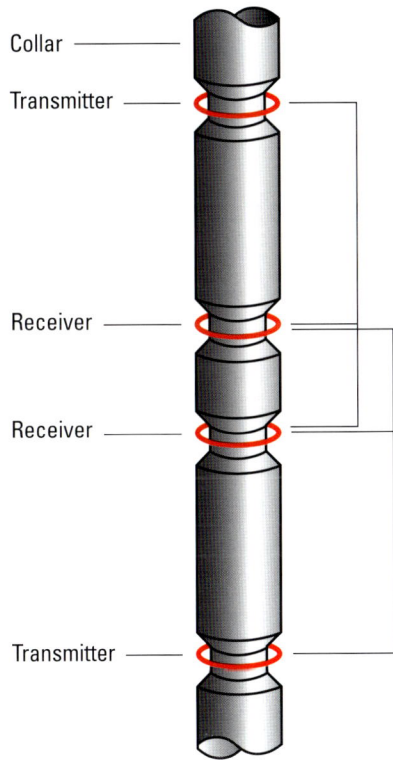

Collar

Transmitter

Receiver

Receiver

Transmitter

Figure 4-6. A standard BHC configuration, combining data from two transmitters placed symmetrically around the receiver pair to provide signal from above and below the receivers, eliminating borehole rugosity and receiver drift effects.

Schlumberger propagation tools dispense with the second physical transmitter, relying instead on the principle of superposition to calculate pseudotransmitter responses, which are then used in the BHC computation.

This linear combination of three sequentially spaced transmitters provides what is called mixed borehole compensation (MBHC), as shown in **Figure 4-7**. The advantage is that tool length and complexity are reduced by eliminating half the transmitters that would otherwise be required. For tools using five transmitter receiver spacings, this eliminates five transmitters.

Figure 4-7. MBHC, using linear combinations of asymmetrically positioned transmitters to compute pseudotransmitters.

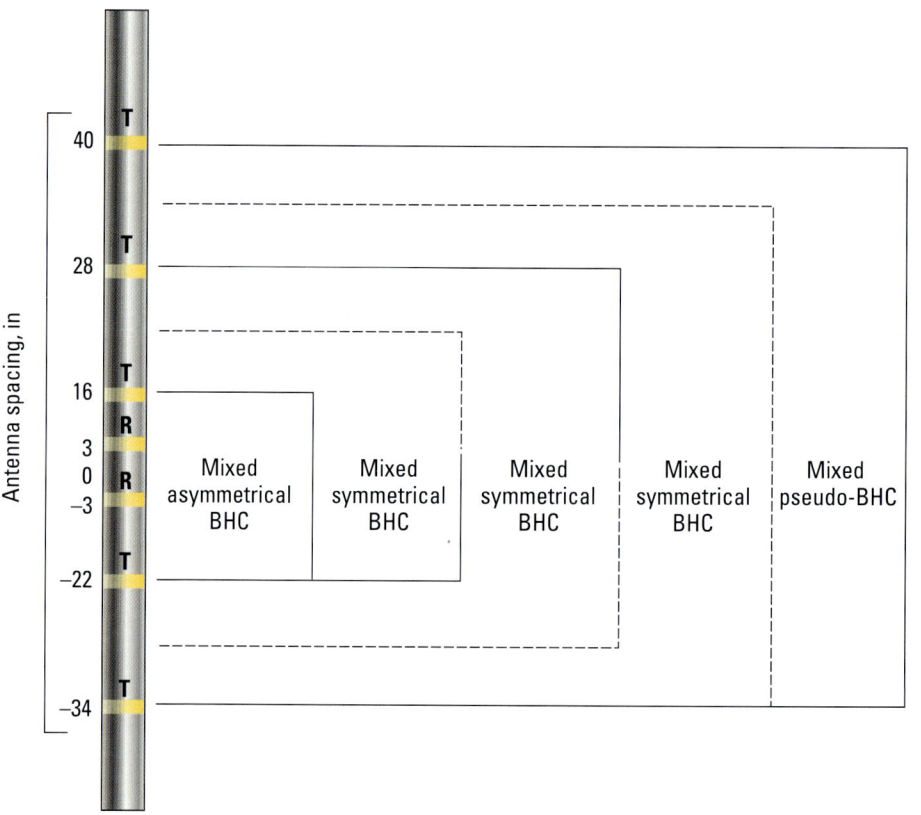

The spiky appearance of the log without MBHC (**Figure 4-8**, top) is caused by overshots in the phase responses induced by borehole washouts and rugosity. These artifacts are canceled out by MBHC (Figure 4-8, bottom).

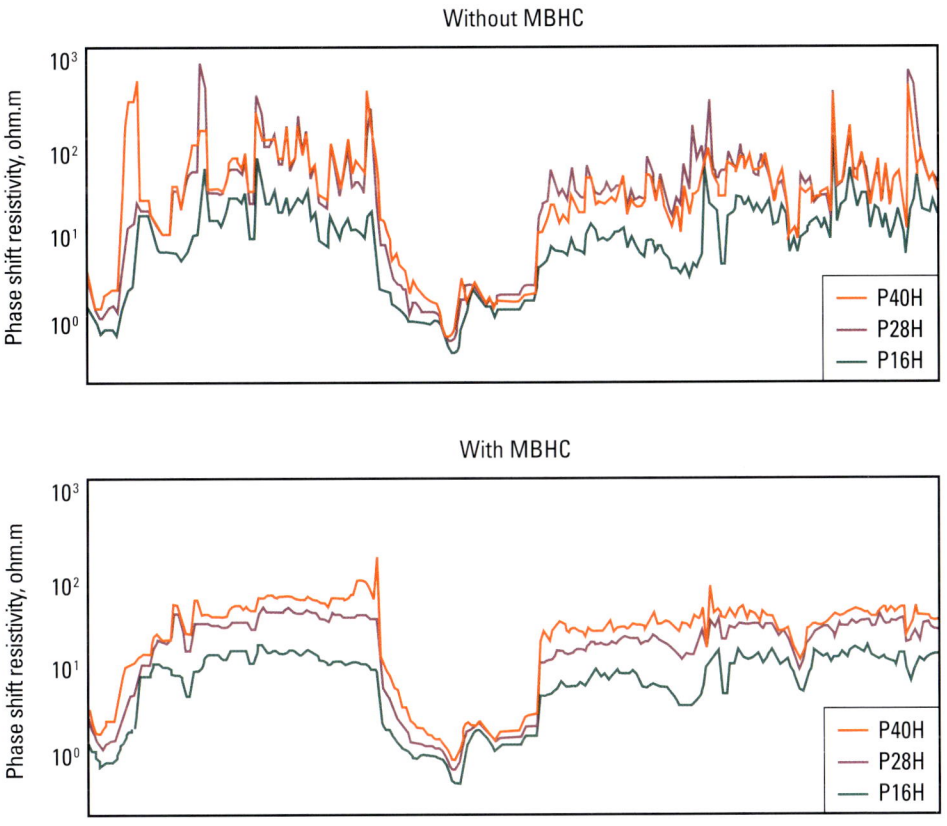

Figure 4-8. Above: Propagation resistivity logs without MBHC. Below: Propagation resistivity logs with MBHC, showing significantly less measurement noise.

Calibration of the propagation tool is performed by measuring the phase shifts and attenuations for the various transmitter-receiver spacings in a nonconductive environment (such as in air far from any conductive material).

The responses obtained in the nonconductive environment are subtracted from subsequent measurements to remove responses associated with the tool. Five MBHC phase shifts and attenuations are then transformed into five calibrated phase shift resistivities and five calibrated attenuation resistivities using a transform similar to that shown in **Figure 4-9**.

Figure 4-9. Transforms converting 2-MHz (top) and 400-kHz (bottom) borehole-compensated and calibrated phase shift (left) and attenuation (right) to formation resistivities.

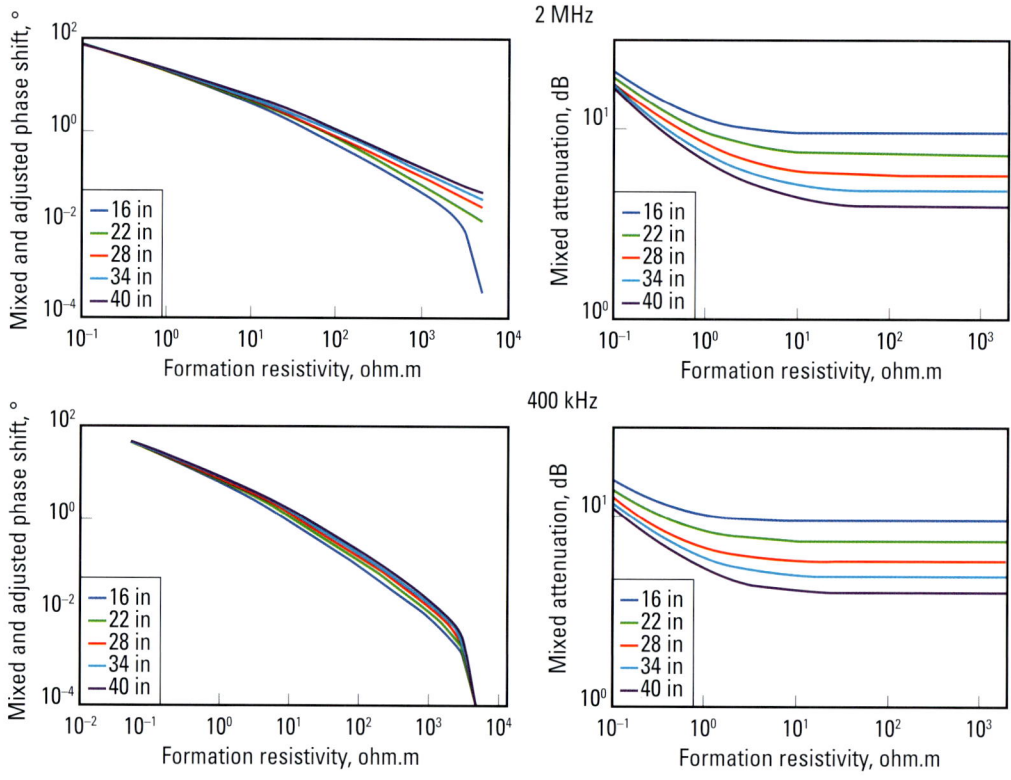

While the transform for phase shift extends well over 1,000 ohm.m, the transform for attenuation is limited to approximately 30 to 70 ohm.m depending on the transmitter-receiver spacing. This limitation exists because the difference in the wave amplitude measured between the two receivers decreases with increasing formation resistivity until the difference becomes smaller than can be reliably measured.

Thus, 2-MHz attenuation resistivities should not be used above approximately 50 ohm.m, and the 400-kHz attenuation resistivities are limited to approximately 10 ohm.m.

4.3.1.2. Depth of investigation

The depth of investigation (DOI) of a propagation resistivity measurement is controlled by

- phase or attenuation measurement of the wave

- transmitter-receiver spacing

- transmitted wave frequency

- formation resistivity.

The first three factors are controlled by tool design. The names of the Schlumberger propagation resistivity measurements encapsulate this information. For example, the resistivity measurement labeled P34H refers to the phase resistivity measured with a 34-in transmitter-receiver spacing at a high frequency (2 MHz). An attenuation resistivity measurement made with a 22-in transmitter-receiver spacing at 400 kHz would be labeled A22L. Despite being measured on the same electromagnetic wave, the phase and attenuation measurements have independent DOIs. As shown in **Figure 4-10**, lines of equal phase are spherical in nature, as the wave travels with equal speed in all directions.

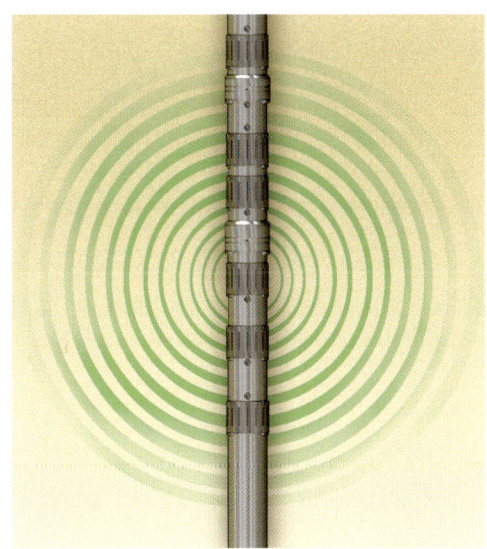

Equal-phase lines

Equal-amplitude lines

Figure 4-10. Left: Lines of equal phase, spherical around a transmitter. Right: Lines of equal amplitude, toroidal.

The corresponding phase resistivity measurement is relatively shallow and axially focused. Lines of equal amplitude form a toroidal shape around the transmitter, as the amplitude is related to the energy of the wave, and the tools are designed to deliver maximum energy in the radial direction. The attenuation measurement is relatively deep but less axially focused, as shown in **Figure 4-11**.

Figure 4-11. Phase and attenuation measurements, which provide two independent volumes of investigation. The phase measurement is shallow and axially focused, and the attenuation is deeper but less axially focused.

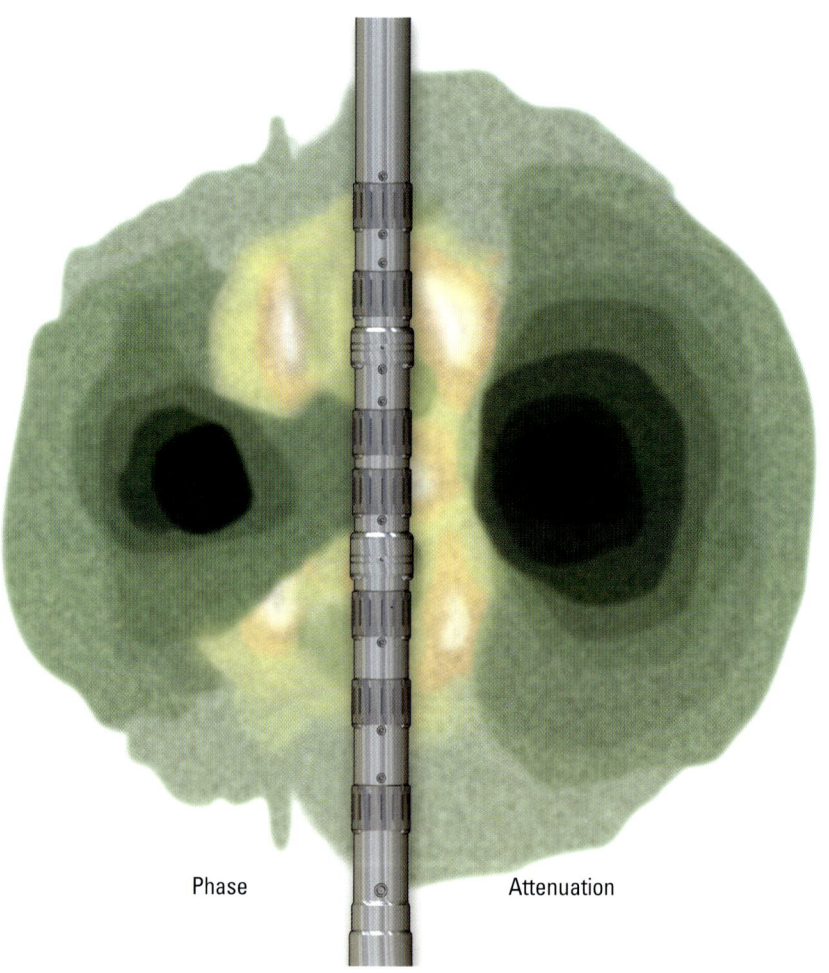

Phase   Attenuation

Because the DOI increases as the transmitter spacing increases, the five phase resistivities represent five different DOIs with nearly identical axial resolutions. Similarly, the five attenuation resistivities represent five deeper-reading measurements.

As shown in **Figure 4-12**, for the range of transmitter-receiver spacings in common use, all the attenuation measurements are deeper than all the phase measurements.

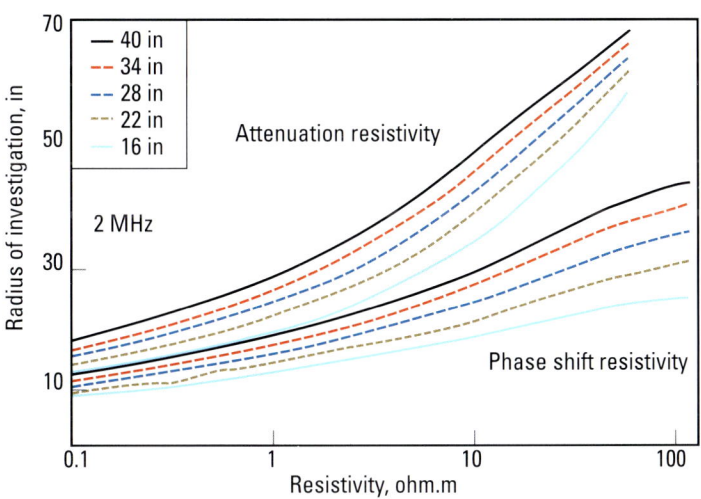

Figure 4-12. 2-MHz measurement radius of investigation plot. The radii of investigation of both phase and attenuation measurements increase with increasing formation resistivity.

For propagation measurements, the radius of investigation is defined as the distance from the borehole axis, at which 50% of the measurement response comes from closer to the borehole, and 50% comes from deeper in the formation. **Figure 4-13** shows the increased radius of investigation obtained when operating at 400 kHz.

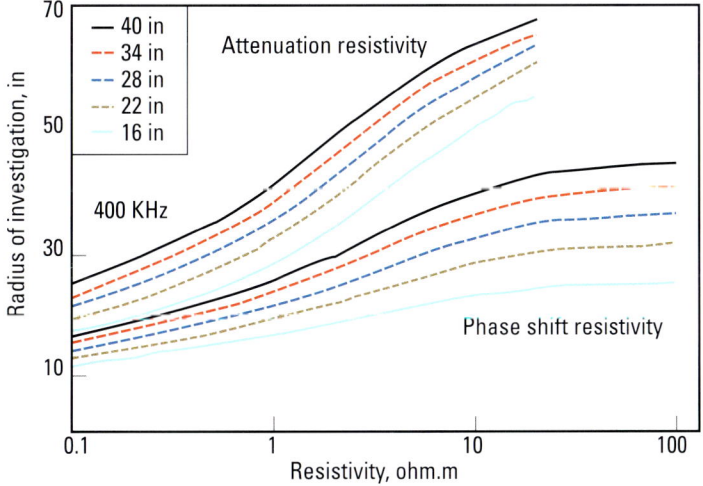

Figure 4-13. 400-kHz measurement radius of investigation plot. The 400-kHz radius of investigation is deeper than the corresponding 2-MHz measurements shown in Figure 4-12.

Note that while the measurement depth increases at lower frequency, the resistivity operating range is reduced at higher resistivities as indicated by the attenuation response truncating at 20 ohm.m for the 400-kHz measurements compared to 50 ohm.m for the 2-MHz response shown in Figure 4-12.

The DOIs of both the 2-MHz and 400-kHz measurements increase with increasing formation resistivity, as shown in Figures 4-12 and 4-13. The measure current induced in the formation by the transmitter will seek the path of least resistance. At low resistivities, the current remains relatively close to the tool. As formation resistivity increases, the current spreads over a larger area. The direct current resistance of a rock is given by

$$r = R\,(L\,/\,A)$$ Equation (4-1),

where

r = resistance seen by the tool (ohm)
L = characteristic measurement length (m)
A = area through which the measurement current passes (m^2).

For high-resistivity formations, the current spreads out over a larger area to reduce the total resistance of the path that it traverses around the tool. This spreading of the induced current results in deeper measurements with increasing formation resistivity as the current spreads deeper in to the formation. It also results in reduced axial resolution with increasing formation resistivity as the current spreads along the tool axis.

In summary,

- attenuation measurements are deeper than phase measurements

- DOI increases with transmitter-receiver spacing

- DOI increases with decreasing transmitter frequency

- DOI increases with formation resistivity.

4.3.1.3. Axial resolution

The axial resolution of a measurement (also known as vertical resolution from the time period when wells were mainly vertical) is a distance that characterizes the ability of the measurement to resolve changes in the formation perpendicular to the tool axis.

The axial resolution of a propagation resistivity measurement is controlled by

- phase or attenuation measurement of the wave

- receiver-receiver spacing

- transmitted wave frequency

- formation resistivity.

Figure 4-14 shows the axial response function of typical phase and attenuation resistivity measurements. The axial response function can be thought of as the "window" along the length of the tool through which the measurement "sees" the formation. The sharper the response function, the thinner the formation layer that the measurement can uniquely resolve. If a layer is thinner than this axial "window," then it will be averaged with the layers above and below it. Note that the phase resistivity response (top) has a thinner "window" (better resolution) than the attenuation measurement (bottom), corresponding to the volume of response for the two measurements shown in Figure 4-11.

As shown in Figure 4-14, there are several different definitions of axial resolution. First, it is the interval within which a large percentage, typically 90%, of the axial response occurs (quantitative resolution). Second, and most commonly quoted numerically, it is the width at the 50% point of the axial response function (width at half maximum). Third, it may refer to the smallest bed thickness for which a significant change can be detected by the measurement (qualitative resolution).

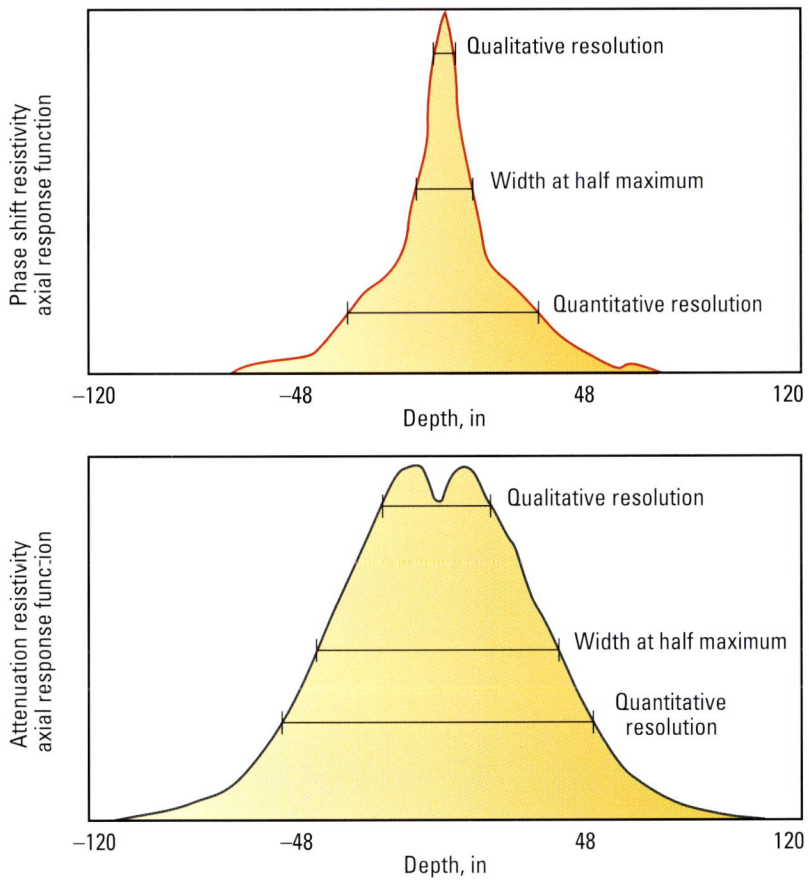

Figure 4-14. Axial resolution of the phase shift measurement (top), which is sharper than the axial resolution of the attenuation measurement (bottom).

Table 4-8 shows the axial resolution, defined as the width of the axial response function at half maximum, for various phase and attenuation measurements in typical formation resistivities. Note that, as shown in Figure 4-14, phase resistivities have sharper axial resolution than attenuation resistivities.

The axial resolution changes only slightly with varying transmitter-receiver spacing because the measurement is taken between the pair of receivers. The axial resolution is strongly dependent on the spacing of the two receivers, but as this is fixed at 6 in for most propagation tools, it is not a factor that needs to be considered for interpretation. The distance to the transmitter has minimal influence on the axial resolution.

Under comparable conditions, the 2-MHz measurements have sharper axial resolutions than the 400-kHz measurements. In general, the 2-MHz measurements can be thought of as sharper and shallower than their 400-kHz counterparts.

The axial resolution degrades with increasing formation resistivity because the current induced in the formation spreads out to reduce the total resistance it experiences as it circulates in the formation around the tool.

Table 4-8. Axial resolution (in ft) for various phase shift and attenuation measurements.

Axial Resolution[†]	Measurement Spacings, in				
Axial resolution values are in ft	**16**	**22**	**28**	**34**	**40**
R = 1 ohm.m					
Phase shift resistivity 2 MHz	0.7	0.7	0.7	0.7	0.7
Phase shift resistivity 400 kHz	1.0	1.0	1.0	1.0	1.0
Attenuation resistivity 2 MHz	1.8	1.8	1.8	1.8	1.8
Attenuation resistivity 400 kHz	3.0	3.5	4.0	4.0	4.0
R = 10 ohm.m					
Phase shift resistivity 2 MHz	1.0	1.0	1.0	1.0	1.0
Attenuation resistivity 2 MHz	4.0	5.0	6.0	6.0	6.0

[†] Width is at half maximum of the response function along the tool axis at the specified formation resistivity.

Figure 4-15 shows the 2-MHz propagation resistivity response to a 4-ft layer of 100 ohm.m and another of 1 ohm.m sandwiched between 10-ohm.m layers. The phase resistivities, with their better axial resolutions, get closer to the true formation resistivity in these thin beds. The shorter transmitter-receiver spacing (16-in phase resistivity) gets closer to true formation resistivity than the 40-in phase resistivity because of its slightly better axial resolution. The attenuation resistivities, having less axial resolution, read considerably lower than true formation resistivity in the 100-ohm.m layer. The axial resolution "window" for the attenuation resistivity includes both the 100-ohm.m layer and the 10-ohm.m layers on either side. The current induced by the tool preferentially flows in the lower-resistivity 10-ohm.m layers on either side of the 100-ohm.m layer, resulting in a lower resistivity reading.

Note that the phase and attenuation measurements get close to true formation resistivity in the 1-ohm.m layer, as this is relatively conductive compared with the surrounding 10-ohm.m layers. The measure current, seeking the path of least resistance, will preferentially flow in this low resistivity layer.

From this example, it can be seen that axial resolution effects and the path of least resistance for the induced currents within the axial resolution "window" must be considered when interpreting resistivity around thin beds.

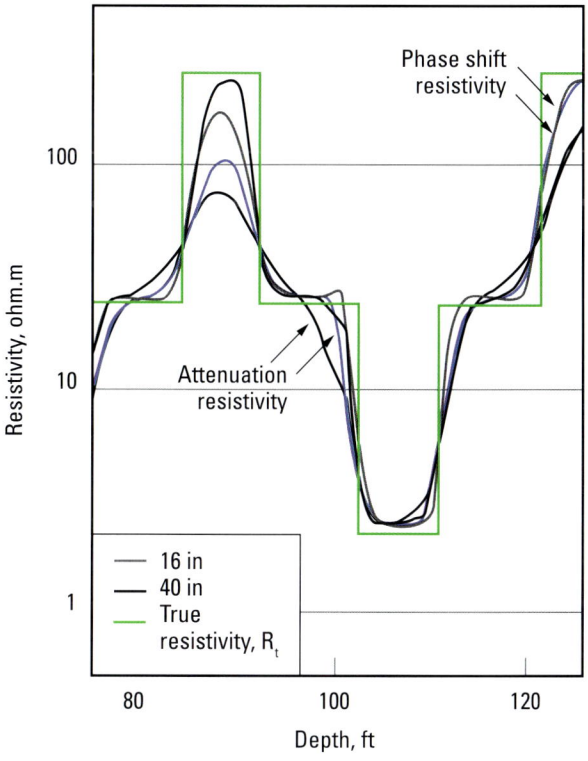

Figure 4-15. Phase and attenuation resistivity responses at 2 MHz to a 4-ft resistive (100-ohm.m) and conductive (1-ohm.m) layer, sandwiched between 10-ohm.m layers.

In summary,

- phase measurements have sharper axial resolutions than the corresponding attenuation measurements

- axial resolution sharpens with increasing transmitter frequency

- axial resolution sharpens with decreasing formation resistivity

- axial resolution sharpens with closer receiver-receiver spacing (this is fixed for a tool)

- axial resolution sharpens slightly with closer transmitter-receiver spacing.

4.3.2. Propagation resistivity environmental effects

The multitransmitter-receiver spacing, multifrequency, phase, and attenuation measurements permit various environmental effects to be identified and, where appropriate, corrected. Where all the measurements read the same formation resistivity, this indicates that the formation is thicker than the axial resolution of the measurements and is homogeneous. Differences among the various resistivity measurements may result from one or more environmental effects.

4.3.2.1. Borehole effect

Conductive mud in the borehole immediately surrounding the tool creates an alternative path for the measure currents induced by the transmitters. The measure currents respond to the borehole and formation resistances in parallel. With increasing contrast between high formation resistivity and low mud resistivity, a greater proportion of the measure current will flow around the tool in the borehole rather than in the formation. Increasing borehole size will also create an easier alternative path and so increase the borehole effect.

The effect may increase or decrease the apparent resistivity response. **Figure 4-16** shows the borehole corrections for the EcoScope phase resistivities as a function of varying borehole diameter, mud resistivity, and formation resistivity. Note that the corrections are not always in the same direction. The magnitude of the correction increases with increasing contrast between the mud and formation resistivities.

Figure 4-16. Borehole corrections for the EcoScope phase resistivities. These corrections vary with borehole size, formation resistivity, and mud resistivity (R_m). The Y axis displays the R_t/R apparent ratio, plotted as a function of true formation resistivity on the X axis. The colored lines correspond to differing mud resistivities in the borehole.

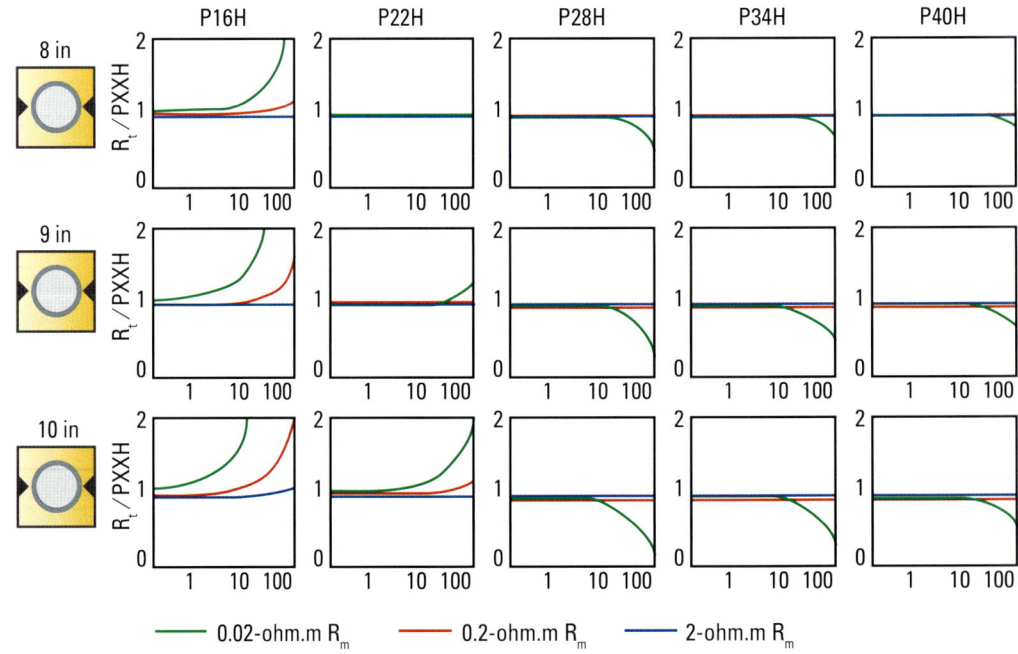

Measurements denoted with a _UNC qualifier (e.g., P40H_UNC) have been borehole-compensated (refer to Section 4.3.1.1) but not borehole-corrected. Borehole-corrected curves do not have the qualifier (e.g., P40H).

In summary,

- a conductive borehole acts as an alternative path for propagation measure currents

- borehole effect may increase or decrease the apparent resistivity response

- borehole-compensated-but-uncorrected resistivities are denoted with a _UNC qualifier

- borehole-corrected resistivities do not have the qualifier (e.g., P40H).

4.3.2.2. Eccentricity effect

Eccentricity effect causes erratic and spiky 2-MHz resistivities when the propagation resistivity tool is run eccentered in a borehole filled with oil-base mud surrounded by low-resistivity formation. As outlined in Section 4.3.1.2, the 2-MHz resistivities are shallower than the 400-kHz. Low formation resistivities result in the induced currents flowing in the formation very close to the tool. If the tool is centered in the borehole, the currents continue to flow in the formation, but if the tool is eccentered, the currents may try to cross the borehole, as shown schematically in **Figure 4-17** (left). If the borehole is filled with nonconductive mud, the current is unable to traverse the borehole, so its path is distorted around the borehole, resulting in erratic resistivity responses as slight changes in the position of the tool in the borehole change the current path.

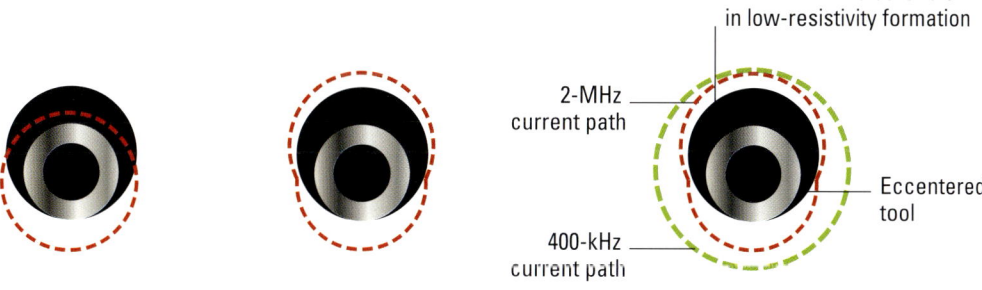

Figure 4-17. Eccentricity effect, occurring when the 2-MHz induced currents try to cross a nonconductive borehole (left), resulting in distortion of the current path (middle), which creates erratic and spiky log responses; the 400-kHz measurement is considerably less sensitive to this effect (right).

Because of their deeper DOIs, the 400-kHz measurements are much less sensitive to the eccentricity effect. In addition, they provide better signal-to-noise response in low-resistivity formations. If eccentricity effect is suspected, then only the 400-kHz resistivities should be used.

In summary, spiky 2-MHz curves may indicate eccentricity effect if

- the propagation tool is not centered in the borehole

- the mud is not conductive (oil-base mud)

- the formation has a resistivity below 2 ohm.m.

If all three are true, the effect may be present, and the 400-kHz resistivities should be used for formation resistivity evaluation.

Blended resistivities combine the best of the 400-kHz responses (at low resistivities) and the 2-MHz responses (at high resistivities) by applying simple threshold logic, as shown in **Figure 4-18**.

Figure 4-18. Blended resistivities, using the 400-kHz resistivities below 1 ohm.m and the 2-MHz resistivities above 2 ohm.m. Between these blending thresholds, a linear combination of the resistivities is used.

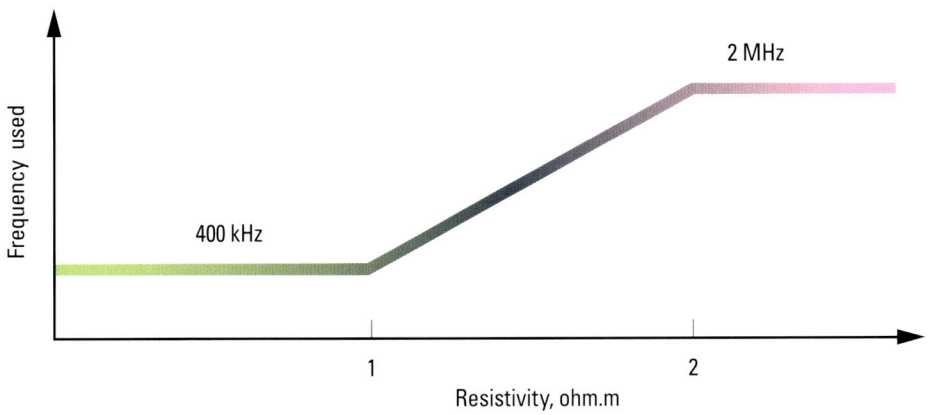

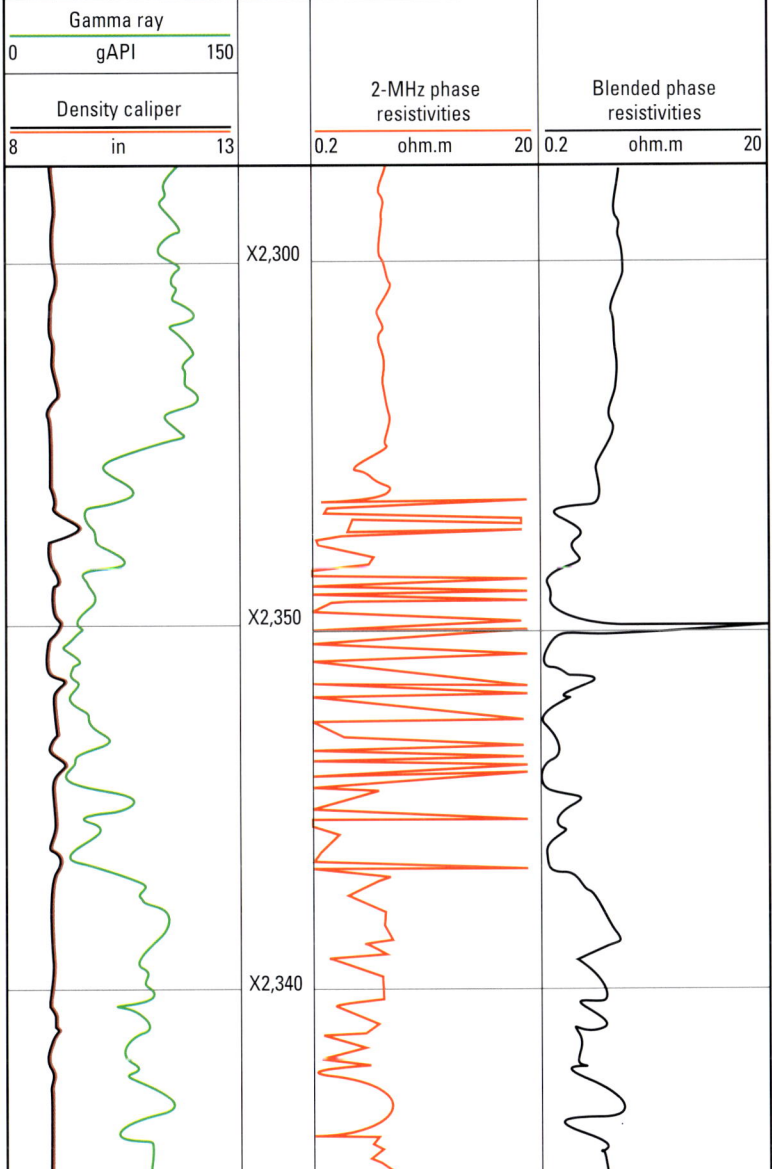

Figure 4-19. Blended resistivities (right), combining the 400-kHz response at low resistivities with the 2-MHz response at high resistivities. This combination eliminates the spikes due to the eccentricity effect seen on the 2-MHz responses (middle track) at low resistivities. The single remaining spike on the blended resistivities suggests that the blending thresholds may need adjustment.

4.3.2.3. Invasion effect

Invasion of mud filtrate into the formation near the wellbore generally results in a change in the resistivity around the well. If the invaded-zone resistivity, R_{xo}, is greater than the uninvaded, or true, formation resistivity, R_t, it is called resistive invasion. ($R_{xo} > R_t$). If invasion reduces the formation resistivity around the wellbore, it is called conductive invasion. ($R_{xo} < R_t$).

Evaluation of the invasion profile is one of the primary reasons for the development of multiple DOI resistivity tools. A greater proportion of the shallow measurement response comes from the invaded zone than for deeper measurements. Hence, conductive invasion will cause a spread of resistivities with the shallow measurements reading lower resistivities than the deep measurements. This creates a conductive invasion resistivity profile, an example of which is shown in **Figure 4-20** (left). If the invasion is very local around the borehole, the separation on the phase resistivities may be significant, with the separation on the deeper attenuation measurements relatively small. Deep invasion will result in significant separation in both the phase and attenuation measurements.

Resistive invasion is characterized by the reverse order of the resistivities. In resistive invasion, the shallowest resistivity reads highest, and the deepest resistivity reads lowest, an example of which is shown in Figure 4-20 (right).

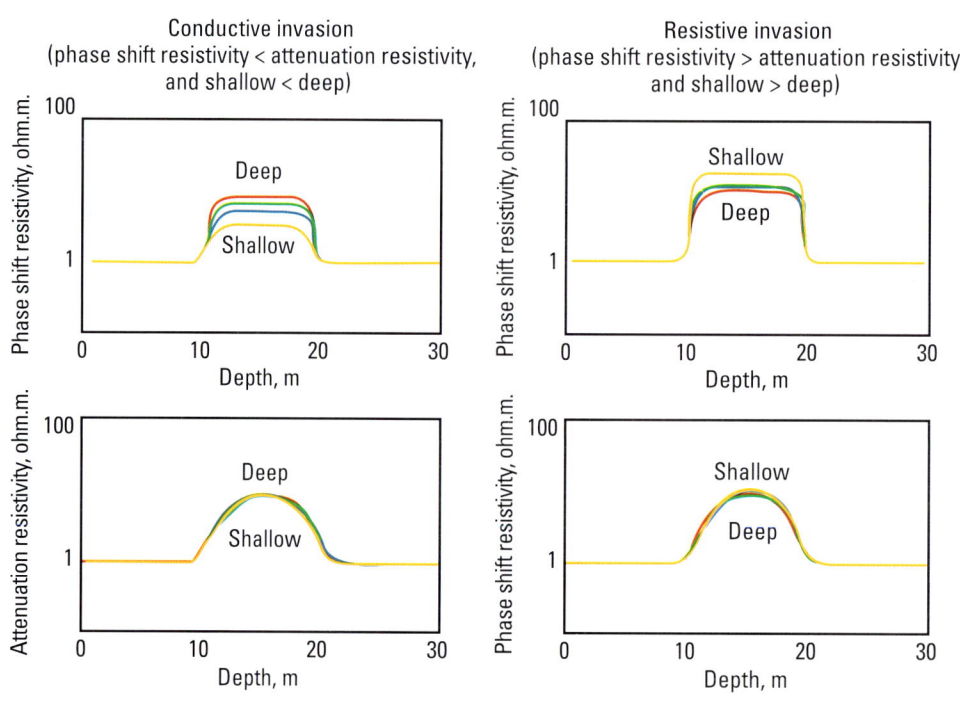

Figure 4-20. Left: A conductive invasion profile, with the shallowest resistivity reading lowest and the deepest resistivity reading highest. Right: A resistive invasion profile, with the shallowest resistivity reading highest and the deepest resistivity reading lowest.

Multiple resistivities are required to solve an invasion model so that R_t can be determined. The simplest invasion model is the piston invasion profile shown in **Figure 4-21** (left). A minimum of three measured resistivities is required to solve for the three unknowns, R_t, R_{xo}, and the invasion radius, r_i.

Propagation resistivity tools induce current loops that circulate in the formation around the tool. Hence, the resistivity they provide is an average from around the borehole. When solving for a piston invasion profile with a propagation tool, it is assumed that r_i, R_{xo}, and R_t are the same in all directions around the borehole (axisymmetric).

The availability of more than three resistivity measurements allows more complex invasion profile models to be solved. The ramp profile shown in Figure 4-21 (right) has four variables (R_t, R_{xo}, r_{i1}, and r_{i2}) and thus requires four resistivity measurements to be able to solve.

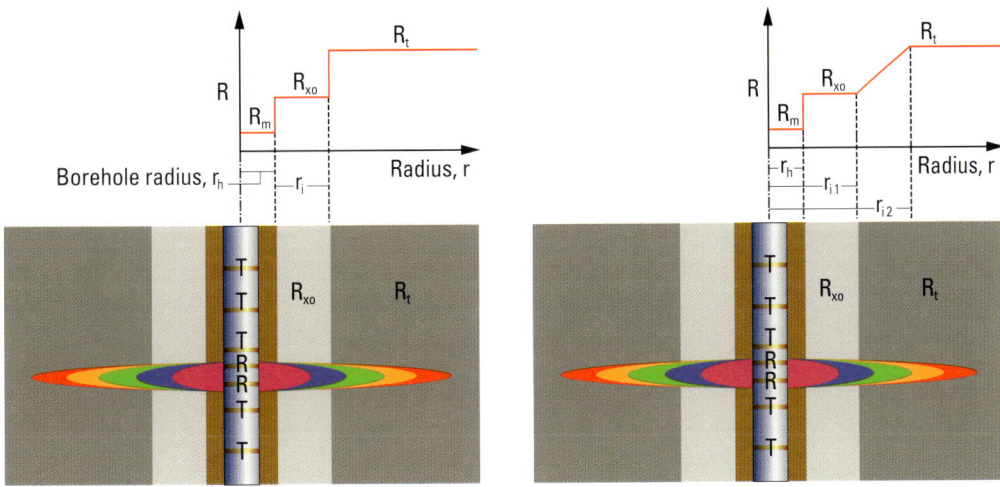

Figure 4-21. Left: The piston, or step invasion, profile. This simplest invasion model can be determined with three resistivity inputs to solve for the three unknowns (R_t, R_{xo}, and r_i). Right: The more sophisticated ramp invasion profile, which requires four resistivity inputs to solve for four unknowns (R_t, R_{xo}, r_{i1}, and r_{i2}). Both examples show conductive invasion, and the approach is also valid for resistive invasion.

The piston invasion profile is the most common model used to interpret resistivity separations. **Figure 4-22** shows modeled responses for three phase and three attenuation resistivities in a piston invasion profile. A uniform 4-in invaded zone around the borehole with an R_{xo} (green line) of 1 ohm.m is modeled in a formation where R_t (pink line) varies from 100 to 0.3 ohm.m. Where R_t is greater than 1 ohm.m, a conductive invasion profile is created. Where R_t is less than 1 ohm.m, a resistive invasion profile is created.

Figure 4-22. Propagation resistivity, responding to conductive and resistive invasion.

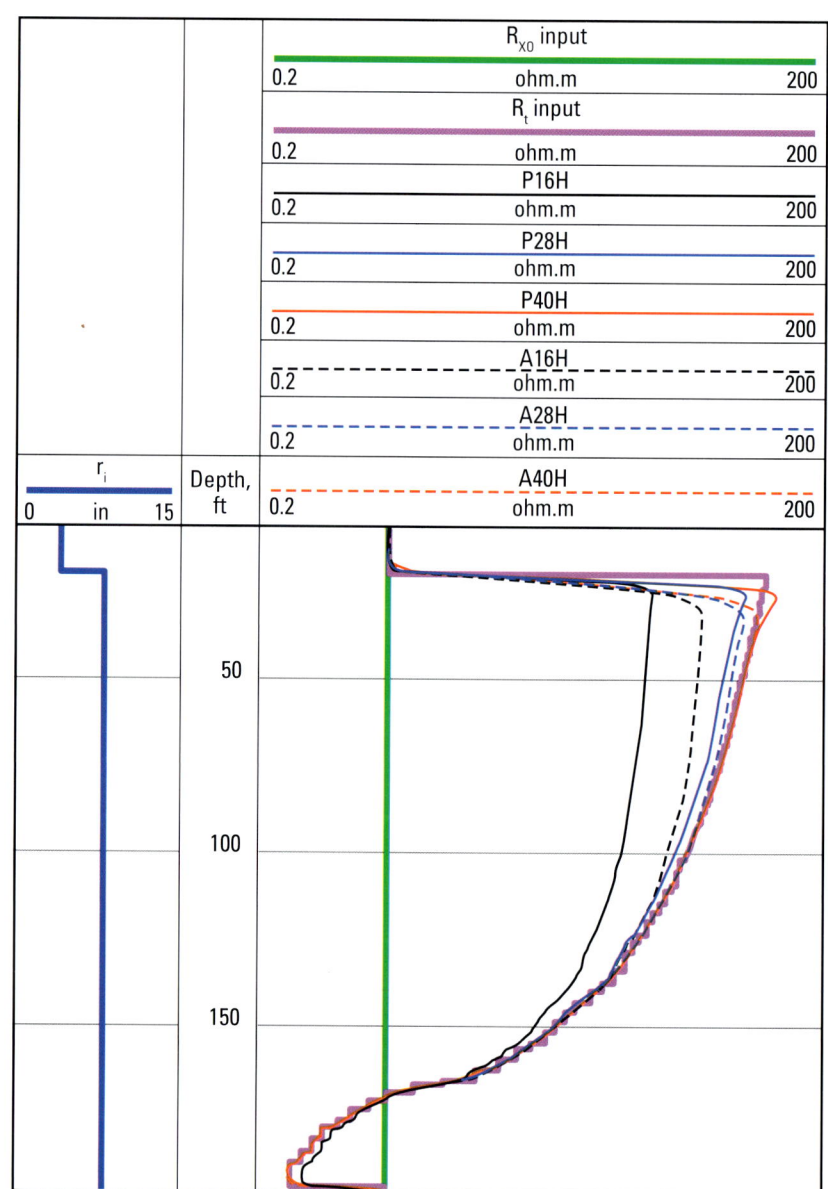

In the conductive invasion profile, the shallower phase and attenuation resistivities read lower than the deeper measurements because the invasion has a greater effect on them. In the resistive profile, the reverse separation is observed. Although r_i remains constant, the resistivity separation increases as the contrast between R_{xo} and R_t increases.

In summary,

- with conductive invasion, shallow phase resistivities decrease (if the invasion is deep or the resistivity contrast high, the shallow attenuation resistivities may also decrease)

- with resistive invasion, shallow phase resistivities increase (if the invasion is deep or the resistivity contrast high, the shallow attenuation resistivities may also increase).

4.3.2.4. Boundary effect

Boundary effect, also known as shoulder effect, on propagation resistivity measurements is controlled by

- resistivity contrast between the layers

- incidence angle between the borehole and layer interface.

The discussion of axial resolution effects in Section 4.3.1.3 considers the effect of thin layers with differing resistivities on the propagation resistivity response when the tool is perpendicular to the layers. As shown in **Figure 4-23**, boundary effects occur when the tool and layering are not perpendicular, so the induced measurement currents circulating around the tool are forced to cross the resistivity contrast rather than run parallel to it. Proximity effects occur where the tool is not crossing the resistivity boundary, but the deeper measurement currents are responding to both the local and proximate layer.

Figure 4-23. Relation of axial resolution, bed boundary, and proximity effects. Axial resolution effects occur when averaging more than one layer within the axial "window" of the measurement, and proximity effects occur when averaging more than one layer within the radial DOI of the measurements. Bed boundary effect is a combination of axial and radial effects as the volumes of investigation of the propagation measurements cross a change in resistivity.

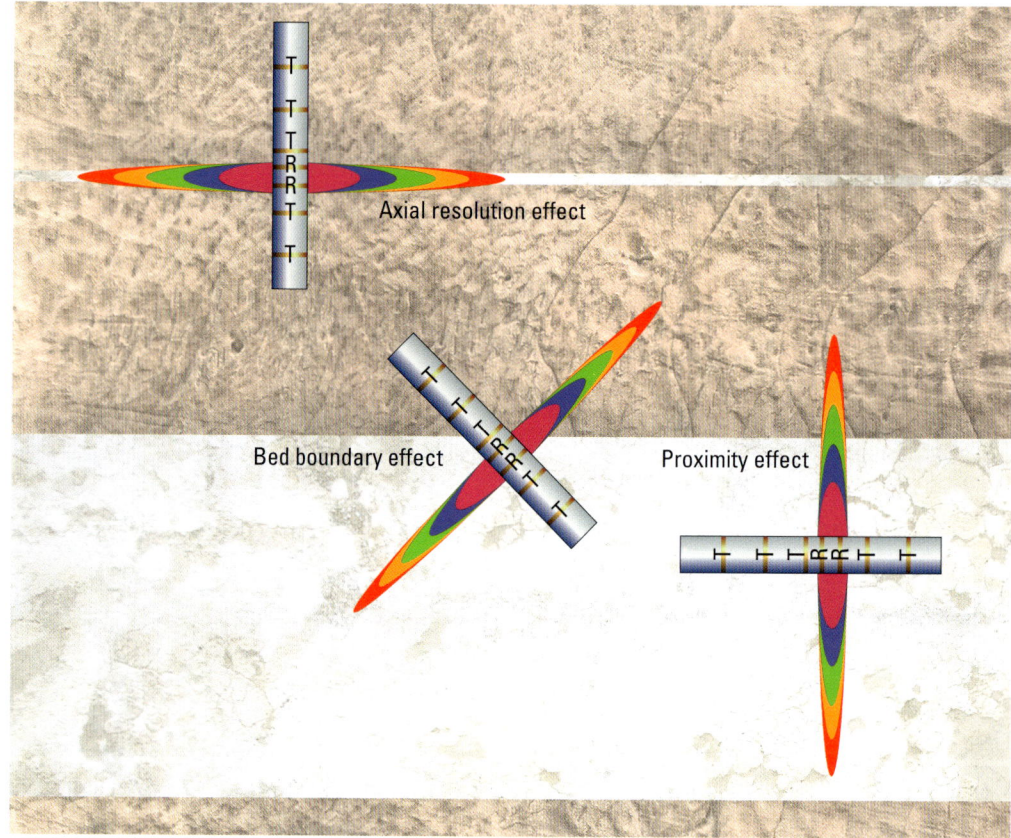

When the propagation tool and formation layering are perpendicular, the induced currents distribute themselves across the layers in inverse proportion to the resistivity. In other words, they respond to the volumetric parallel resistivity of the two layers, as shown in **Figure 4-24**.

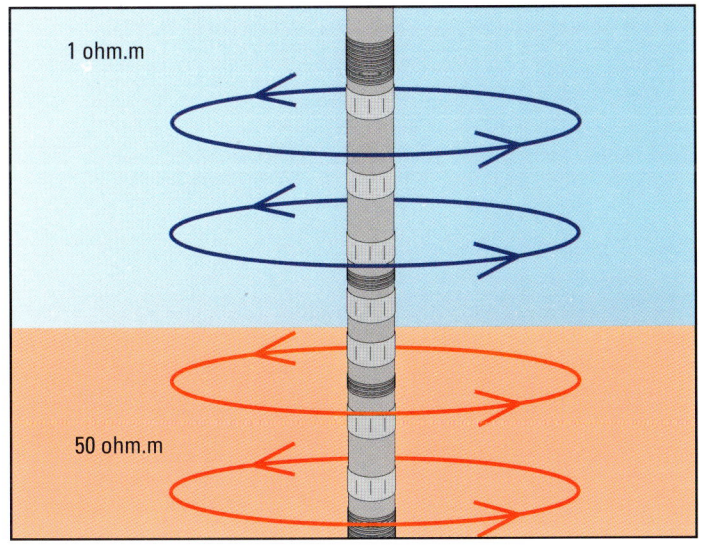

Figure 4-24. Measure currents parallel to formation resistivity boundaries.

In a well drilled perpendicular to the layering, the propogation resistivity response is the volumetric parallel sum of the layer resistivities:

$$\frac{1}{R_{measurement}} = \frac{V_{upper}}{R_{upper}} + \frac{V_{lower}}{R_{lower}}$$

(Equation 4-2),

where

$R_{measurement}$ = measured propagation resistivity
V_{upper} = volumetric proportion of the upper layer in the volume of measurement
R_{upper} = resistivity of the upper layer in the volume of measurement
V_{lower} = volumetric proportion of the lower layer in the volume of measurement
R_{lower} = resistivity of the lower layer in the volume of measurement.

Thus,

$$R_{measurement} = \frac{1}{\dfrac{0.5}{1} + \dfrac{0.5}{50}} = 1.96 \text{ ohm.m.}$$

When a propagation resistivity tool approaches a change in formation resistivity with a low incidence angle between the tool and layering, the measure currents are forced to cross both layers, so the resistivity response is the series sum of the resistivities, as shown in **Figure 4-25**.

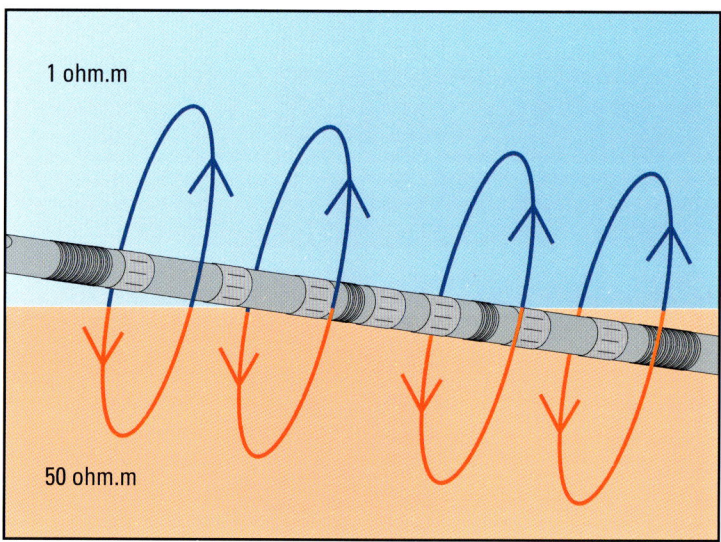

Figure 4-25. Measure currents forced to cross formation resistivity boundaries.

1 ohm.m

50 ohm.m

In a well drilled parallel to the layering, the propagation resistivity response is the volumetric series sum of the layer resistivities:

$$R_{measurement} = V_{upper} \times R_{upper} + V_{lower} \times R_{lower}$$ (Equation 4-3).

Thus,

$$R_{measurement} = 0.5 \times 1 + 0.5 \times 50 = 25.5 \text{ ohm.m}.$$

There is a significant difference in the resistivity response to the same layers as a function of the incidence angle between the borehole and layering.

For propagation resistivities, the measurement response is further complicated by polarization horns due to charge buildup at the resistivity interface. As shown in **Figure 4-26**, the current induced in the formation by the propagation resistivity tool must traverse both the upper and lower layers. Kirchoff's law states that the sum of the voltages around a circuit must be 0. Given that the current is the same in the two layers, but the resistance seen by the current is different, the voltage drop in the two layers will differ. To balance this difference, charges accumulate at the boundaries. For a 2-MHz measurement, these charges are oscillating two million times per second with a slight delay to the measure current, resulting in interference with the measurement signal. This interference reduces both the measured phase shift and attenuation, which results in an increase in the computed resistivities (refer to Figure 4-9). The apparent resistivity increase as the tool approaches and crosses the resistivity contrast boundary is called a polarization horn.

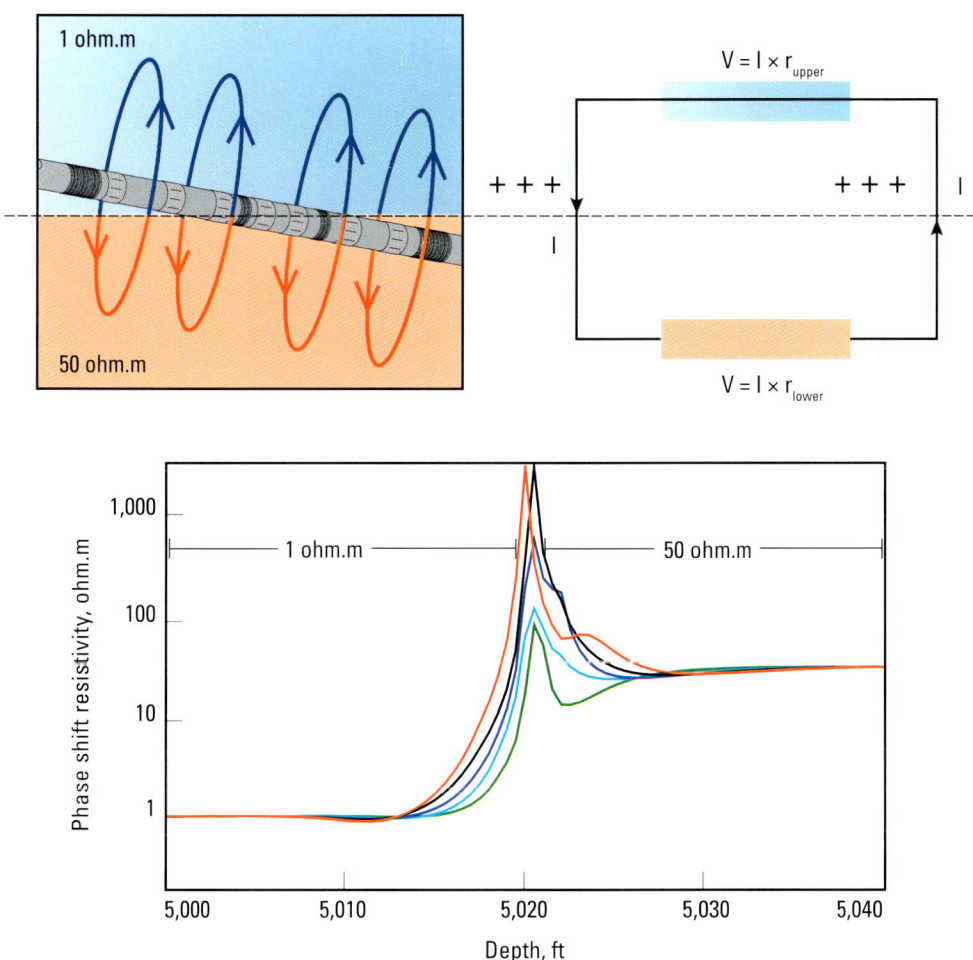

Figure 4-26. Polarization horns, caused by charge buildup at the resistivity interface.

r_{upper} = resistance experienced by current in upper layer
r_{lower} = resistance experienced by current in lower layer

The magnitude of the polarization horn depends on the resistivity contrast between the layers and the incidence angle between the tool and layers. As shown in **Figure 4-27**, assuming horizontal layering, the polarization horn magnitude increases to a maximum as the well inclination approaches 90° (parallel to the layers).

Figure 4-27. Polarization effect on phase (upper) and attenuation (middle) resistivities, which increases as the tool and layering come closer to being parallel—the angles shown are wellbore inclinations assuming that the layers are horizontal. Polarization effect on phase resistivities (lower) is more pronounced than that observed on attenuation resistivities.

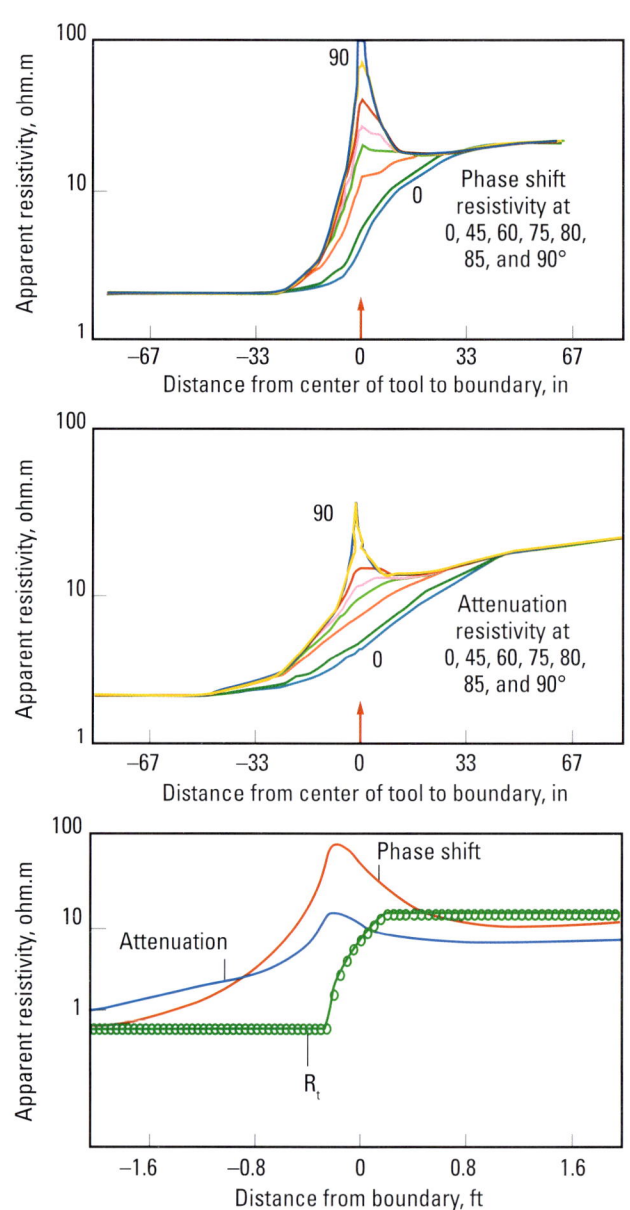

The polarization effect is greater on phase resistivities than on attenuation resistivities and greater on the deeper measurements than the shallower. This difference results in a spread of the responses from shallow-reading-low to deep-reading-high, with the phase resistivities reading higher than the attenuation resistivities. Note that this is distinct from a conductive invasion profile where the shallow measurements also read lower than the deeper, but the phase resistivities read lower than the attenuation resistivities.

In summary, bed boundary effect results in resistivity curve separation due to the measure current crossing more than one layer. Polarization horns may form if

- the incidence angle between the tool and layering forces current to cross a resistivity contrast (approximately 45°).

- the resistivity contrast is sufficient to cause significant charge buildup on the interface.

4.3.2.5. Anisotropy effect

Resistivity anisotropy is the phenomenon where the measured formation resistivity varies depending on the direction it is measured. It is common where formations are thinly layered, such as that shown in **Figure 4-28**. The resistivity in layered formations is sometimes called transverse isotropic because the resistivity does not vary when measuring in any direction parallel to the layering. Only when measuring across layers does the anisotropy become apparent.

Figure 4-28. Resistivity anisotropy, potentially caused by alternating formation layers with differing resistivities.

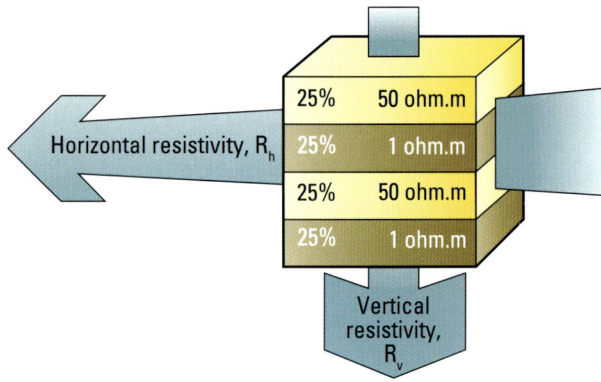

In the simple bimodal layer model shown in Figure 4-28, R_h is the volumetric parallel sum of the layer resistivities:

$$\frac{1}{R_h} = \frac{V_1}{R_1} + \frac{V_2}{R_2}$$

(Equation 4-4),

where

V_1 = volumetric proportion of the lower-resistivity layer in the volume of measurement
V_2 = volumetric proportion of the higher-resistivity layer in the volume of measurement
R_1 = isotropic resistivity value of the lower-resistivity layer.
R_2 = isotropic resistivity value of the higher-resistivity layer.

Thus,

$$R_h = \frac{1}{\dfrac{(0.25+0.25)}{1} + \dfrac{(0.25+0.25)}{50}} = 1.96 \text{ ohm.m.}$$

R_v is the volumetric series sum

$$R_v = V_1 \times R_1 + V_2 \times R_2 \qquad \text{(Equation 4-5).}$$

Thus,

$$R_v = (0.25 + 0.25) \times 1 + (0.25 + 0.25) \times 50 = 25.5 \text{ ohm.m.}$$

Note that the terms horizontal resistivity and vertical resistivity are somewhat misleading. The quantities of interest are the resistivities measured parallel and perpendicular to the layering. The horizontal and parallel terminology is based on the assumption that the layering is horizontal, which may not be the case. To be more accurate, horizontal resistivity should be called parallel resistivity, and vertical resistivity should be called perpendicular resistivity.

The propagation resistivity response to multiple thin layers can be considered the sum of numerous shoulder bed effects. Rather than a single polarization horn associated with traversing a single resistivity change, anisotropy causes the phase resistivity to read continuously higher than the attenuation resistivity. As with a polarization horn at a single boundary, there must be sufficient resistivity contrast between the layers, and the angle between the borehole and layering must be sufficiently small to force the measure currents to traverse multiple layers, as shown in **Figure 4-29** (right).

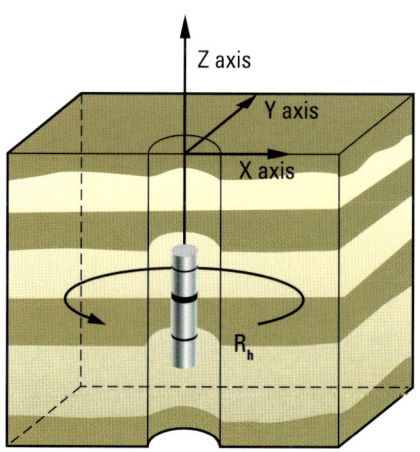

Perpendicular to layers;
propagation tools measure R_h

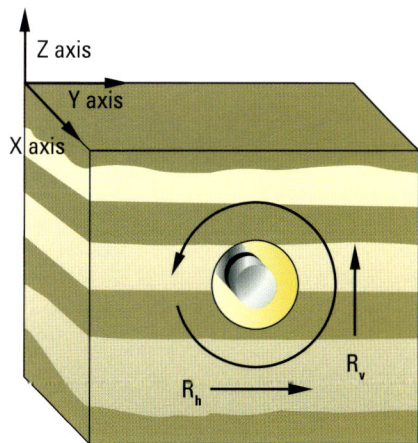

Parallel to layers;
propagation tools measure a combination
of R_h and R_v

Figure 4-29. Left: Propagation resistivity tool, oriented perpendicular to formation layering. The induced current loops circulate parallel to the layers providing a measurement of the horizontal resistivity. Right: A tool close to parallel with the layering. The induced current loops traverse the layers and run parallel to them, providing a measurement sensitive to a combination of horizontal and vertical resistivities.

The separation between the phase and attenuation resistivities increases with increasing resistivity contrast between the formation layers (the anisotropy ratio) and with decreasing incidence angle between the borehole and layering (**Figure 4-30**).

Figure 4-30. Left: Resistivity log, showing the anisotropy signature of phase resistivities increasing from shallowest to deepest and reading higher than attenuation resistivity. R_h and R_v, computed from the data, are also presented. Right: Phase shift and attenuation resistivity response as a function of incidence angle. As with polarization horns, sensitivity to anisotropy increases as the incidence angle between the borehole and layering decreases.

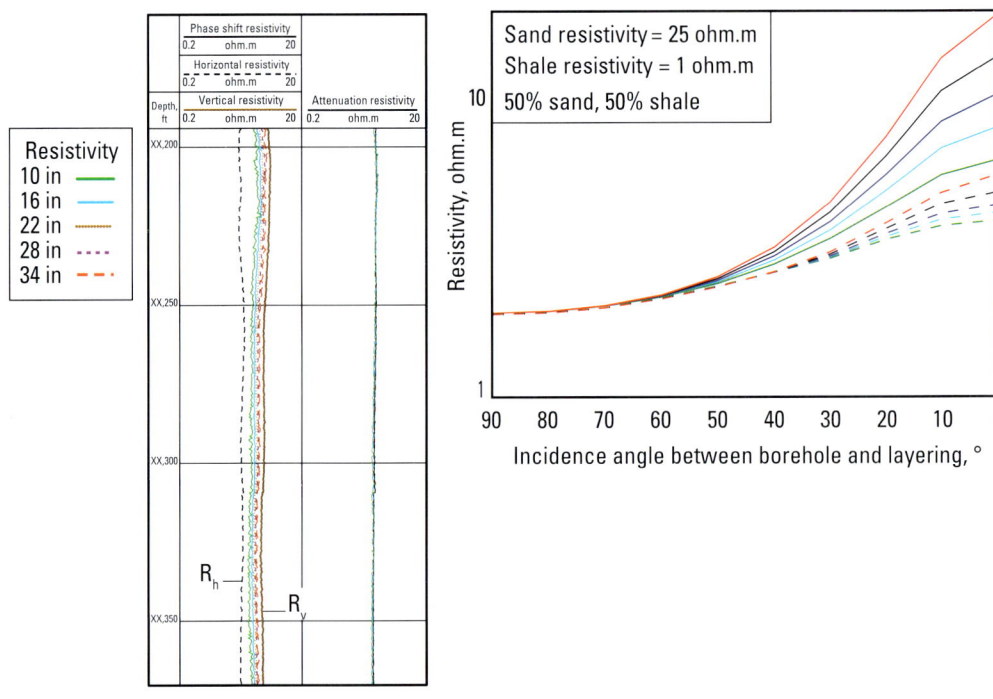

Shown in **Figure 4-31**, resistivity anisotropy causes the phase resistivities to increase from shallowest to deepest and read higher than the attenuation resistivities.

Anisotropy
(phase shift resistivity > attenuation resistivity, and shallow < deep)

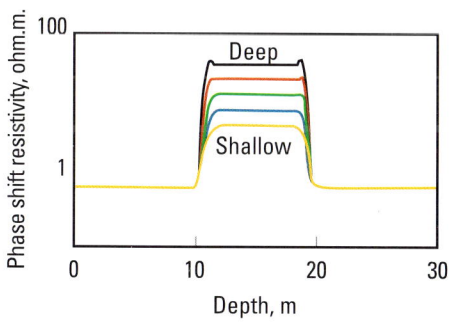
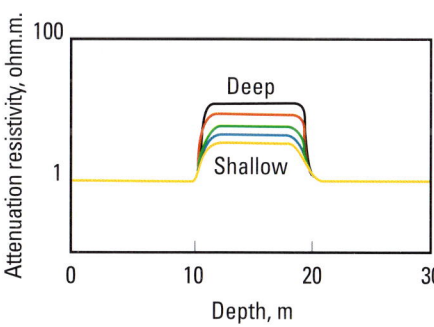

Figure 4-31. Resistivity anisotropy, causing the phase resistivities to increase from shallowest to deepest and read higher than the attenuation resistivities.

In summary,

- the tool must be at a sufficiently low incidence angle to the layering (less than approximately 45°) to force induced currents to cut through multiple layers in order to be sensitive to the anisotropy

- anisotropy results in a log response similar to a polarization horn extended along the borehole

- both phase and attenuation resistivities increase from shallowest to deepest, but the phase resistivities read higher than the corresponding attenuation resistivities.

4.3.2.6. Proximity effect

Proximity effects occur when the deeper propagation measurements respond to more than one layer, as shown schematically in Figure 4-23 (right). The proximity of a conductive layer is particularly apparent because the propagation measurement depth of investigation increases with increasing local formation resistivity, and the induced current seeks the path of least resistance. A conductive layer gives the induced current an alternate path with lower resistance.

In the example shown in **Figure 4-32**, the well crosses from a 100-ohm.m layer into a 1-ohm.m layer. The phase resistivities are unaffected by the proximity of the low-resistivity layer until the tool is very close to the layer. The deeper attenuation measurements begin decreasing from the local layer resistivity of 100 ohm.m while the lower layer is still 10 ft of true vertical depth (TVD) away. As the well approaches the 1-ohm.m layer, the attenuation resistivities decrease further as more of their volume of investigation falls within the low-resistivity layer. The high resistivity contrast and low incidence angle between the borehole and layering produce a large polarization horn.

Figure 4-32. Proximity effect on propagation resistivities. The proximity of a low-resistivity layer reduces the deeper attenuation resistivities (bottom) before the shallower phase resistivities respond (middle) as the well (green line top) traverses from a high-resistivity (brown) to a low-resistivity layer (blue).

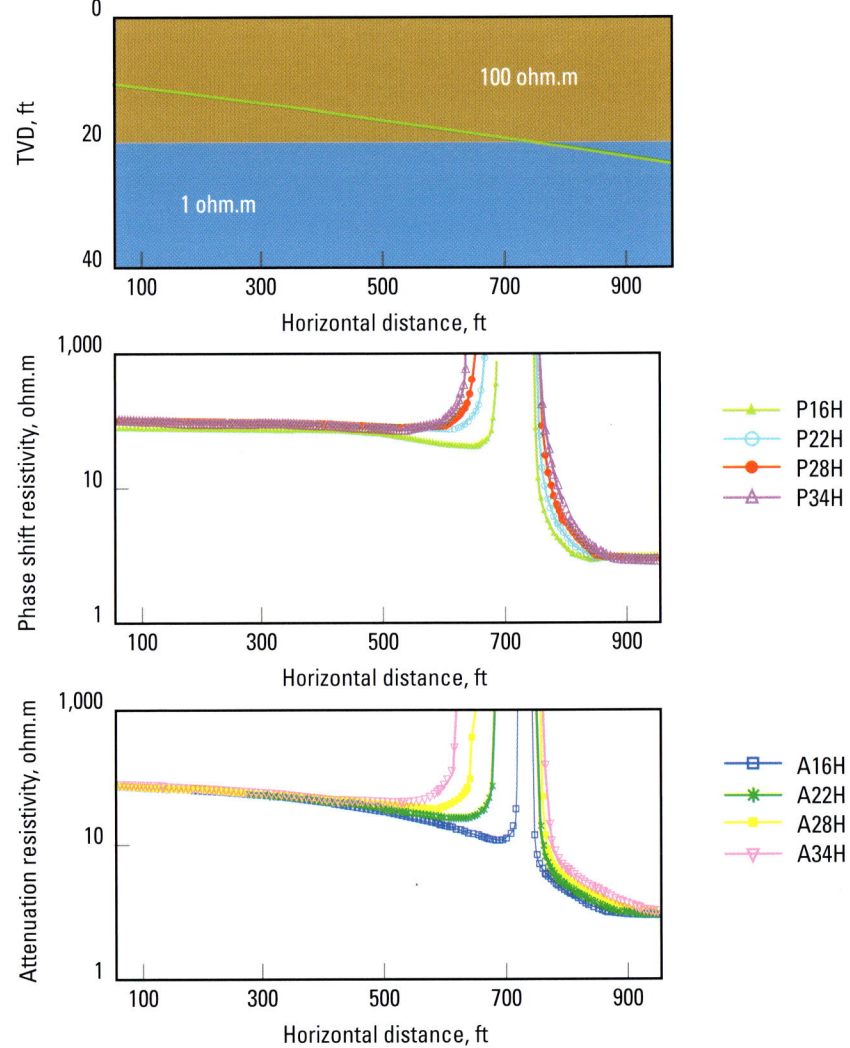

As the well enters the 1-ohm.m layer, the phase resistivities rapidly drop to 1 ohm.m. The attenuation resistivities remain higher as these deeper measurements continue to have a proportion of their response from the 100-ohm.m layer above. When the well is approximately 3-ft-TVD from the high-resistivity layer, all the phase and attenuation measurements read the local resistivity of 1 ohm.m. This rapid return to 1 ohm.m. is because the DOI of the propagation measurement decreases as the local resistivity decreases and because the high-resistivity layer above does not offer a path of lower resistance to the measure currents, so they remain in the 1-ohm.m layer.

In summary, responses on the deep measurements not seen on the shallower measurements may occur because of proximity effects from a nearby layer.

4.3.2.7. Dielectric effect

At propagation measurement frequencies, the total impedance of a rock has a direct current resistance component and a frequency-dependent component that is a function of the formation dielectric constant. Common practice is to transform attenuation and phase shift independently to resistivity, assuming a certain transform between the relative dielectric constant and resistivity. This relationship, shown as the dielectric assumption in **Figure 4-33**, loses accuracy at high resistivities. By using the measured phase shifts and attenuations, the formation resistivity and dielectric constant can be solved for simultaneously without requiring an assumed relationship between them, as shown graphically in Figure 4-33. The resistivity determined in this manner extends the range of measurement, typically up to 3,000 ohm.m.

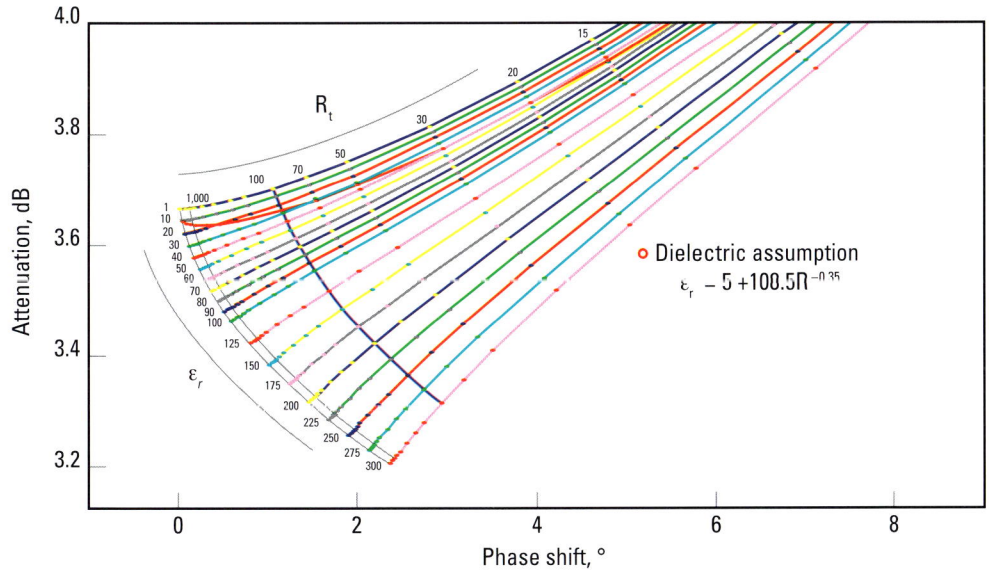

Figure 4-33. Formation resistivity and dielectric constant, which can be determined using both the measured attenuation and phase shift as inputs rather than depending on the dielectric assumption. This chart is for the 2-MHz response at 40-in transmitter-receiver spacing on a 6.75-in-diameter tool.

The attenuation measurement is more sensitive than the phase resistivities to the formation dielectric constant being higher than calculated from the empirical dielectric assumption. This sensitivity tends to occur in high-resistivity formations such as volcanic rocks. In these cases, the shallow attenuation measurements show a higher resistivity than the deeper attenuation measurements. The phase resistivities show little or no separation, as shown in **Figure 4-34**.

Figure 4-34. Dielectric effect, causing the shallow attenuation resistivities to read higher than the deep attenuation resistivities. The phase resistivities show little or no separation.

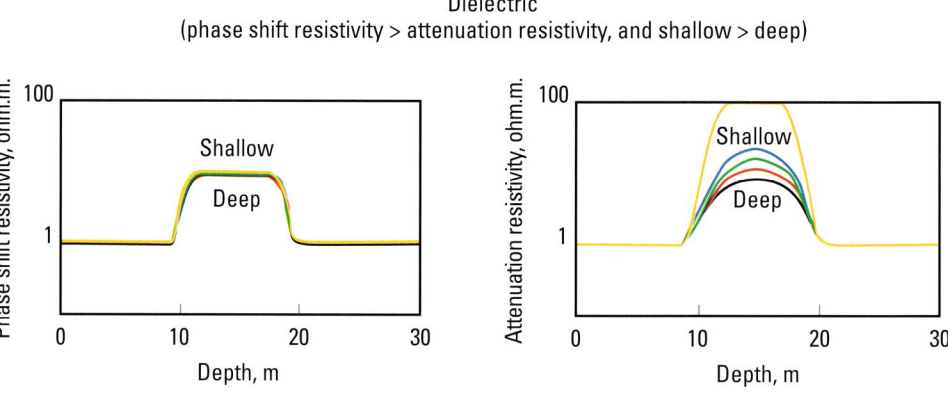

In summary, unusually high shallow attenuation response in high-resistivity formations may be caused by dielectric effect.

4.3.3. ARCWizard processing

As outlined in Section 4.3.2, there are several environmental effects that may cause separations on the propagation resistivity responses. While there may be more than one effect present at any one time, often there is one dominant effect. ARCWizard processing has been developed to identify the dominant effect and, where appropriate, deliver a formation resistivity compensated for the effect.

ARCWizard processing (**Figure 4-35**) checks the consistency of the raw phase shift and attenuation measurements to quality check (QC) the transmitter and receiver responses. If a repairable anomaly, such as a single transmitter problem is identified, a flag is raised requesting permission to rectify the problem. If an unrepairable anomaly, such as a receiver or multiple transmitter failure, is identified, a flag is set indicating that the resistivities cannot be repaired.

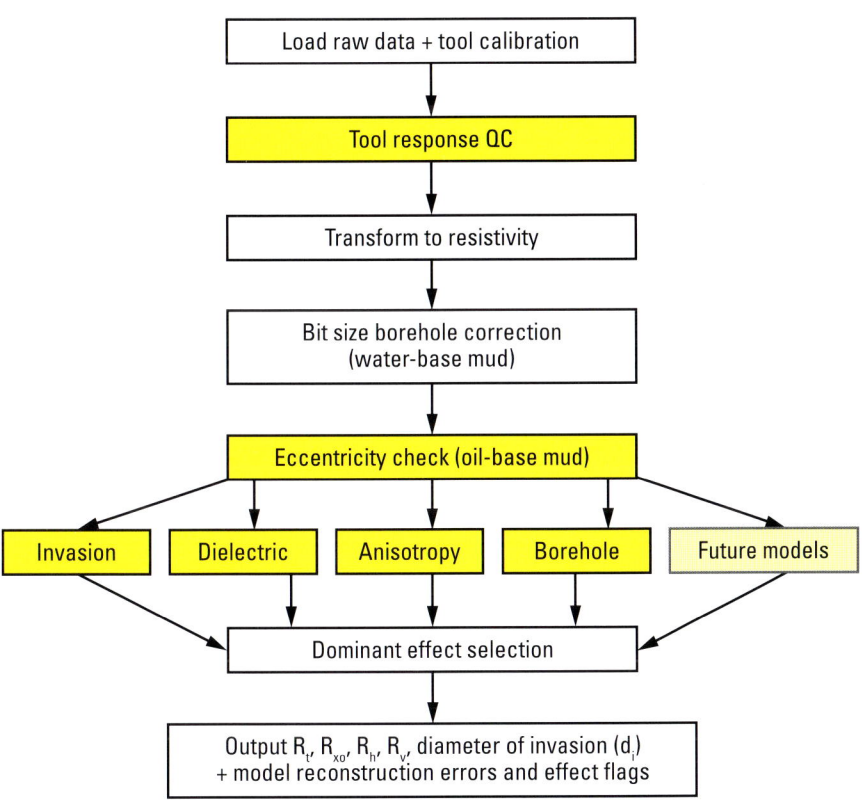

Figure 4-35. ARCWizard processing, which solves for each of the major effects that cause resistivity separation, selects the dominant effect, and delivers formation resistivity compensated for the effect (a more detailed ARCWizard processing flowchart is provided in Figure 6-5).

The quality-controlled resistivities are borehole-corrected and checked for eccentricity effect. If this effect is identified, the 2-MHz responses are ignored, and the 400-kHz measurements are provided as representative of the formation resistivity. No further processing is applied to the 400-kHz measurements.

Where no eccentricity effect is identified, the 2-MHz resistivities are inverted in parallel through single effect models that account for anisotropy, borehole enlargement, dielectric and invasion effects. In each case, the output is a model-fitting error indicating how well the model was able to explain the resistivity response separations and the formation resistivities corrected for the effect.

Outputs include

- anisotropy—vertical and horizontal resistivities
- borehole enlargement—hole diameter and corrected formation resistivity
- dielectric effect—true resistivity and dielectric constant
- invasion effect—true resistivity, invaded-zone resistivity, and invasion diameter.

The dominant effect is selected based on the model-fitting error of all of the models. Where none of the models is able to create a good match with the measured data, a multieffect flag is raised, indicating that the processing is unable to identify a single dominant effect. This indication may be because of the presence of multiple effects at the same depth or because of an effect for which a numerical model is not currently implemented, such as shoulder bed effect. When multiple effects are detected, R_t is set equal to one of the measured propagation resistivity curves as selected by the user.

Where one model shows low error in reconstructing the measured resistivities and all other models show large errors (a poor match to the measured data), the interval is flagged with high confidence as being influenced by the selected model. The ability of one model to explain the resistivity separation, combined with the poor match from the other models, gives confidence that the separation is because of a single dominant effect.

Where all the models show an ability to explain the resistivity response separation, the interval is flagged as a uniform formation, which generally occurs where the resistivity curves overlay. In this case, each of the models will be able to explain the response, as there is a trivial case for each model. For example, an isotropic formation—in which $R_v = R_h$—and an invaded formation—in which $R_{xo} = R_t$—will both give a good match with the measured data. In a uniform formation, $R_t = R_{xo} = R_h = R_v$.

The left panel of **Figure 4-36** shows the measured resistivity responses in a 60°-inclined, 8.5-in borehole drilled with oil-base mud. Resistivity curve separation can be seen on the 2-MHz responses in both the upper and lower intervals. The 400-kHz responses show separation only in the lower interval. In the lower interval, the phase resistivities read higher than the corresponding attenuation resistivities (which are presented on a scale shifted by one decade).

The right panel shows the results of ARCWizard processing. The flag track displays colored flags identifying the selected dominant effect. The width of the shaded area indicates the confidence in the selected effect. In this example, the eccentricity and anisotropy effects have been selected with high confidence while the interval showing uniform formation has lower confidence. The flag track also displays the reconstruction error associated with each model. The vertical gray line in the flag track indicates the selection threshold. Where all the reconstruction errors fall to the right of the selection threshold, the formation is flagged as uniform.

In the lower interval, only the anisotropy flag falls to the right of the selection threshold while all other models show reconstruction error greater than the threshold. Consequently, the anisotropy model is selected, with high confidence that it is the dominant effect.

The formation resistivity track displays the effect-corrected resistivities. Where anisotropy is indicated, R_t and R_{xo} are set equal to R_h. This allocation is arbitrary because R_t and R_{xo} have no meaning in an uninvaded anisotropic formation. Similarly, in an invaded formation, R_h and R_v values are set equal to R_t because they have no meaning in an invaded isotropic formation.

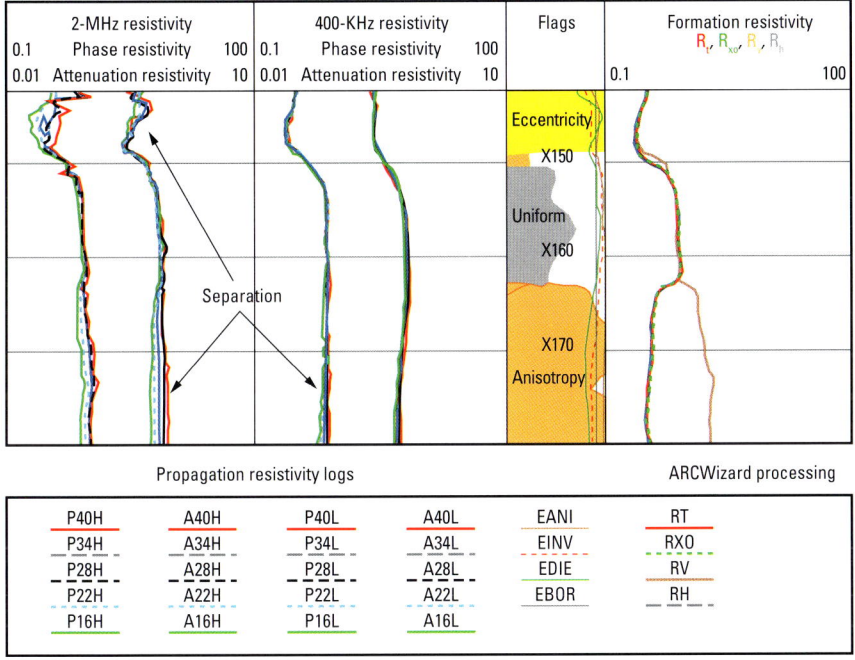

Figure 4-36. ARCWizard processing in 60°-inclined, 8.5-in borehole drilled with oil-base mud. The processing identifies eccentricity effect as the cause of the resistivity curve separation in the upper interval and resistivity anisotropy as the cause of the resistivity separation in the lower interval. It identifies the uniform formation of the section where the resistivity curves overlay.

In cases where ARCWizard processing is unable to distinguish a single dominant effect, more sophisticated modeling may be necessary to identify the influences on the measured log responses. Such modeling currently involves manual trial-and-error processing using sophisticated electromagnetic forward modeling software that requires expert user supervision.

4.3.4. Propagation resistivity applications

Propagation resistivity measurements are generally acquired for determination of the uninvaded formation resistivity, which is subsequently used for formation fluid saturation calculations. Multiple DOI resistivities allow a range of environmental effects to be detected, quantified, and corrected, including

- borehole effect

- tool eccentering effect

- invasion—conductive and resistive

- boundary effect

- anisotropy

- dielectric effect.

Resistivity measurements are also used for correlation and well placement applications. As the deepest of the triple-combo measurements, propagation resistivity often provides the earliest indication of a high-angle well approaching another layer.

4.3.5. Propagation resistivity log quality control

The shop calibration must have been performed less than 3 months before the acquisition date and be in tolerance. Refer to the Schlumberger LWD Quality Control Manual for details.

One of the most powerful QCs is to visually verify that the resistivity responses either overlay or separate in a coherent manner. The EcoScope hardware incorporates checks to verify the electrical integrity of the transmitter loops and the correct performance of the acquisition electronics. The LQC logic is outlined in **Table 4-9**.

Table 4-9. Propagation resistivity QC flag logic.

QC_RES	Hardware	Remarks/User Actions
Green	Resistivity status word, RESSTAT = 0	P34H ≤ 3,000 ohm.m and P34L ≥ 0.1 ohm.m and A34L ≥ 0.1 ohm.m and 0.5 < (corrected P34H / P34H_UNC) < 1.5
Yellow	Hardware error indicated by one of the following hardware status words being set Memory error or CAN bus receiver buffer overflow (lost data) or 2-MHz or 400-kHz synthesizer error or Error in communications with the resistivity analog-digital converter or Error in communications with the slow channel analog-digital converter or 3.3-V supply out of tolerance or 20-V supply out of tolerance or Configuration parameters not received	3,000 ohm.m < P34H < 5,000 ohm.m or P34L < 0.1 ohm.m or A34L < 0.1 ohm.m or (Corrected P34H / P34H_UNC) ≥ 1.5 or (Corrected P34H / P34H_UNC) ≤ 0.5
Red	Board error or Memory usage indicator = full	P34H ≥ 5,000 ohm.m
White		No resistivity ticks for more than 2 ft

4.3.6. Propagation resistivity channels and parameters

4.3.6.1. Real time

Table 4-10. Propagation resistivity real-time parameters.

Parameter Name	Description	Unit
RES_ENVCOR_OPT_RT	Enabled resistivity environmental correction	Yes/no
GRAD	Real-time formation temperature gradient	degF/ft
RTST	Real-time surface temperature	degF
T_IN	Mud temperature at surface	degF
MRIN	Mud resistivity at surface	ohm.m
EcoScope tool OD	Tool OD from current BHA	in
BS_RT	Bit size from current BHA	in

Table 4-11. Propagation resistivity real-time transmitted channels.

Transmitted Channel	Description
A40[†]B_DH_ECO_RT	Array resistivity-compensated (ARC) blended attenuation resistivity, 16-in spacing, real time, computed downhole
A40[†]H_UNC_DH_ECO_RT	ARC uncorrected attenuation resistivity, 16-in spacing at 2 MHz, real time, computed downhole
A40[†]L_UNC_DH_ECO_RT	ARC uncorrected attenuation resistivity, 16-in spacing at 400 kHz, real time, computed downhole
P40[‡]B_DH_ECO_RT	ARC blended phase shift resistivity, 16-in spacing, real time, computed downhole
P40[‡]H_UNC_DH_ECO_RT	ARC uncorrected phase shift resistivity, 16-in spacing at 2 MHz, real time, computed downhole
P40[‡]L_UNC_DH_ECO_RT	ARC uncorrected phase shift resistivity, 16-in spacing at 400 kHz, real time, computed downhole
PHNRH_DH_ECO_RT	Phase at near receiver at 2 MHz, real time, computed downhole
PHNRL_DH_ECO_RT	Phase at near receiver at 400 kHz, real time, computed downhole
QC_RES_DH_ECO_RT	Resistivity quality indicator, real time, computed downhole

[†] Attenuation resistivity channels with 16-, 22-, 28-, and 34-in intransmitter-receiver spacings can also be transmitted.

[‡] Phase shift resistivity channels with 16-, 22-, 28-, and 34-in transmitter-receiver spacings can also be transmitted.

Table 4-12. Propagation resistivity real-time computed channels.

Computed Channel	Description
A40[†]H_ECO_RT	ARC attenuation resistivity, 16-in spacing at 2 MHz, environmentally corrected, real time
A40[†]L_ECO_RT	ARC attenuation resistivity, 16-in spacing at 400 kHz, environmentally corrected, real time
P40[‡]H_ECO_RT	ARC phase shift resistivity, 16-in spacing at 2 MHz, environmentally corrected, real time
P40[‡]L_ECO_RT	ARC phase shift resistivity, 16-in spacing at 400 kHz, environmentally corrected, real time
A40[§]H_COND_UNC_ECO_RT	Uncorrected attenuation conductivity, 16-in spacing at 2 MHz, real time, computed downhole
A40[§]L_COND_UNC_ECO_RT	Uncorrected attenuation conductivity, 16-in spacing at 400 kHz, real time, computed downhole
P40[††]H_COND_UNC_ECO_RT	Uncorrected phase shift conductivity 16-in spacing at 2 MHz, real time, computed downhole
P40[††]L_COND_UNC_ECO_RT	Uncorrected phase shift conductivity 16-in [spacing at 400 kHz, real time, computed downhole

[†] Environmentally corrected attenuation resistivity channels with 16-, 22-, 28-, and 34-in transmitter-receiver spacings are also computed provided that the corresponding downhole channel has been transmitted.

[†] Environmentally corrected phase shift resistivity channels with 16-, 22-, 28-, and 34-in transmitter-receiver spacings are also computed provided that the corresponding downhole channel has been transmitted.

[§] Attenuation conductivity channels with 16-, 22-, 28-, and 34-in transmitter-receiver spacings are also computed provided that the corresponding downhole channel has been transmitted.

[††] Phase shift conductivity channels with 16-, 22-, 28-, and 34-in transmitter-receiver spacings are also computed provided that the corresponding downhole channel has been transmitted.

4.3.6.2. Recorded mode

Table 4-13. Propagation resistivity recorded-mode parameters.

Parameter Name	Description	Unit
OBMF_RM	Oil-base mud flag	yes/no
RMS_RM	Resistivity of mud sample	ohm.m
MST_RM	Mud sample temperature	degF
BS_RM	Bit sizc	in
ATMP_ARC	ARC select temperature channel	(Tool_Temp, Annulus_Temp, Interpolated_Temp)
Blended Resistivity Processing Parameters		
HIGH_BLEND	High-resistivity threshold for blending	ohm.m
LOW_BLEND	Low-resistivity threshold for blending	ohm.m
ARCWizard Processing Parameters		
EN_WIZARD	Enabled ARCWizard processing	yes/no
SDPTH	Wizard process start depth	ft
EDPTH	Wizard process stop depth	ft
MSWS	ARCWizard model switch window	ft
ERRCT	Percentage error cutoff	%
GRSH	Gamma ray shale (invasion computation cutoff)	gAPI
INVAS_COMPUTE	Invasion computation option	NA[†]
DIELEC_COMPUTE	Dielectric computation option	NA
ANISO_COMPUTE	Anisotropy computation option	NA
BH_COMPUTE	BHC computation option	NA
UNIFORM_COMPUTE	Uniform rock computation option	NA
MULTIEFFECT_COMPUTE	Multieffect computation option	NA
PRTD	Preferred resistivity log for real-time display when multiple effects detected	ohm.m
DIPF[‡]	Formation dip angle	°
AZMF[‡]	Formation dip azimuth	°

[†] Not applicable.

[‡] Used for anisotropy processing in conjunction with the well trajectory to compute the incidence angle between the tool and formation layering.

Table 4-14. Propagation resistivity recorded-mode output channels.

Output Channel	Description
A40†H_UNC	ARC uncorrected attenuation resistivity, 40-in spacing at 2 MHz
A40†L_UNC	ARC uncorrected attenuation resistivity, 40-in spacing at 400 KHz
AT40‡H	ARC compensated attenuation, 40-in spacing at 2 MHz
AT40‡L	ARC compensated attenuation, 40-in spacing at 400 KHz
P40§H_UNC	ARC uncorrected phase shift resistivity, 40-in spacing at 2 MHz
P40§L_UNC	ARC uncorrected phase shift resistivity, 40-in spacing at 400 KHz
PS40††H	ARC compensated phase shift, 40-in spacing at 2 MHz
PS40††L	ARC compensated phase shift, 40-in spacing at 400 KHz
TAB_ARC_RES	ARC resistivity time after bit
TICK_ARC_RES	ARC resistivity samples
QC_FRT	Uninvaded-zone resistivity quality indicator
QC_RES	Resistivity quality indicator
QC_RXO	Invaded-zone resistivity quality indicator
A40‡‡H	ARC attenuation resistivity, 40-in spacing at 2 MHz, environmentally corrected
A40‡‡L	ARC attenuation resistivity, 40-in spacing at 400 KHz, environmentally corrected
A40‡‡B	ARC blended attenuation resistivity, 40-in spacing, environmentally corrected
P40§§H	ARC phase shift resistivity, 40-in spacing at 2 MHz, environmentally corrected
P40§§L	ARC phase shift resistivity, 40-in spacing at 400 KHz, environmentally corrected
P40§§B	ARC blended phase shift resistivity, 40-in spacing, environmentally corrected
A40†††H_COND	ARC attenuation conductivity, 40-in spacing at 2 MHz, environmentally corrected
A40†††L_COND	ARC attenuation conductivity, 40-in spacing at 400 KHz, environmentally corrected
A40†††B_COND	ARC blended attenuation conductivity, 40-in [101.6-cm] spacing, environmentally corrected
P40‡‡‡H_COND	ARC phase shift conductivity, 40-in spacing at 2 MHz, environmentally corrected
P40‡‡‡L_COND	ARC phase shift conductivity, 40-in spacing at 400 KHz, environmentally corrected
P40‡‡‡B_COND	ARC blended phase shift conductivity, 40-in spacing, environmentally corrected

† Uncorrected attenuation resistivity channels with 16-, 22-, 28-, and 34-in transmitter-receiver spacings are also provided.

‡ Compensated attenuation channels with 16-, 22-, 28-, and 34-in transmitter-receiver spacings are also provided.

§ Uncorrected phase shift resistivity channels with 16-, 22-, 28-, and 34-in transmitter-receiver spacings are also provided.

†† Compensated phase shift channels with 16-, 22-, 28-, and 34-in transmitter-receiver spacings are also provided.

‡‡ Environmentally corrected attenuation resistivity channels with 16-, 22-, 28-, and 34-in transmitter-receiver spacings are also provided.

§§ Environmentally corrected phase shift resistivity channels with 16-, 22-, 28-, and 34-in transmitter-receiver spacings are also provided.

††† Environmentally corrected attenuation conductivity channels with 16-, 22-, 28-, and 34-in transmitter-receiver spacings are also provided.

‡‡‡ Environmentally corrected phase shift conductivity channels with 16-, 22-, 28-, and 34-in transmitter-receiver spacings are also provided.

Table 4-15. ARCWizard individual model processing output channels.

Output Channel	Description
FTCK	ARC tool status check flag = 0, none abnormal = 1, T1 is abnormal, repairable = 2, T2 is abnormal, repairable = 3, T3 is abnormal, repairable = 4, T4 is abnormal, repairable = 5, T5 is abnormal, repairable = 6, R1 abnormal, nonrepairable = 7, R2 is abnormal, nonrepairable = 8, all other nonrepairable cases (any two or more transmitters and/or any one or more receivers are abnormal)
Eccentricity Check	
FECC	ARC tool eccentricity check flag = 0, no tool eccentricity effect = 1, tool eccentricity effect exists
Invasion Model	
Rt_Inv_QC	True resistivity QC log from invasion inversion model
Rxo_Inv_QC	Flushed-zone resistivity QC log from invasion inversion model
DI_Inv_QC	Invasion diameter QC log from invasion inversion model
DI_ARC	ARC invasion diameter
EINV	ARCWizard invasion inversion percentage error
Anisotropy Model	
RH_INV_QC	Horizontal resistivity QC log from anisotropy inversion model
RV_ARC	ARC vertical resistivity
EANI	ARCWizard anisotropy inversion percentage error
Dielectric Model	
RT_Die_QC	Formation resistivity QC log from dielectric inversion model
EPS_Die_QC	Relative dielectric constant QC log from dielectric inversion model
EDIE	ARCWizard dielectric inversion percentage error
Borehole Model	
RT_Bor_QC	Formation resistivity QC log from borehole inversion model
HD_INV_QC	Borehole diameter QC log from borehole inversion model
EBOR	Borehole model inversion percentage error

Table 4-16. ARCWizard processing output channels.

Output Channel	Description
RT_ARC	ARC true formation resistivity
RXO_ARC	ARC flushed-zone resistivity
DI_ARC	ARC invasion diameter
RH_ARC	ARC horizontal resistivity
RV_ARC	ARC vertical resistivity
CALE_ARC	Borehole diameter from electrical caliper
Rh_Inv_Flt	Horizontal resistivity from anisotropy model inversion
Rv_Inv_Flt	Vertical resistivity from anisotropy model inversion
Rt_Inv_Flt	True resistivity from invasion model inversion
Rxo_Inv_Flt	Flushed-zone resistivity from invasion model inversion
DI_Inv_Flt	Diameter of invasion from invasion model inversion
RT_Bor_Flt	Borehole-effect-free resistivity from borehole model inversion
HD_Inv_Flt	Hole diameter from borehole model inversion
RT_Die_Flt	Dielectric effect-free resistivity from dielectric model inversion
EPS_Die_Flt	Relative dielectric constant from dielectric model inversion
FERR_ANI	Anisotropy model inversion percentage error
FERR_INV	Invasion model inversion percentage error
FERR_DIE	Dielectric model inversion percentage error
FERR_BOR	Borehole model inversion percentage error
FANI	ARC anisotropy effect flag
FINV	ARC invasion effect flag
FDIE	ARC dielectric effect flag
FBOR	ARC borehole effect flag
FECC	ARC tool eccentricity check flag
FUNI	ARC uniform rock flag
FMUL	ARC multieffect flag

4.3.7. Propagation resistivity measurement specifications

Table 4-17. Resistivity measurement specifications—range and accuracy.

Measurement	Accuracy	Range
Phase shift resistivity, 2 MHz	2%	0.2 to 60 ohm.m
	0.3 mS/m	60 to 3,000 ohm.m
Phase shift resistivity, 400 kHz	2%	0.1 to 10 ohm.m
	2 mS/m	10 to 100 ohm.m
Attenuation resistivity, 2 MHz	3%	0.2 to 25 ohm.m
	1.5 mS/m	25 to 50 ohm.m
Attenuation resistivity, 400 kHz	3%	0.1 to 3 ohm.m
	10 mS/m	3 to 10 ohm.m

Table 4-18. Resistivity measurement specifications—axial resolution.

Axial resolution[†], ft	Measurement spacings, in				
	16	22	28	34	40
R = 1 ohm.m					
Phase shift resistivity, 2 MHz	0.7	0.7	0.7	0.7	0.7
Phase shift resistivity, 400 kHz	1.0	1.0	1.0	1.0	1.0
Attenuation resistivity, 2 MHz	1.8	1.8	1.8	1.8	1.8
Attenuation resistivity, 400 kHz	3.0	3.5	4.0	4.0	4.0
R = 10 ohm.m					
Phase shift resistivity, 2 MHz	1.0	1.0	1.0	1.0	1.0
Attenuation resistivity, 2 MHz	4.0	5.0	6.0	6.0	6.0

[†] Width at half maximum of the response function along the tool axis at the specified formation resistivity.

Table 4-19. Resistivity measurement specification—radius of investigation.

Radius of investigation[†], ft	Measurement spacings, in				
	16	22	28	34	40
R = 1 ohm.m					
Phase shift resistivity, 2 MHz	13	14	15	17	18
Phase shift resistivity, 400 kHz	17	19	22	25	27
Attenuation resistivity, 2 MHz	19	22	24	26	29
Attenuation resistivity, 400 kHz	27	30	33	36	38
R = 20 ohm.m					
Phase shift resistivity, 2 MHz	18	22	26	30	34
Attenuation resistivity, 2 MHz	44	48	52	56	60

[†] Radius at which the integrated radial geometrical factor reaches 0.5 in the specified formation resistivity.

4.4. Azimuthal bulk density

The azimuthal bulk density measurement uses a side-loaded chemical cesium (Cs)-137 gamma ray source, which is positioned in the collar closer to the formation wall than in conventional LWD tools. This proximity to the formation results in better statistics, improved measurement DOIs, and reduced borehole sensitivity. The density measurement is updated every 10 seconds. The outgoing and returning gamma rays can be focused to create an azimuthal measurement of the formation. This density measurement is referred to as the azimuthal or gamma-gamma bulk density to distinguish it from the sourceless, or neutron-gamma, density measurement outlined in Section 4.9.

4.4.1. Azimuthal bulk density theory of measurement

Formation density is determined based on the down-scattering of gamma ray energies as they interact with the electrons of the atoms in the formation. This Compton scattering creates a reduction in the number of gamma rays of a specific energy range with increasing electron density of the material they pass through. Based on a correlation between the number of electrons and atomic mass, the measured electron density is converted to bulk density.

The EcoScope gamma-gamma density measurement differs from the previous-generation density measurement in that the gamma ray source is side-loaded into the tool. While this means that the source cannot be fished if the tool becomes stuck downhole, it also means that the source can be positioned in the wall of the collar closer to the formation. The resultant increase in gamma ray counts in the detectors has allowed the detectors to be moved farther from the source, increasing the measurement DOI and reducing borehole sensitivity. The increased statistics also improve the measurement precision.

While formation density can be determined with a single gamma ray source and detector, in most logging tools two detectors are used to create a compensated density measurement, which corrects for parallel standoff between the detectors and borehole wall. The detector closer to the source is called the short-spacing detector; the farther is called the long-spacing detector, as shown in **Figure 4-37**.

Figure 4-37. Dual detector density measurement, which compensates for the effect of a limited standoff between the detectors and borehole wall.

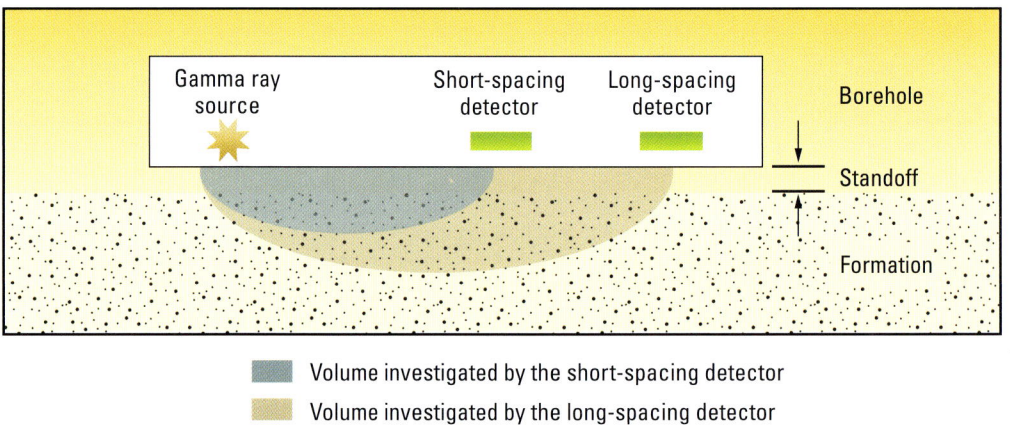

4.4.2. Azimuthal bulk density environmental corrections

4.4.2.1. Spine-and-ribs standoff compensation

The limited penetrating ability of gamma rays means that the gamma-gamma density measurement requires good contact between the detectors and the borehole wall. Any standoff will result in the material between the detectors and the borehole wall being included in the measurement volume. In light muds, the spine-and-ribs algorithm compensates for parallel standoff of up to 1 in (less in heavier muds).

The spine-and-ribs correction compares the density measured using the short- and long-spacing detectors, as shown in **Figure 4-38**. If they read the same density, they plot on the spine, and no correction is applied. As standoff increases in light mud, the density measured in the short-spacing detector will decrease more rapidly than the long-spacing, as it has a shallower DOI and is therefore more affected by mud in front of the detector. As the short-spacing density falls faster than the long-spacing, the data falls on a rib. The shape of the rib is characterized during tool design, so it is possible to determine the density correction, $\Delta\rho$, that must be applied to the long-spacing density to recover the true formation density corrected for the standoff. The density correction is an indication of standoff between the detectors and formation. Large density correction values (greater than 0.2 g/cm^3) indicate excessive standoff, and the corresponding density data should be disregarded.

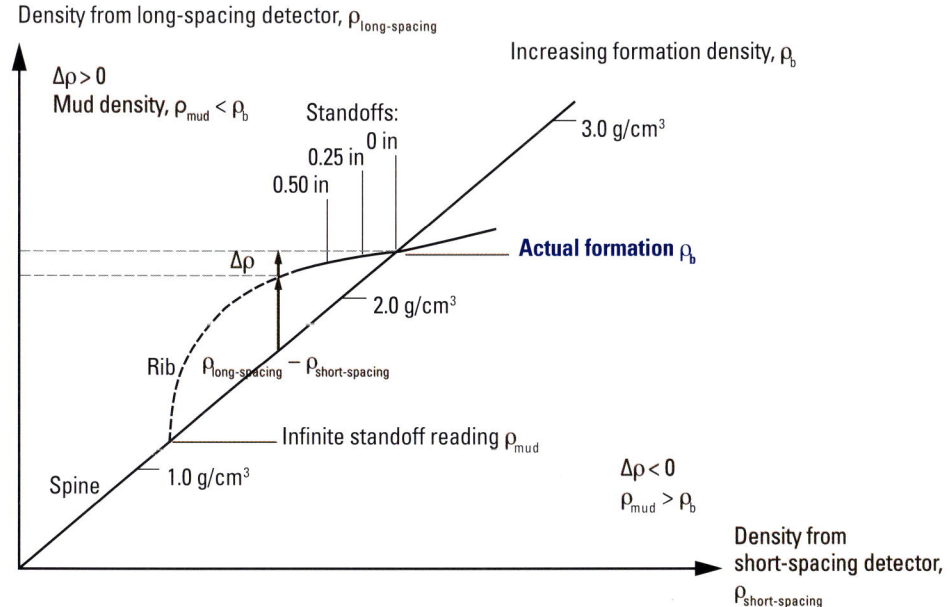

Figure 4-38. Density spine-and-ribs technique, correcting for standoff between the detectors and the borehole wall.

Because gamma rays are relatively easily stopped, the density measurement is shallow (2 to 3 in). It can be well focused, making it suitable for azimuthal measurements and imaging. Contact with the formation is enhanced by the presence of a stabilizer with special low-density windows installed to exclude drilling mud and improve gamma ray transport to and from the formation. However, stabilizers can affect the drilling performance of the BHA, so the decision to run stabilized or slick (without the density stabilizer) is a decision that should be taken in conjunction with the directional driller.

To facilitate a compromise between drilling and data quality requirements, the EcoScope density has been engineered to have a deeper DOI and hence, less sensitivity to the borehole than previous-generation LWD density measurements. In addition, the slide-on density stabilizer design allows the diameter of the stabilizer to be changed without requiring a new tool. Stabilizers with diameters of 7.875, 8.25, and 9.375 in are currently available.

One of the consequences of the slide-on stabilizer design is that there is a small gap between the stabilizer sleeve and the tool collar so that the stabilizer sleeve is able to slide on and off when required. This gap creates a gamma ray path between the source and detectors in much the same way as standoff from the borehole wall. The spine-and-ribs correction automatically compensates for this small gap and any standoff. Consequently, the EcoScope azimuthal density correction often reads slightly higher than previous-generation tools, which did not have the slide-on stabilizer. Traditionally, one of the azimuthal density QCs has been to ensure that the density correction is very close to 0. The EcoScope density correction may read slightly higher than 0 despite good contact between the tool and borehole wall because of the presence of the gap between the stabilizer sleeve and the tool collar, but the data is still valid.

4.4.2.2. Image-derived density processing for path-of-best-contact determination

Image derived density (IDD) processing extracts the best formation density and photoelectric factor (PEF) from the array of 16-sector image data acquired by the EcoScope gamma-gamma density detectors by determining the path of best contact between the detectors and the formation.

When the EcoScope tool is run with a stabilizer of similar diameter to the bit, good formation data can be acquired around the borehole as the tool rotates, as shown in **Figure 4-39** (left). In deviated and horizontal boreholes, it is generally assumed that the best contact between the density detectors and the borehole wall occurs when the detectors are facing the bottom of the hole. This may not be the case where an undergauge stabilizer allows the tool to move about in the borehole during rotation, as shown in Figure 4-39, or where a BHA rubs against the opposite side of the borehole at a change in the trajectory. For example, as the trajectory drops, the point of best contact may be the top of the borehole.

In these circumstances, the bottom quadrant density, ROBB, may suffer more standoff than the spine-and-ribs algorithm can compensate. In most mud systems, this standoff leads to a low-density reading, which gives a misleading gas separation when compared with the neutron log. The image in the lower right of Figure 4-39 shows a bright stripe corresponding to formation density surrounded by dark coloring associated with low-density mud.

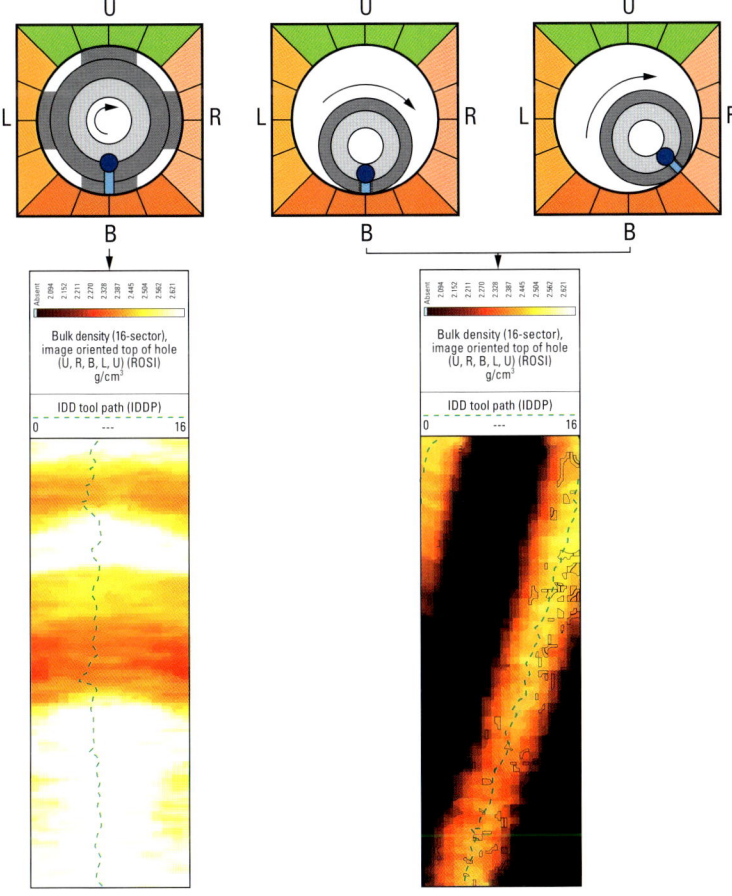

Figure 4-39. The cross sections of a gauge-stabilized EcoScope tool (left) and undergauge-stabilized tool (right) in a deviated borehole, with corresponding density images. The undergauge-stabilized tool may move in the borehole during drilling, resulting in the bottom quadrant density not corresponding to the point of best contact.

The bright stripe corresponds to the path of best contact between the density detectors and borehole wall. IDD processing uses a logic that looks for reasonable values of long-spacing density, the difference between long- and short-spacing densities, and the volumetric PEF (U). The expected ranges of these measurements are shown in **Figure 4-40**.

Figure 4-40. Expected ranges for the long-spacing density (A), the difference between long and short-spacing densities (B), and U (C). The quality of the detector contact with the borehole wall is determined by mutliplying the three confidence factors.

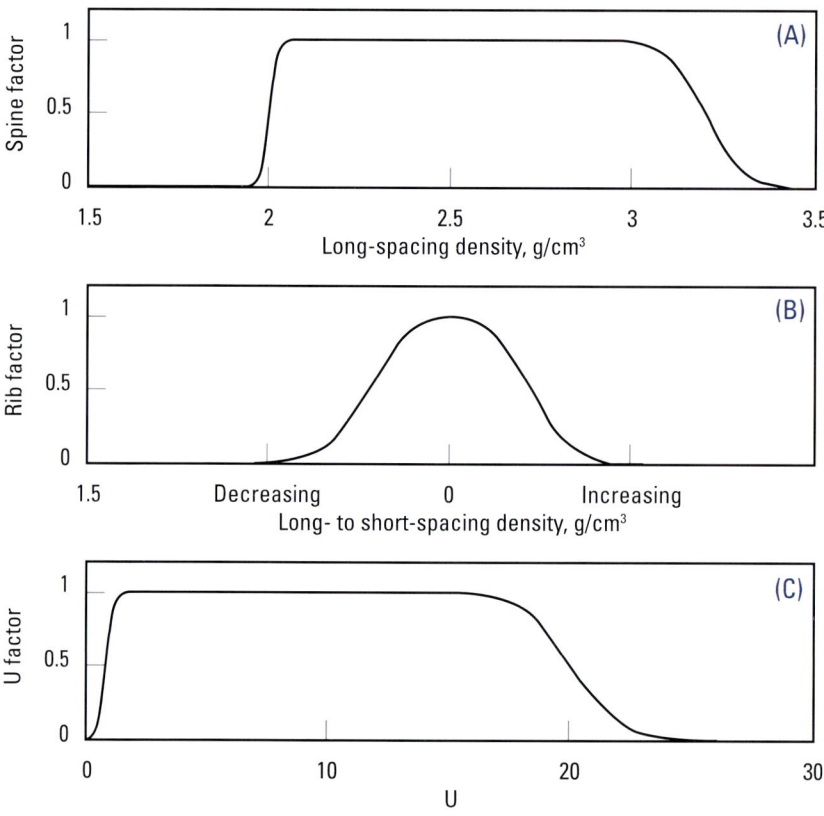

An IDD quality factor, IDDQ, is computed in each sector at each depth by multiplying the confidence factors from these three measurements. An IDDQ close to 1 indicates good contact with the formation and hence, more representative formation density and photoelectric measurements. At each depth, the centroid of the 16 IDDQ factors (one for each of the sectors) is identified as the IDD tool path, IDDP, as shown in **Figure 4-41**.

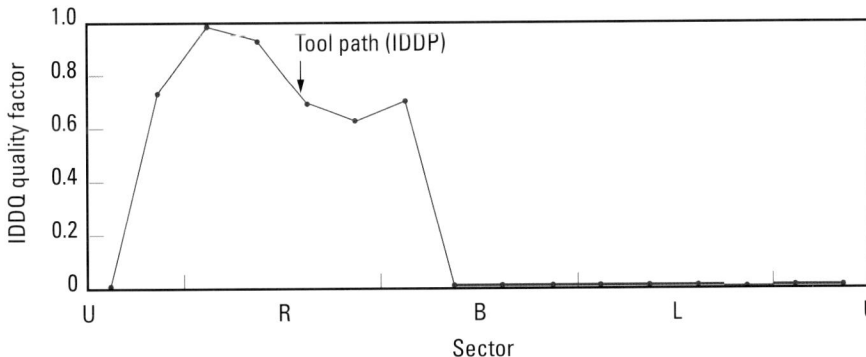

Figure 4-41. The point of best contact with the borehole wall, IDDP, at a given depth, determined by computing the centroid of the quality factors from each of the 16 sectors.

To ensure sufficient statistical precision of the azimuthal measurements, count rates from four sectors are generally averaged. A quality factor, QIDD, is computed over the four sectors centered on the tool path, and another quality factor, QRBB, is computed over the four sectors centered on the bottom of the hole. A quality ratio, IDQR, is then computed according to the equation

$$IDQR = \frac{QRBB}{QRBB + QIDD}$$
(Equation 4-6).

IDQR will be close to 0 when the IDD processing yields a better tool path than assuming the bottom four sectors give the path of best contact. IDQR will be closer to 1 when IDD processing does not provide a better tool path, as may be the case with a stabilized tool.

A user-defined threshold, IDQT, is applied to IDQR to determine whether the IDD or bottom measurements are provided in the IDD channels for density, IDRO, density correction, IDDR, PEF, IDPE, and volumetric PEF, IDU.

Table 4-20. Threshold logic for IDD processing.

IDQT	Image-Derived Output			
	IDRO =	IDDR =	IDPE =	IDU =
IDQT = 2 (DEFAULT)				
Stabilized tool	ROBB	DRBB	PEB	UB
Slick tool	IDD output	IDD output	IDD output	IDD output
0 < IDQT < 1				
IDQR lower than IDQT	IDD density	IDD density correction	IDD PEF	IDD volumetric PEF
IDQR equal or higher than IDQT	ROBB	DRBB	PEB	UB

Figure 4-42 shows how the user, by setting the quality threshold IDQT between 0 and 1, can control whether the bottom quadrant or image-derived outputs are used in the channels IDRO, IDDR, IDPE, and IDU.

Figure 4-42. Threshold logic, used to select either bottom quadrant or image-derived outputs for the channels IDRO, IDDR, IDPE, and IDU.

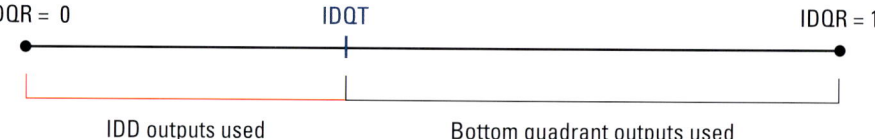

Figure 4-43 shows an example of IDD processing. The following images are shown.

- Track 3—The density image shows the light colors associated with high density meandering from the bottom of the hole (center of the track) surrounded by dark coloring associated with low-density mud.

- Track 4—The quality image, computed from the of long-spacing density, the difference between long and short-spacing densities, and the volumetric PEF, U, in each sector, shows a dark stripe of high contact quality surrounded by bright colors, indicating poor contact between the detectors and borehole.

The centroid of the high-quality region (IDDP, dotted green line on both image tracks) indicates the tool path along the borehole wall. The right track shows the thermal neutron porosity (TNPH, blue dashed line) compared with the bottom density (ROBB, black line) and IDD (IDRO, red line). The large separation between TNPH and ROBB at approximately X0,000 ft suggests the presence of gas in the formation, but this was not expected in this clean, oil-filled limestone reservoir. The density image shows that at this depth, the detectors are not in good contact with the borehole wall in the bottom sectors. IDD processing identifies that the tool is riding up one side of the borehole. The density computed using the four sectors centered on the computed tool path lies much closer to the TNPH with a hydrocarbon separation similar to that seen at the top and bottom of the interval. Note that the ROBB and IDRO overlay where the point of best contact is close to the bottom of the hole. The IDD correction (IDDR, red dotted line) indicates a small standoff in the affected interval that is compensated by the spine-and-ribs processing.

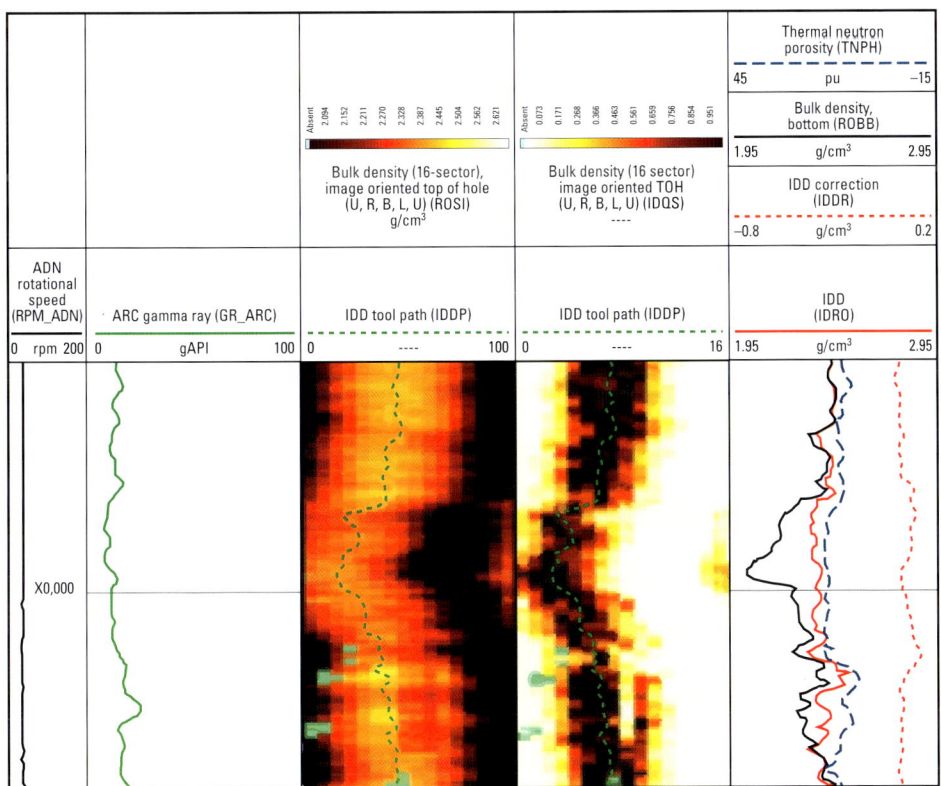

Figure 4-43. IDD processing, showing the bulk density image (left image track), the quality image (right image track) computed from the density, density correction, and U in each sector at each depth. The IDD, (IDRO, solid red line, right track), density correction, (IDDR, dotted red line, right track), and PEF (IDPE) are computed from the sectors centered around the tool path (green dotted line on the images). Bottom density (ROBB) shows significant separation from the neutron response, suggesting the presence of gas, while IDRO is more consistent with the neutron, showing that the formation is actually oil bearing.

The assumption that the highest density is the true formation density is flawed because this method tends to track the density of the least porous layer, resulting in underestimation of formation porosity when crossing multiple layers with a highly deviated or horizontal hole. The assumption that the point of best contact is always at the bottom of the hole can also give misleading results. IDD processing uses the full suite of density-related image data to derive the best possible formation density and PEF information.

4.4.2.3. Threaded or spiraled borehole processing for reduced rugosity effect

Threaded or spiraled rugosity is a periodic drilling-induced oscillation of the borehole that affects many LWD measurements. Density, because of the relatively shallow DOI, may be severely affected by this condition. An example of spiraled hole and its effect on log measurements is provided in **Figure 4-44**.

Figure 4-44. Periodic oscillations in the borehole wall, resulting in oscillations in the density measurements due to standoff—particularly evident on the density image (bottom).

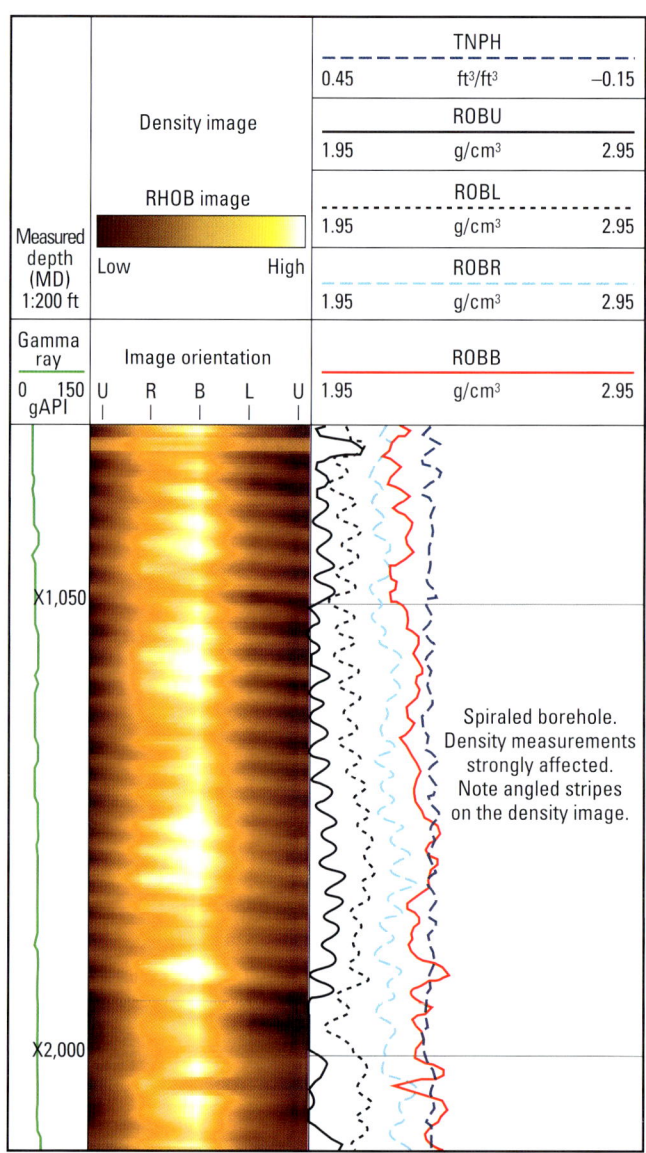

Advanced processing has been developed to extract formation density from spiral hole–affected measurements. The correction algorithm scans the long- and short-spacing density data determining the spatial frequency of the spiral induced noise, after which an adaptive filter removes it. The filtered arrays are processed using the traditional spine-and-ribs technique to compensate for any remaining standoff effect.

An example of the processing output is provided below. **Figure 4-45** (top) shows a set of quadrant density curves severely affected by spiraling. Figure 4-45 (bottom) shows the same data processed to remove threaded borehole effect. Notice that only the periodic noise has been filtered out. The signal coming from the laminated formation is preserved.

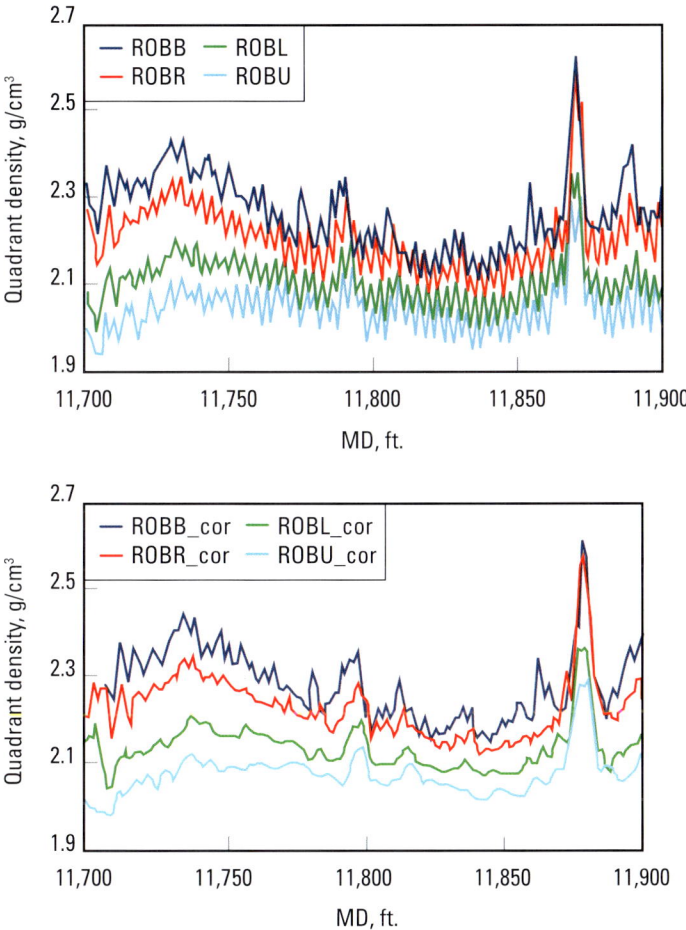

Figure 4-45. Borehole spiraling, resulting in periodic oscillation on the density data (top), which can make formation features difficult to recognize. After spiral hole processing, the periodic noise is removed, leaving the formation layers clearly recognizable.

4.4.2.4. High-resolution (alpha-processed) density

The spine-and-ribs standoff compensation technique outlined in Section 4.4.2.1 uses the short-spacing density to correct the long-spacing density for standoff effect. The resulting compensated density measurement has an axial resolution defined by the spacing between the source and long-spacing detector.

In good borehole conditions where the standoff between the detectors and borehole wall is small and consistent, an alternative technique, called alpha processing, can be applied. In alpha processing, the long-spacing density is used to correct the short-spacing density such that the resulting alpha-processed density has improved axial resolution, as it is defined by the spacing between the source and the short-spacing detector.

The processing calculates the average multiplying factor (alpha factor) required on the short-spacing density to have it match the long-spacing density over a user-specified depth interval, typically around 5 ft. This factor is then applied to the short-spacing response over the corresponding interval, resulting in a high-resolution density measurement, as shown in **Figure 4-46**.

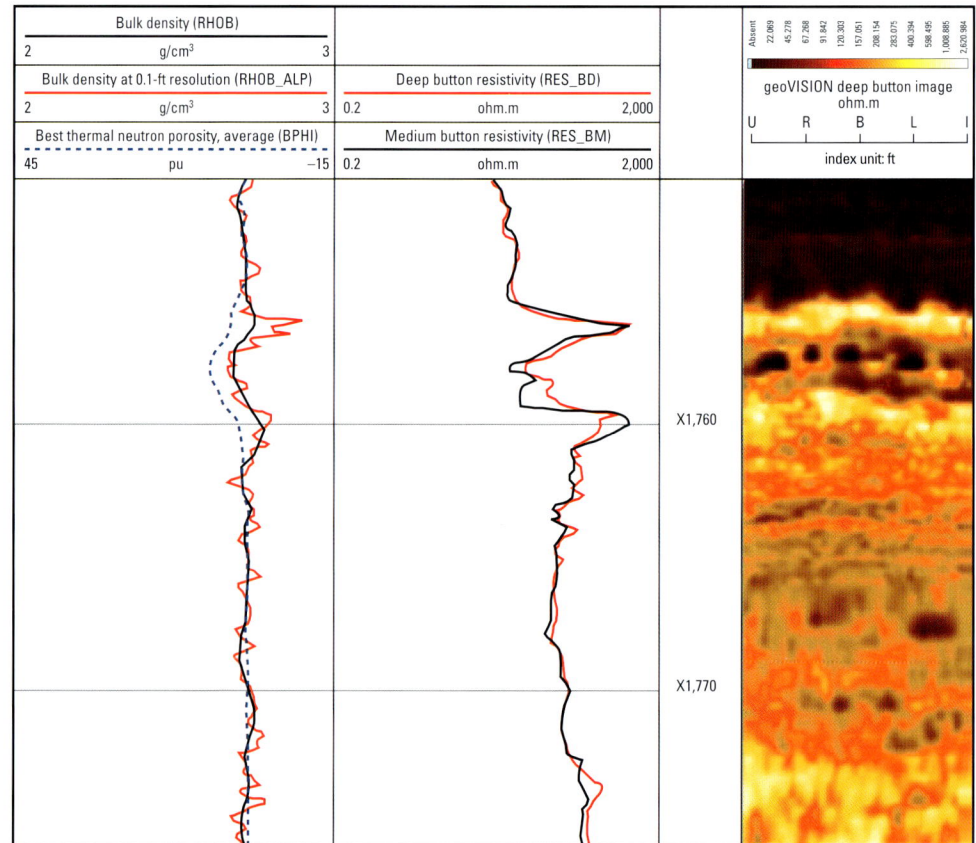

Figure 4-46. Alpha-processed density (RHOB_ALP, red line, left track), showing improved resolution compared with the normal resolution density (RHOB, black line, left track) in good borehole conditions. Alpha processing uses the long-spacing density to correct the short-spacing density rather than the other way around, as applied in conventional spine-and-ribs processing. High-resolution geoVISION laterolog resistivities (middle track) and image (right track) confirm the presence of thin, high-resistivity features that correlate with the high-resolution density response.

4.4.3. Azimuthal bulk density applications

Density is a volumetric formation property. It responds to the volumetric sum of the components in the formation. Hence, the density of a formation can be summarized by the equation

$$\rho_{bulk} = \phi \left(\rho_{fluid} \right) + \rho_{grain} \left(1 - \phi \right)$$ (Equation 4-7),

where

ρ_{bulk} = bulk density of the formation
ϕ = formation porosity in volume fraction (v/v)
ρ_{fluid} = density of the fluid in the pores
ρ_{grain} = density of the formation matrix.

When ρ_{fluid} and ρ_{grain} are known, ρ_{bulk} allows ϕ to be calculated according to the equation

$$\phi = \frac{\rho_{grain} - \rho_{bulk}}{\rho_{grain} - \rho_{fluid}}$$ (Equation 4-8).

The density measurement is generally the primary means of evaluating formation porosity. **Figure 4-47** illustrates the significant difference in density between common formation lithologies and the fluids that fill the pore spaces. This contrast in densities makes the density measurement sensitive to formation porosity.

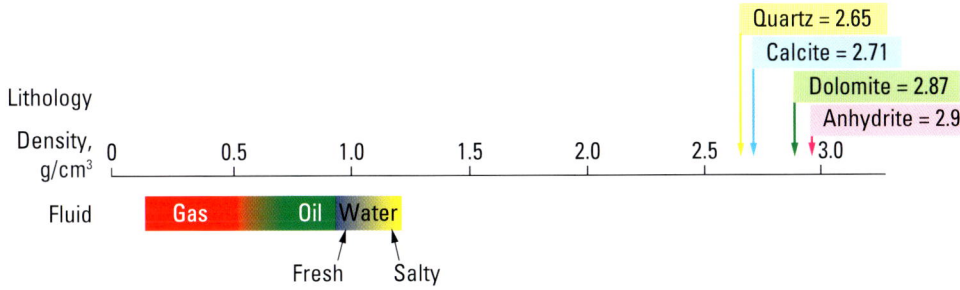

Figure 4-47. A density scale, showing typical fluid and matrix densities.

Normally, additional information is required to assist in identifying the fluids and formation lithology so that the appropriate fluid and formation grain densities can be determined and the formation porosity calculated. The neutron measurement is frequently used in conjunction with the density measurement to help identify formation fluid and lithology. For further details on comparison of the density and neutron responses for formation fluid and lithology identification, please refer to Section 4.8.3. Lithology information can also be extracted from the photoelectric measurement and spectroscopy.

The density image is often used to identify features crossing the borehole and to evaluate the incidence angle between the borehole and formation layering. This is generally sufficient information to allow well placement decisions for positioning the well in the desired location within the geological sequence.

The incidence angle and azimuth information extracted from the images can be converted to true formation dip and azimuth by accounting for the well trajectory inclination and azimuth from the directional surveys.

4.4.4. Azimuthal bulk density log quality control

The shop calibration must have been performed less than 1 month before the acquisition date and must be in tolerance. The calibration must be performed in the calibration blocks appropriate for the stabilizer size. Refer to the Schlumberger LWD Quality Control Manual for details.

If the electronic chassis or stabilizer has been removed from the tool, a new calibration must be performed.

The source (identified by serial number) used during logging must match the source with which the tool was calibrated.

Because the major effect that degrades the bulk density measurement is standoff from the borehole wall, the density correction should be checked to ensure that it reads close to 0. Because of the internal gap between the collar and stabilizer sleeve, the density correction may read slightly higher than 0 because the effect of the gap is included in the density correction.

When plotted on a lithology-compatible scale, the density and neutron responses should overlay in a clean, freshwater-filled formation of the corresponding lithology after all environmental corrections have been applied. Refer to Section 4.8.3 for details.

If azimuthal density information is to be used, then rotation of the tool over the interval of interest must be verified.

The EcoScope hardware incorporates checks to verify correct performance of the acquisition detectors and electronics. Additional checks to ensure that the measurements fall within the range expected for downhole formation are outlined in the QC logic shown in Table **4-21**.

Table 4-21. Azimuthal bulk density QC flag logic.

QC_RHOB	Hardware	Measurement
Green	Short- and long-spacing gamma high-voltage control loop status is FINE and Short- and long-spacing gamma high-voltage control is AUTO and Short- and long-spacing gamma pulse shape compensation control loop status is FINE and Short- and long-spacing gamma pulse shape compensation control is AUTO	$1.7\ \text{g/cm}^3 \leq \text{ROBB} \leq 3.05\ \text{g/cm}^3$ and $-0.1\ \text{g/cm}^3 \leq \text{DRHB} \leq 0.3\ \text{g/cm}^3$ and $0.5 \leq \text{UB} \leq 300$ and $\text{CRPM} > 20$ and $\text{ROP} < 450\ \text{ft/h}$
Yellow	Short- or long-spacing gamma high-voltage control loop status is COARSE or Short- or long-spacing gamma pulse shape compensation control loop status is COARSE	$1\ \text{g/cm}^3 < \text{ROBB} < 1.7\ \text{g/cm}^3$ or $3.05\ \text{g/cm}^3 < \text{ROBB} < 3.3\ \text{g/cm}^3$ or $-0.2\ \text{g/cm}^3 < \text{DRHB} < -0.1\ \text{g/cm}^3$ or $0.3\ \text{g/cm}^3 < \text{DRHB} < 0.5\ \text{g/cm}^3$ or $\text{UB} < 0.5$ or $300 < \text{UB} < 1000$ $\text{CRPM} < 20$ or $\text{ROP} > 450\ \text{ft/h}$
Red	Short- or long-spacing gamma high-voltage control loop status is SEARCH or Short- or long-spacing gamma pulse shape compensation control loop status is SEARCH, WAIT or Short- or long-spacing gamma pulse shape compensation control is MANUAL or HOLD or Short- or long-spacing gamma high-voltage control is MANUAL or HOLD or Density configuration and calibration parameters have not been received or have bad values or Density windows out of range: CES_SSW1A < 200 or > 30,000 or CES_SSW3A < 1,000 or > 50,000 or CES_LSW3A < 100 or > 30,000 Gamma-gamma density hardware flag set or Recorded mode only: Memory usage indicator = full	$\text{ROBB} \leq 1\ \text{g/cm}^3$ or $\text{ROBB} \geq 3.3\ \text{g/cm}^3$ or $\text{DRHB} \leq -0.2\ \text{g/cm}^3$ or $\text{DRHB} \geq 0.5\ \text{g/cm}^3$ or $\text{UB} > 1000$
White		No new density tick for more than 2 ft

4.4.5. Azimuthal bulk density channels and parameters

4.4.5.1. Real time

Table 4-22. Azimuthal bulk density real-time parameters.

Parameter Name	Description	Unit
MDEN	Matrix density for density porosity	g/cm^3
MFD	Mud filtrate density	g/cm^3
IDD Processing Parameters		
IDQT	Image-derived quality threshold	NA[†]

[†] Not applicable.

Table 4-23. Azimuthal bulk density real-time transmitted channels.

Transmitted Channel	Description
RHOB_DH_ECO_RT[†]	Bulk density average, real time, computed downhole
ROBB_DH_ECO_RT[†]	Bulk density, bottom, real time, computed downhole
ROBL_DH_ECO_RT[†]	Bulk density, left, real time, computed downhole
ROBU_DH_ECO_RT[†]	Bulk density, up, real time, computed downhole
ROBR_DH_ECO_RT[†]	Bulk density, right, real time, computed downhole
DRHO_DH_ECO_RT	Bulk density correction, average, real time, computed downhole
DRHB_DH_ECO_RT	Bulk density correction, bottom, real time, computed downhole
DRHL_DH_ECO_RT	Bulk density correction, left, real time, computed downhole
DRHU_DH_ECO_RT	Bulk density correction, up, real time, computed downhole
DRHR_DH_ECO_RT	Bulk density correction, right, real time, computed downhole
ROSI_STA_TOH_ECO_RT	Bulk density (16-sector), image oriented top of hole (U,R,B,L,U), real time
QC_RHOB_DH_ECO_RT	Density quality indicator, real time, computed downhole
SRPM_DH_ECO_RT	rpm standard deviation, real time, computed downhole
VRPM_DH_ECO_RT	Rotational speed variation, real time
IDD Transmitted Channels	
IDDP_DH_ECO_RT	IDD tool path, real time, computed downhole
IDDR_DH_ECO_RT	IDD correction, real time, computed downhole
IDRO_DH_ECO_RT	IDD, real time, computed downhole

[†] Corresponding short- and long-spacing densities can also be transmitted in real time.

Table 4-24. Azimuthal bulk density real-time computed channels.

Computed Channel	Description
DPHI_ECO_RT	Density porosity from RHOB_RT, real time
DPHB_ECO_RT	Density porosity from ROBB_RT, real time
DPHL_ECO_RT	Density porosity from ROBL_RT, real time
DPHU_ECO_RT	Density porosity from ROBU_RT, real time
DPHR_ECO_RT	Density poroslty from ROBR_RT, real tlme
ROIQ_ECO_RT	Bulk density image from quadrants, real time
ROSI_DYN_TOH_ECO_RT	Bulk density (16-sector), dynamic image oriented top of hole (U,R,B,L,U), real time
RHOB_ECO_RT[†]	Bulk density, average, real time
ROBB_ECO_RT[†]	Bulk density, bottom, real time
ROBL_ECO_RT[†]	Bulk density, left, real time
ROBU_ECO_RT[†]	Bulk density, up, real time
ROBR_ECO_RT[†]	Bulk density, right, real time
DRHB_ECO_RT[†]	Bulk density correction, bottom, real time
DRHL_ECO_RT[†]	Bulk density correction, left, real time
DRHO_ECO_RT[†]	Bulk density correction, average, real time
DRHU_ECO_RT[†]	Bulk density correction, up, real time
DRHR_ECO_RT[†]	Bulk density correction, right, real time

[†] The compensated density and density correction can be computed at surface if the corresponding short- and long-spacing detector densities are transmitted.

4.4.5.2. Recorded mode

Table 4-25. Azimuthal bulk density recorded-mode parameters.

Parameter Name	Description	Unit
RHOM_RM	Matrix density, recorded mode	g/cm^3
RHOF_RM	Mud filtrate density, recorded mode	g/cm^3
BS_RM	Bit size, recorded mode	in
STAB_SIZE	Stabilizer size	in
ALPHA_DEN_OPT	Density-enhanced vertical resolution processing switch	On/off (default is off)
EVRL	Enhanced vertical resolution processing averaging number of samples, recorded mode	NA[†]
DYN_IMAGE_OPT	Generated dynamic normalized image	Yes/no (default is yes)
WSDI	Window size of dynamic normalization image	NA
STOH	Top of hole sector	NA
IMAGE_MAX_SRHOB	Image RHOB (segment) right scale	NA
IMAGE_MIN_SRHOB	Image RHOB (segment) left scale	NA
IDD Processing Parameters		
IDQT	Image-derived quality threshold	NA
IMAGE_MAX_IDDQ	Image density quality right scale	NA
IMAGE_MIN_IDDQ	Image density quality left scale	NA

[†] Not applicable.

Table 4-26. Azimuthal bulk density recorded-mode channels.

Output Channel	Description
Azimuthal Average, 6-in Sampling	
RHOB	Bulk density, average
DRHO	Bulk density correction, average
DPHI	Density porosity from RHOB
RHOL	Long-spacing bulk density, average
RHOS	Short-spacing bulk density, average
QC_RHOB	Density quality indicator
TAB_DEN	Density time after bit
TICK_DEN	Density samples
Quadrants, 6-in Sampling	
ROBB	Bulk density, bottom
ROBL	Bulk density, left
ROBU	Bulk density, up
ROBR	Bulk density, right
DRHB	Bulk density correction, bottom
DRHL	Bulk density correction, left
DRHU	Bulk density correction, up
DRHR	Bulk density correction, right
DPHB	Density porosity from ROBB
DPHL	Density porosity from ROBL
DPHU	Density porosity from ROBU
DPHR	Density porosity from ROBR
ROLB	Long-spacing bulk density, bottom
ROLL	Long-spacing bulk density, left
ROLU	Long-spacing bulk density, up
ROLR	Long-spacing bulk density, right
ROSB	Short-spacing bulk density, bottom
ROSL	Short-spacing bulk density, left
ROSU	Short-spacing bulk density, up
ROSR	Short-spacing bulk density, right
16-Sector, 6-in Sampling	
RLSC	Long-spacing bulk density, 16-sector
RSSC	Short-spacing bulk density, 16-sector
ROSC	Bulk density, 16-sector
DRSC	Bulk density correction, 16-sector

IDD Channels	
IDRO	IDD
IDDR	IDD correction
IDDP	IDD tool path in sectors
IDDQ	Quality image from IDD, 16-sector
IDDR	IDD quality ratio
IDQS	Scaled quality image from IDD, 16-sector
QIDD	Tool path quality reference
QRBB	ROBB quality reference
Azimuthal Average, 0.1-ft Sampling	
RHOB_ALP	Bulk density at 0.1-ft resolution
RHOS_ALP	Short-spacing bulk density at 0.1-ft resolution
NRHO	Bulk density, enhanced vertical resolution
DPHN	Density porosity from NRHO
Quadrants, 0.1-ft Sampling	
ROBB_ALP	Bulk density at 0.1-ft resolution, bottom
ROSB_ALP	Short-spacing bulk density at 0.1-ft resolution, bottom
NROB	Bulk density, enhanced vertical resolution, bottom
DPNB	Density porosity from NROB
16-Sector, 0.1-ft Sampling	
ROSI	Bulk density image, 0.1 ft, 16-sector
ROSI_DYN	Dynamically normalized bulk density image, 0.1 ft, 16-sector
DRSI	Bulk density correction image, 0.1 ft, 16-sector
RLSI	Long-spacing bulk density image, 0.1 ft, 16-sector
RSSI	Short-spacing bulk density image, 0.1 ft, 16-sector

4.4.6. Azimuthal bulk density measurement specifications

Table 4-27. Azimuthal bulk density measurement specification at 200-ft/h ROP, assuming a three-level spatial average at zero standoff.

Item	Value	Remarks
Range	1 to 3.05 g/cm^3	
Axial resolution[†]	6 in	
Compensated azimuthal average density accuracy[‡]	0.015 g/cm^3	
Compensated azimuthal average density precision		
Stabilizer size = 7.875 in	0.006 g/cm^3	At 2.5 g/cm^3 and 200 ft/h
Stabilizer size = 8.25 in	0.006 g/cm^3	At 2.5 g/cm^3 and 200 ft/h
Stabilizer size = 9.375 in	0.009 g/cm^3	At 2.5 g/cm^3 and 200 ft/h

[†] With enhanced resolution processing.

[‡] Values provided in good hole conditions.

4.5. Azimuthal photoelectric factor

The low-energy gamma rays used for the azimuthal density measurement also induce a photoelectric response from the formation. This shallow, focused measurement is acquired every 10 seconds and is generally used for formation lithology identification and imaging.

4.5.1. Azimuthal photoelectric factor theory of measurement

After several scatterings, source gamma rays have lost a large fraction of their energy. At low energies, the photoelectric effect becomes important. In this process, a low-energy gamma ray collides with an electron in the outer shell of an atom and imparts its energy to the electron, which is ejected from the atom's electric field. The gamma ray is absorbed into the atom and disappears completely.

The probability of photoelectric absorption increases rapidly with an increasing atomic number (number of protons in the nucleus) as the outer shell of the atom becomes larger and the electrons in it more loosely bound than in atoms with a lower atomic number. The effect decreases with increasing gamma ray energy as scattering becomes dominant at higher energies. Because the photoelectric effect is strongly related to the atomic number of the atom that absorbs the gamma ray, determining the PEF helps identify formation lithology.

Formation lithologies containing elements with a large atomic number absorb some of the low-energy gamma rays before they can be detected. Conversely, fewer low-energy gamma rays are absorbed in lithologies containing materials with a low atomic number, so more low-energy gamma rays are detected. Consequently, the volumetric photoelectric factor, U, is inversely proportional to the low-energy gamma ray count rate.

U is related to the PEF and electron density, ρ_e, according to the equation

$$U \text{ (b/cm}^3\text{)} = PEF \text{ (b/electron)} \times \rho_e \text{(electrons/cm}^3\text{)} \qquad \text{(Equation 4-9)}.$$

Typical values of these parameters for common minerals and liquids are shown in **Table 4-28**.

Table 4-28. Density and photoelectric parameters of common minerals and liquids.

Name	PEF, b/electron	Bulk density, ρ_{bulk}, g/cm^3	ρ_e, electrons/cm^3	U = PEF × ρ_e, b/cm^3
Anhydrite	5.055	2.960	2.957	14.95
Barite	266.8	4.500	4.011	1070.0
Calcite	5.084	2.710	2.708	13.77
Dolomite	3.142	2.870	2.864	9.00
Quartz	1.806	2.654	2.65	4.79
Water	0.358	1.000	1.110	0.40
Saltwater (120 ppk)	0.807	1.086	1.185	0.96
Oil[†]	0.125	0.850	0.970	0.12
"Clean" sandstone[†]	1.745	2.308	2.330	4.07
"Dirty" sandstone[†]	2.70	2.394	2.414	6.52
Average shale[†]	3.42	2.650	2.645	9.05

[†] The values given are illustrative only.

4.5.2. Azimuthal photoelectric factor environmental corrections

The low-energy gamma rays involved in photoelectric interactions do not travel far. The photoelectric response measured in a gamma ray detector originates in an area in the immediate proximity of the detector. While this means that the measurement has good axial resolution, it also means that it is a very shallow measurement. Also, because the volume investigated by the two detectors does not overlap significantly, standoff compensation is not applied. Hence, the photoelectric measurement is a shallow, uncompensated measurement that is sensitive to borehole rugosity and heavy muds. Standoff in front of the detector will invalidate the photoelectric measurement. Even if not required for the deeper, compensated density measurement, IDD processing (refer to Section 4.4.2.2) is often useful for improving the photoelectric response by determining the point of best contact between the photoelectric-measuring detector and borehole wall. Determining the path of best contact and measuring the PEF along that path is the only environmental correction applied to the photoelectric measurement.

4.5.3. Azimuthal photoelectric factor applications

PEF is commonly displayed on logs for lithology identification purposes. Typical PEF values for common fluids and lithologies are shown in **Figure 4-48**.

Figure 4-48. Typical PEF values for common fluids and lithologies.

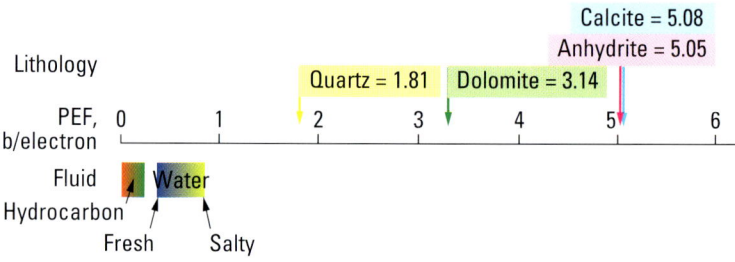

Note that the PEF response is not volumetric. It should only be used for qualitative lithology identification.

Quantitative interpretation should be performed with U, as its response conforms to the simple volumetric equation

$$U_{bulk} = \phi \times U_{fluid} + (1 - \phi) \times U_{grain}$$ (Equation 4-10),

where

U_{bulk} = bulk volumetric PEF
U_{fluid} = volumetric PEF of the fluid in the pores
U_{grain} = volumetric PEF of the formation grain.

Typical U values for common fluids and lithologies are shown in **Figure 4-49**.

Figure 4-49. Typical volumetric PEF values for common fluids and lithologies.

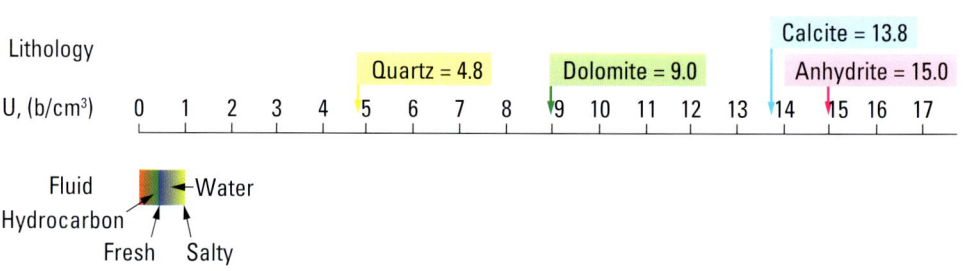

Consider the example of a clean formation, as outlined in **Table 4-29**. Note that the change in U and PEF is greater when the formation lithology changes from limestone to sandstone than when either the porosity or fluid type changes. This strong sensitivity to the formation grain and relative insensitivity to the fluid makes the PEF measurement useful for formation lithology identification.

Table 4-29. Volumetric PEF and density responses, calculated for changes in formation porosity, lithology, and fluid.

	U, b/cm^3	PEF, b/electron	Density, g/cm^3
Input			
Water	0.4	0.358	1
Gas	0.03	0.12	0.25
Limestone	13.8	5.08	2.71
Sandstone	4.8	1.81	2.65
	U, b/cm^3	PEF, b/electron	Density, g/cm^3
Output			
10-pu water-filled limestone	12.46	4.91	2.54
34-pu water-filled limestone	9.24	4.27	2.13
% change due to porosity	**34.8%**	**14.9%**	**19.3%**
34-pu water-filled limestone	9.24	4.27	2.13
34-pu water-filled sandstone	3.30	1.58	2.09
% change due to lithology	**64.3%**	**63.0%**	**1.9%**
34-pu water-filled limestone	9.24	4.27	2.13
34-pu gas-filled limestone	9.12	4.73	1.87
% change due to fluid	**1.4%**	**10.8%**	**12.0%**

The PEF measurement is relatively insensitive to porosity because of the low PEF values of the fluids that fill the pore spaces. PEF is very sensitive to changes in lithology.

Multiplying the PEF measurement by ρ_e yields U. As the formation electron density is sensitive to the porosity, the U measurement is sensitive to both porosity and lithology.

The density measurement is primarily sensitive to porosity and fluid type, but also has some sensitivity to lithology because of differences in their grain densities.

4.5.4. Azimuthal photoelectric factor log quality control

The shop calibration must have been performed less than 1 month before the acquisition date and must be in tolerance. The calibration must be performed in the calibration blocks appropriate for the stabilizer size. Refer to the Schlumberger LWD Quality Control Manual for details.

The source (identified by serial number) used during logging must match the source with which the tool was calibrated.

The PEF measurement is very shallow and often suffers from wellbore influences such as borehole rugosity, standoff, and heavy mud effects, which rapidly make the log unusable.

Barite mud is of particular concern because small quantities of barite near the detector render the photoelectric measurements unusable because of the extremely high photoelectric response of barite (its U is 1,090 b/cm^3).

Good contact with the borehole wall is essential for a valid measurement. IDD processing may improve the photoelectric measurement by calculating the PEF along the path of best contact between the tool and borehole wall.

If azimuthal PEF information is to be used, then rotation of the tool over the interval of interest must be verified.

The EcoScope hardware incorporates checks to verify correct performance of the acquisition detectors and electronics. Additional checks to ensure that the measurements fall within the range expected for downhole formation are outlined in the QC logic shown in **Table 4-30**.

Table 4-30. PEF QC flag logic.

QC_PEF	Hardware	Measurement
Green	Short-spacing gamma high-voltage control loop status is FINE and Short-spacing gamma high-voltage control is AUTO and Short-spacing gamma pulse shape compensation control loop status is FINE and Short-spacing gamma pulse shape compensation control is AUTO	$1.5 \leq PEB \leq 6$ and Collar rpm > 20 and ROP < 450 ft/h
Yellow	Short-spacing gamma high-voltage control loop status is COARSE or Short-spacing gamma pulse shape compensation control loop status is COARSE	$0.2 < PEB < 1.50$ or $6 < PEB < 10$ or Collar rpm < 20 or ROP > 450 ft/h
Red	Short-spacing gamma high-voltage control loop status is SEARCH or Short-spacing gamma pulse shape compensation control loop status is SEARCH, WAIT or Short-spacing gamma pulse shape compensation control is MANUAL or HOLD or Short-spacing gamma high-voltage control is MANUAL or HOLD or Density configuration and calibration parameters have not been received or have bad values or Density windows out of range CES_SSW1A < 200 or > 30,000 or CES_SSW3A < 1,000 or > 50,000 or CES_LSW3A < 100 or > 30,000 or Gamma-gamma density hardware flag set or Memory usage indicator = full	$PEB \leq 0.2$ or $PEB \geq 10$
White		No new density tick for more than 2 ft

4.5.5. Azimuthal photoelectric factor channels and parameters

4.5.5.1. Real time

Table 4-31. Azimuthal PEF real-time parameters.

Parameter Name	Description	Unit
IDD Processing Parameters		
IDQT	Image-derived quality threshold	

Table 4-32. Azimuthal photoelectric real-time transmitted channels.[†]

Transmitted Channel	Description
PEF_DH_ECO_RT	PEF average, real time, computed downhole
PEB_DH_ECO_RT	PEF bottom, real time, computed downhole
PEL_DH_ECO_RT	PEF left, real time, computed downhole
PEU_DH_ECO_RT	PEF up, real time, computed downhole
PER_DH_ECO_RT	PEF right, real time, computed downhole
PESI_STA_TOH_ECO_RT	PEF (16-sector), image oriented top of hole (U,R,B,L,U), real time
IDD Transmitted Channels	
IDDP_DH_ECO_RT	IDD tool path, real time, computed downhole
IDRO_DH_ECO_RT	Image-derived PEF, real time, computed downhole

[†] The real-time PEF QC indicator is incorporated in the QC_RHOB_DH_ECO_RT channel.

Table 4-33. Azimuthal PEF real-time computed channels.

Computed Channel	Description
PESI_DYN_TOH_ECO_RT	PEF (16-sector) dynamic image oriented top of hole (U,R,B,L,U), real time

4.5.5.2. Recorded mode

Table 4-34. Azimuthal PEF recorded-mode parameters.

Parameter Name	Description	Unit
STOH	Top of hole sector	NA[†]
WSDI	Window size of dynamic normalization image	NA
IMAGE_MAX_SPEF	Image PEF (segment) right scale	NA
IMAGE_MIN_SPEF	Image PEF (segment) left scale	NA
IDD Processing Parameters		
IDQT	Image-derived quality threshold	NA

[†] Not applicable.

Table 4-35. Azimuthal PEF recorded-mode channels.

Output Channel	Description
Azimuthal Average, 6-in Sampling	
PEF	Photoelectric factor
U	Volumetric PEF
QC_PEF	Photoelectric effect quality indicator
Quadrants, 8-in Sampling	
PEB	PEF, bottom
PEL	PEF, left
PEU	PEF, up
PER	PEF, right
UB	U, bottom
UL	U, left
UU	U, up
UR	U, right
16-Sector, 6-in Sampling	
PESC	PEF, 16-sector
USC	U, 16-sector
IDD Channels	
IDU	Image-derived volumetric PEF
IDPE	Image-derived PEF
IDDP	IDD tool path in sectors
IDDQ	Quality image from IDD, 16-sector
IDQR	IDD quality ratio
IDQS	Scaled quality image from IDD, 16-sector
QIDD	Tool path quality reference
QRBB	ROBB quality reference
16-Sector, 0.1-ft Sampling	
PESI	PEF image, 0.1 ft, 16-sector
PESI_DYN	Dynamically normalized PEF image, 0.1 ft, 16-sector
USI	U image, 0.1 ft, 16-sector

4.5.6. Azimuthal photoelectric factor measurement specifications

Table 4-36. Azimuthal PEF specifications at 200-ft/h ROP assuming a three-level spatial average at zero standoff.

Item	Value	Remarks
Range	1 to 100 units	
Axial resolution	2 in	
Azimuthal average accuracy[†]	5%	
Azimuthal average precision[†]		
Stabilizer size = 7.875 in	0.1 unit	For PEF = 3 and 200 ft/h
Stabilizer size = 8.25 in	0.15 unit	For PEF = 3 and 200 ft/h
Stabilizer size = 9.375 in	0.2 unit	For PEF = 3 and 200 ft/h

[†] Values provided apply in good hole conditions.

4.6. Neutron fundamentals

Neutron generation, interaction with the formation, and detection, are fundamental to the EcoScope measurements of

- hydrogen index (HI)

- thermal neutron porosity

- neutron-gamma density (NGD)

- elemental spectroscopy

- thermal neutron capture cross section (sigma)

The following outlines the common physics. Sections 4.7 to 4.11 should be read in conjunction with this material.

4.6.1. Pulsed neutron generator

The EcoScope tool uses a PNG to provide neutrons, thus removing the need for a chemical neutron source. This electronically controlled source of high-energy neutrons is composed of a high-voltage ladder and a minitron, as shown in **Figure 4-50**.

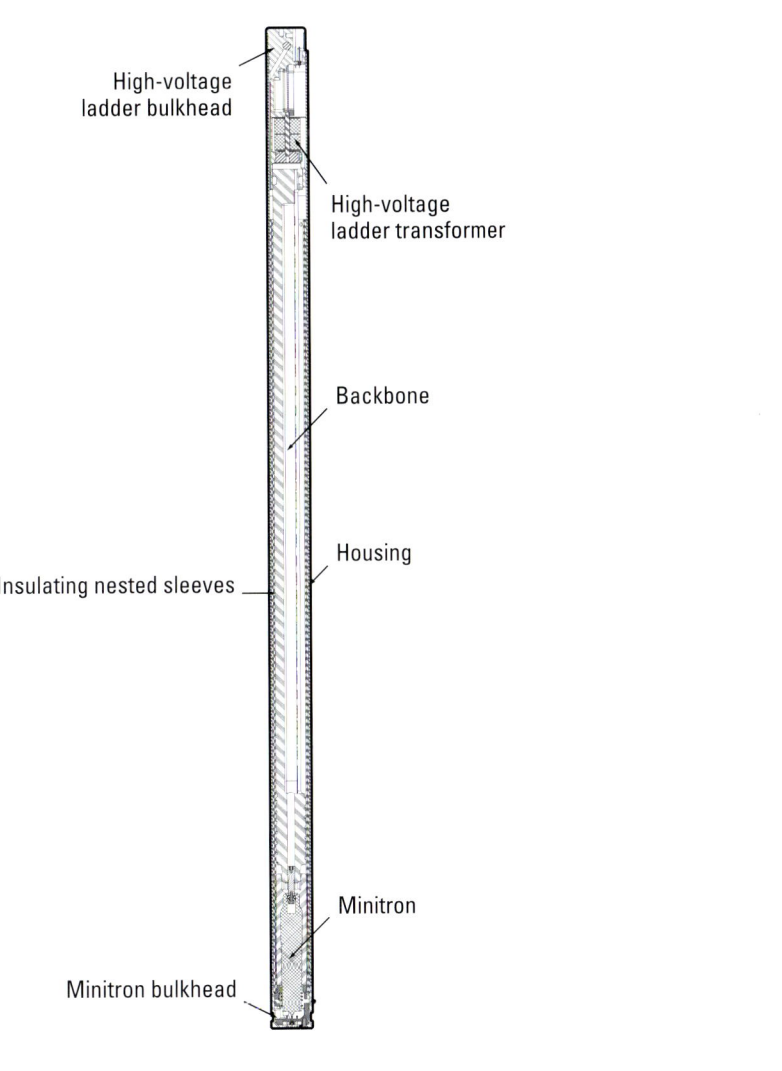

High-voltage
ladder bulkhead

High-voltage
ladder transformer

Backbone

Housing

Insulating nested sleeves

Minitron

Minitron bulkhead

Figure 4-50. The PNG, composed of a high-voltage ladder and a minitron.

The high-voltage ladder generates extremely high voltages (80 to 100 kV) that are used to accelerate hydrogen (H) ions across the minitron where they collide with other hydrogen atoms. Deuterium (hydrogen with one neutron) and tritium (hydrogen with two neutrons) atoms fuse to form helium (He) and eject a neutron in the process, shown schematically in **Figure 4-51**.

Figure 4-51. Deuterium and tritium fuse in a high-energy collision to form helium and expel a neutron with 14.1 MeV of kinetic energy.

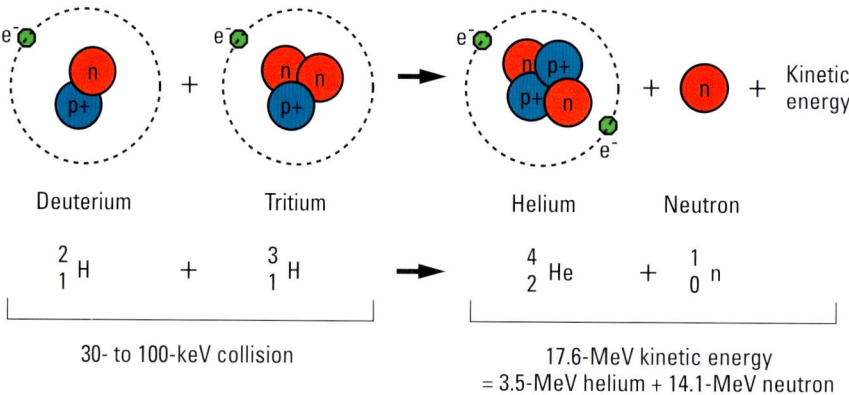

Deuterium   Tritium   Helium   Neutron

$$^{2}_{1}H \;+\; ^{3}_{1}H \;\longrightarrow\; ^{4}_{2}He \;+\; ^{1}_{0}n$$

30- to 100-keV collision   17.6-MeV kinetic energy
= 3.5-MeV helium + 14.1-MeV neutron

The supply of neutrons can be controlled by the voltages applied to the PNG. The ability to generate pulses of neutrons rather than the continuous stream supplied by a chemical neutron source facilitates a number of measurements, including capture spectroscopy, thermal neutron capture cross section (sigma), and NGD, each of which is outlined in subsequent sections. The HI and thermal neutron measurements do not require the pulsing feature of the PNG, but use average neutron count rates in near and far neutron detectors over an accumulation period of 10 seconds.

When operating, the PNG emits millions of high-energy neutrons each second. To ensure operational safety, the PNG cannot be operated outside of a logging tool. When installed in an EcoScope tool, a sequence of safety interlock conditions must be met before neutrons can be generated (**Figure 4-52**).

- User input—the field engineer must enter a password while initializing the tool to enable neutron generation.

- Physical key—a field neutron plug (FNP) must be installed in the readout port of the tool to enable neutron generation. During tool transportation, a conventional readout port plug is installed in the readout port. The FNP is kept in a separate locked container during transportation.

- Power supply—the EcoScope PNG can only be operated when run with a TeleScope measurement-while-drilling (MWD) system. When mud circulation spins the TeleScope power turbine, a command is sent to the EcoScope tool, giving permission for the PNG to fire. Other external voltages to the EcoScope will not fire the PNG. Only power generated by the TeleScope tool during mud circulation will enable the PNG to generate neutrons.

- Environment check—sensors in the tool must register a temperature greater than 0 degC and sufficient fluid pressure around the tool to confirm that the tool is below the surface before the PNG will fire.

On removal of the FNP from the tool readout port at surface to retrieve the data from the memory of the tool, the password is reset, disabling the PNG.

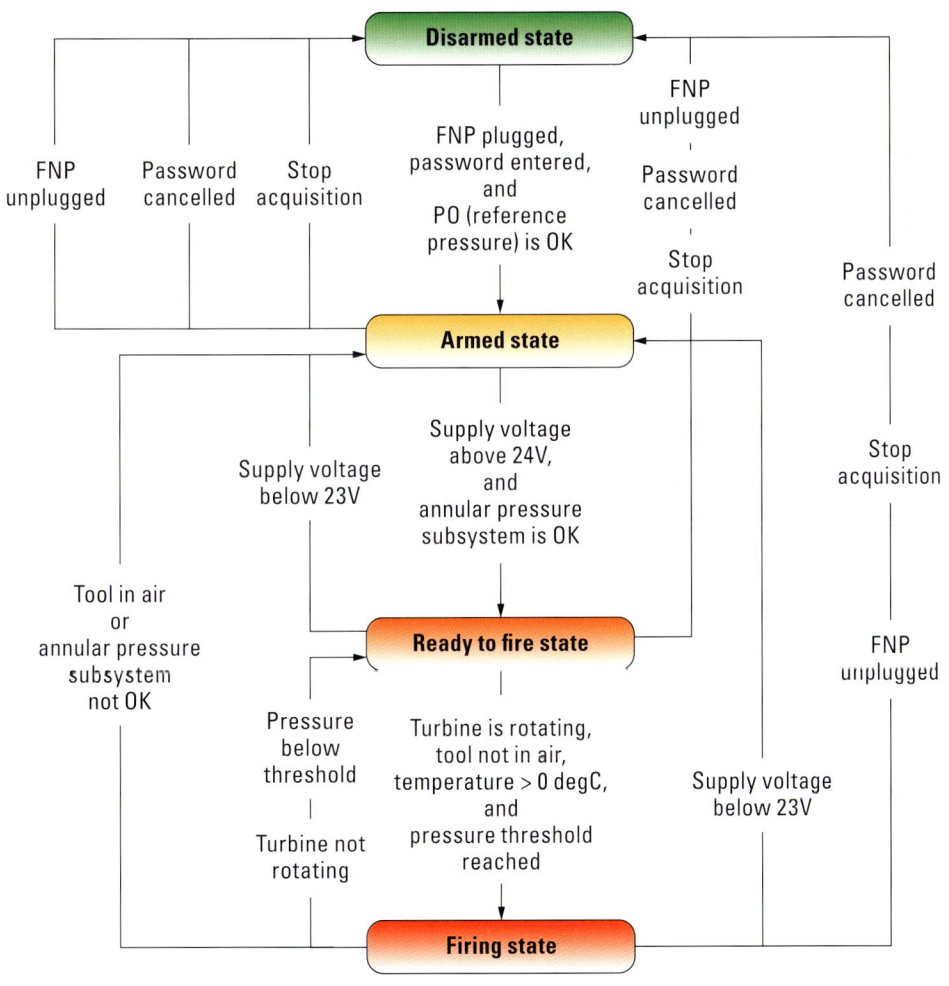

Figure 4-52. PNG safety interlock system flowchart.

4.6.2. Neutron interactions

Several types of neutron interactions lead to the slowing down and the absorption of neutrons.

Figure 4-53 shows the typical time scale and the associated neutron energies involved in the slowing down of 14-MeV neutrons from a PNG to thermal energy and their eventual capture.

Figure 4-53. Neutron slowing down, thermal neutron life, and final capture.

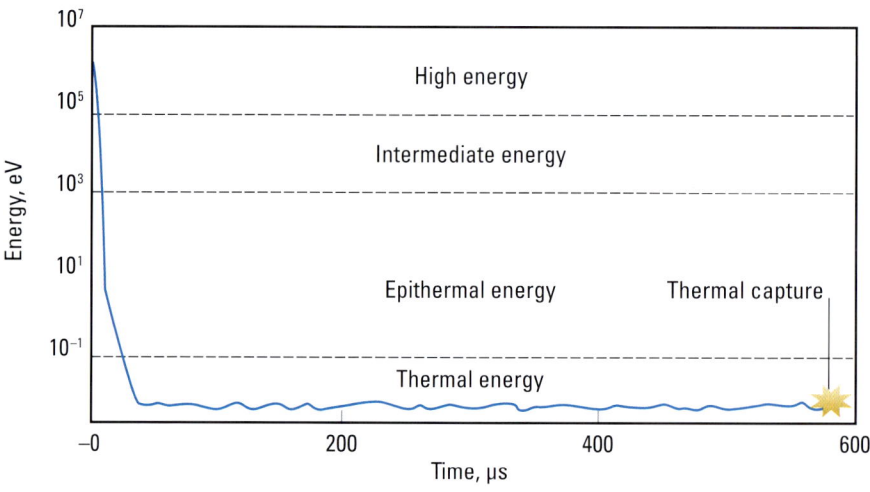

Neutron energies are broadly classed as

- high-energy   > 100 keV
- intermediate   100 eV to 100 keV
- epithermal   0.4 to 100 eV
- thermal   0.025 eV average at 77 degF.

After emission, neutron energy decreases rapidly (in approximately 40 μs) to thermal energy, at which point the neutron bounces around randomly until it is captured by a nucleus. The average energy of a thermal neutron is solely a function of the temperature. At thermal energy, the neutron loses and gains energy while it is colliding with the nuclei in the material through which it travels. The times shown in Figure 4-53 are for a single typical neutron. Others may slow down more quickly or more slowly or be consumed in a nuclear reaction before they reach thermal energies.

The interactions that result in the slowing down and absorption of neutrons include

- neutron elastic scattering
- neutron inelastic scattering
- neutron reactions
- neutron capture.

In the first two types of interactions, the neutron loses energy (i.e., it gets slowed down). In many of the nuclear reactions, the neutron will be consumed. The capture of a neutron (absorption of the neutron) by a nucleus is an important reaction, particularly for thermal neutrons. For neutron porosity measurements, elastic scattering (particularly by hydrogen) is by far the most important interaction.

4.6.2.1. Neutron elastic scattering

Neutrons interact exclusively with nuclei in the material through which they pass. They are not affected by the presence of electrons. Elastic scattering occurs when a neutron bounces off a nucleus. Such a collision is similar to the collision between two steel balls. The incoming neutron loses some of its kinetic energy to the nucleus in the collision. The energy lost by the neutron equals the kinetic energy gained by the collision partner nucleus. Total kinetic energy and momentum are conserved in an elastic collision. If the nucleus is much larger than the neutron, only a small fraction of the neutron's energy is imparted to the nucleus (**Figure 4-54**).

Figure 4-54. Elastic scattering on a heavier nucleus. The incoming neutron (A) only loses a small fraction of its energy (B).

The largest energy transfer occurs in a head-on collision with a particle of the same mass. **Figure 4-55** shows a head-on collision of a neutron with a proton (hydrogen nucleus). In this case, the neutron loses all of its kinetic energy (it stops), and the collision partner recoils with the same energy (and thus speed) as the incoming neutron.

Figure 4-55. A head-on collision between neutron and H nucleus (proton). The incoming neutron (A) loses all its energy in a single collision (B).

Because the neutron mass is virtually equal to the proton mass (mass of the hydrogen nucleus), hydrogen is particularly effective in slowing down neutrons.

4.6.2.2. Neutron inelastic scattering

At high energies (>1 MeV) collisions between a neutron and nucleus can put the nucleus in an excited state, which is normally short lived. The neutron leaves the collision with an energy loss equal to the energy needed to move the nucleus into its excited state plus the kinetic energy (recoil) imparted to the nucleus. In most cases, the excited state of the nucleus decays by the emission of one or more gamma rays in less than 1 ns. **Figure 4-56** shows the excitation and decay of a nucleus by the inelastic scattering of a high-energy neutron and the subsequent emission of a gamma ray.

Figure 4-56. Inelastic scattering of a neutron, with a target nucleus. The incoming neutron (A) loses a large fraction of its energy in exciting the nucleus (B), and shortly after the collision (10^{-15} to 10^{-9} seconds), the excited nucleus deexcites through the emission of one (or more) gamma rays (C).

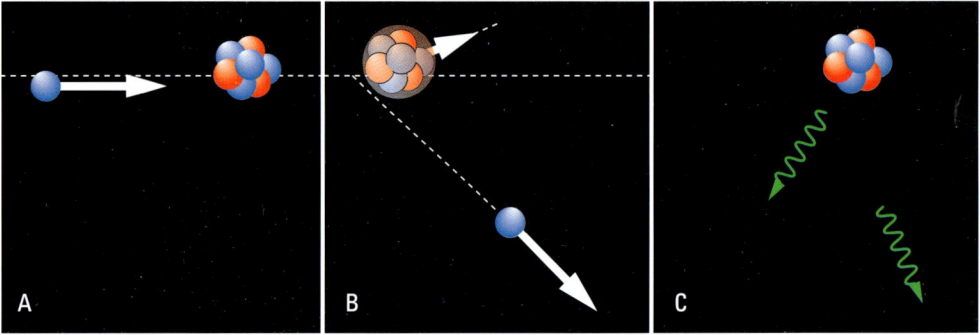

4.6.2.3. Neutron reactions

In a neutron reaction, the incoming neutron collides with the target nucleus by knocking out one or more nucleons. In many cases, the neutron is not reemitted after the interaction. An example of such an interaction is

neutron + $^{16}O \rightarrow$ ^{16}N + proton.

This interaction leads to oxygen-activation because the nucleus produced is radioactive with a half-life of approximately 7.1 seconds. An alternative reaction between the neutron and oxygen is

neutron + $^{16}O \rightarrow$ ^{13}C + ^{4}He.

This reaction, illustrated in **Figure 4-57**, leads to an excited state of carbon (C)-13, which decays to the stable ground state through gamma ray emission. In both these cases, the neutron is consumed as a consequence of the interaction, and gamma rays of characteristic energies are released.

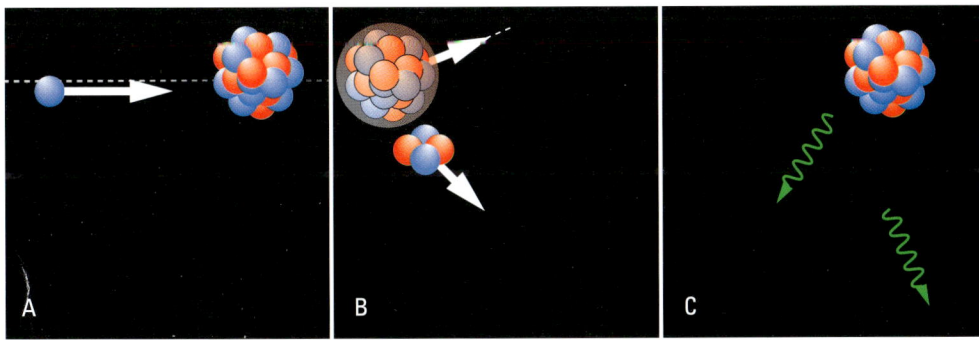

Figure 4-57. An inelastic reaction of a neutron with a heavier nucleus. As a result of the reaction of the incoming nucleus (A), the helium nucleus (alpha particle) gets knocked out of the target nucleus, the neutron is absorbed (B), and the highly excited nucleus emits one or more gamma rays (C).

4.6.2.4. Neutron capture

Neutron capture leads to the absorption of the neutron by the target nucleus. This reaction is important for thermal- and, to a much lesser extent, higher-energy neutrons. The sequence of events during neutron capture is illustrated in **Figure 4-58**. The target nucleus absorbs the thermal neutron. The newly formed nucleus is in a highly excited state. This excited state decays by the emission of one or more gamma rays to the ground state. The emission of the gamma rays takes typically 10^{-14} to 10^{-9} seconds.

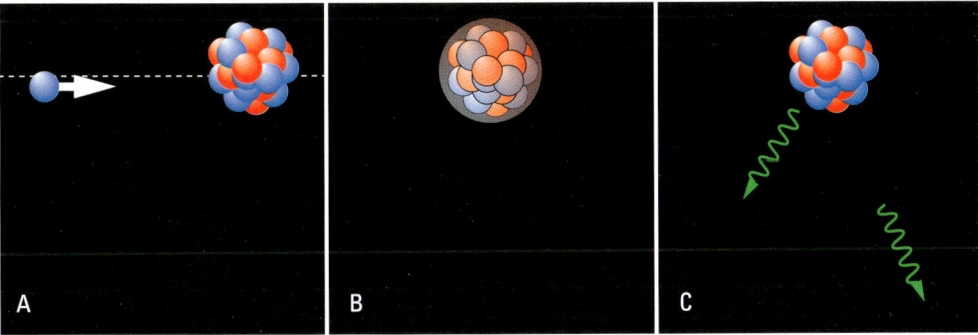

Figure 4-58. Neutron capture, where the incoming slow neutron (A) gets absorbed by a nucleus, and the resulting new nucleus is highly excited (B) and deexcites through gamma ray emission to its ground state within 10^{-14} to approximately 10^{-9} seconds.

For example, the capture of a neutron by silicon (Si)-28,

neutron + $^{28}Si \rightarrow$ ^{29}Si,

results in the creation of an excited silicon-29 nucleus, which deexcites by the emission of several gamma rays. These gamma rays, which are specific to this isotope, can be used to identify the element that emits them. Therefore, they can be used to detect and quantify the presence of silicon in well logging. Silicon-29 is a stable isotope and does not decay further after the emission of the gamma rays. However, in many cases, the nucleus resulting from the capture reaction is not stable and will decay further with a half-life ranging from seconds to many years.

The probability of a thermal neutron being captured by a nucleus is described by the thermal neutron capture cross section. This cross section varies widely from element to element and from isotope to isotope. The cross section can be pictured as the effective cross-sectional area of a given nucleus for a capture reaction with a neutron. This is also called the microscopic capture cross section. It is typically measured in units of barn (1 barn (b) = 10^{-24} cm^2).

For well logging, the macroscopic capture cross section (sigma) is used. The macroscopic cross section, expressed in capture units, cu, is the total cross section seen by a neutron passing through a piece of matter with a thickness of 10^{-3} cm and an area of 1 cm^2. A cu therefore equals 10^{-3} cm/cm^2 = 10^{-3} cm^{-1}. A simplified picture is shown in **Figure 4-59**. The part on the left shows a schematic view through a thin layer (10^{-3} cm) of material. The circular areas represent the effective cross section of the individual nuclei. For the purpose of the definition, it is assumed that there is no overlap of the cross-sectional areas.

Figure 4-59. A simplified depiction of the macroscopic cross section. For the purpose of the definition, it is assumed that there is no overlap among the circles, which represent the cross sections of the individual nuclei.

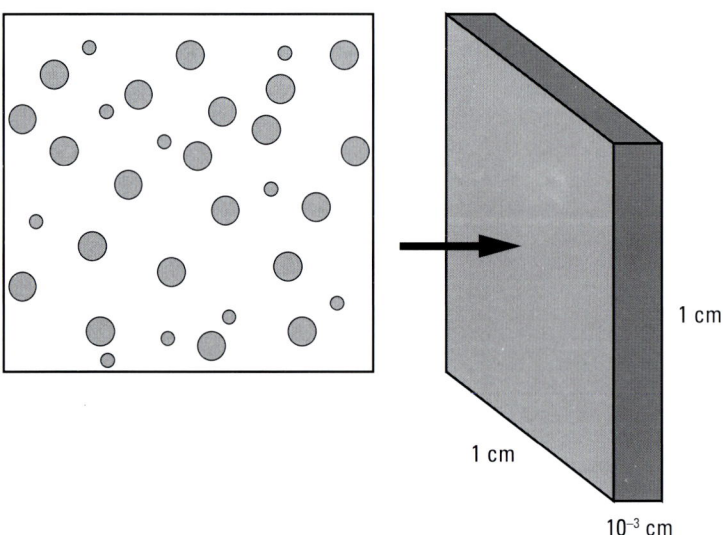

1 cm

1 cm

10^{-3} cm

The neutron capture cross section of the formation can be computed from the decay of the thermal neutrons as a function of time. This can be determined either by measuring the neutron die-away directly or by measuring the die-away of capture gamma rays. The decay time of the thermal neutron population is related to sigma by

$$\Sigma = \frac{4{,}550}{\tau} \text{ (cu)}$$

(Equation 4-11).

where

Σ = sigma

τ = decay constant of the quasiexponential decay of the neutrons or the associated gamma rays in us.

Sigma logging makes use of the fact that chlorine (Cl) has a much larger capture cross section than hydrogen. Saturated saltwater will therefore manifest itself by a large sigma (approximately 123 cu) while oil or freshwater show a lower sigma (approximately 22 cu).

4.6.3. Neutron detectors

All Schlumberger neutron tools use helium-3 detectors to detect thermal and epithermal neutrons. This detector type is chosen because of its high sensitivity to neutrons and its virtual immunity to gamma rays.

The basic design of a helium-3 tube is shown in **Figure 4-60**. The tube consists of a cylinder, which is filled with helium-3 gas, typically at a pressure of 3 to 20 atm. A thin wire passes through the middle of the cylinder. A positive high voltage, between 1,500 and 2,500 V, is applied to the wire, creating a high electric field close to the center of the tube.

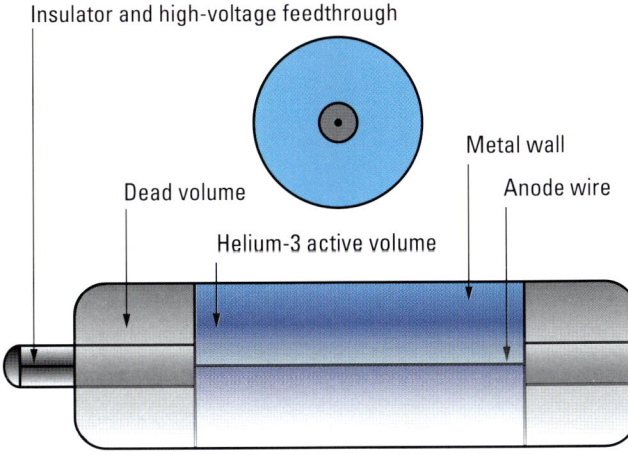

Insulator and high-voltage feedthrough

Metal wall

Anode wire

Dead volume

Helium-3 active volume

Figure 4-60. Basic construction of a helium-3 thermal neutron detector.

Because neutrons do not carry an electrical charge, their detection is performed indirectly through their interaction with the nuclei of the helium-3 gas. The following interaction takes place.

Neutron + ^{3}He $\rightarrow$ ^{3}H + proton + 750 keV.

The resulting tritium nucleus (hydrogen-3) and proton are charged particles that are stopped in the detector. In the process, they ionize the gas, creating positive ions and free electrons. The electric potential accelerates the electrons, which then ionize more atoms along the path and generate more free electrons. The cascade of electrons (and ions) creates a current, which is detected and further amplified in the detector electronics.

The ^{3}He reaction allows detection of both thermal and epithermal neutrons. To create a detector that only responds to epithermal neutrons, a helium-3 tube is wrapped in cadmium (Cd) foil. Cadmium absorbs thermal neutrons with energies up to 0.4 eV very well, thereby preventing them from entering the detector, while it is almost completely transparent to epithermal neutrons with energies greater than 0.4 eV.

4.7. Hydrogen index

Enhanced neutron energy, count rate, and a suite of detectors enable determination of formation HI. This measurement is less sensitive than the corresponding thermal neutron porosity to environmental conditions such as borehole effects, salty formation water, or shale, thereby simplifying interpretation.

4.7.1. Hydrogen index theory of measurement

Please refer to Section 4.6 for details on neutron generation, interactions, and detection.

The neutron measurements use count rates in near and far helium-3 tubes. High-energy neutrons emitted by the PNG interact with atoms in the surrounding formation, borehole, and tool. Because of the dominant role hydrogen plays in moderating (slowing down) the neutrons to the low energies at which they are usually measured, elastic scattering with hydrogen usually dominates the neutron response.

HI, the quantity of hydrogen per unit of volume relative to water at standard temperature and pressure, can be related to porosity because hydrogen is mostly present in the fluid (with relatively little hydrogen in the formation matrix).

$$HI = \frac{\text{Hydrogen concentration}}{\text{Hydrogen concentration in water at standard temperature and pressure}} \qquad \text{(Equation 4-12).}$$

An HI of 1 indicates that the amount of hydrogen is equal to that if the same volume were entirely filled with water at standard temperature and pressure.

By measuring the number of neutrons returning to the near and far helium-3 detectors the formation, HI can be inferred. The formation HI is subsequently used to derive ϕ according to the equation

$$HI_{bulk} = \phi \times HI_{fluid} + \text{environmental effects} \qquad \text{(Equation 4-13)},$$

where

HI_{bulk} = HI of the formation

HI_{fluid} = HI of the fluid filling the pore space.

Formation porosity is either expressed as a volume to volume ratio (v/v), such that it takes a value between 0 and 1, or as a percentage, such that it takes a value between 0 and 100. The volume fraction form must be used in calculations.

While elastic scattering with hydrogen dominates the neutron response, it should be remembered that all the elements in a material and each of the neutron interactions described in Section 4.6.2 contribute to the observed response. These environmental effects need to be taken in to account when using neutron data.

The cumulative effect of the neutron interactions is a reduction in the number of neutrons passing from the PNG through the formation to the detector. This can be described semiquantitatively in terms of three fundamental formation properties: HI, density, and sigma.

$$N(x) \approx N_0 e^{-\alpha.HI.x} e^{-\beta.\rho.x} e^{-\Sigma x} \qquad \text{(Equation 4-14)},$$

where

$N(x)$ = number of neutrons per second after traveling a distance x

N_0 = number of source neutrons per second

x = distance traveled in the material

α = neutron microscopic cross section (except capture) for hydrogen

β = neutron microscopic cross section (except capture) for elements other than hydrogen

ρ = material density (atom concentration of other elements).

The quantities α and β are constants for a given detector spacing and neutron source energy. They can be interpreted as HI and density sensitivities.

The first factor in Equation 4-14, $e^{-\alpha.HI.x}$, approximates the ideal response of an HI measurement; one which responds only to the HI of the formation and has no dependence on the other elements of the material (or the capture cross section of hydrogen).

The second factor in Equation 4-14, $e^{-\beta.\rho.x}$, represents the dependence of the response on the elements other than hydrogen. As these are generally much heavier than hydrogen, they tend to elastically scatter the neutrons with minimal reduction in neutron energy. Because these heavy elements account for most of the density of the material, this factor represents the density dependence of the response.

The third factor in Equation 4-14, $e^{-\Sigma x}$, approximates the dependence of the response on the thermal neutron capture cross section of all the elements in the material, including hydrogen. Thermal neutron capture reduces the number of neutrons that reach the detector by absorbing them. This can be significant in saline water because of the large thermal neutron capture cross section (sigma) of chlorine in the saltwater.

The thermal neutron capture effect can be dramatically reduced by detecting epithermal neutrons rather than thermal neutrons. Rather than wrap the EcoScope neutron detectors in cadmium to absorb the thermal neutrons impinging on the detectors, which would significantly reduce the number of neutrons detected, the natural thermal neutron absorption of the thick EcoScope collar walls is used. **Figure 4-61** shows how increasing wall thickness increases the proportion of epithermal to thermal neutrons detected in the tool. This increase is because the iron (Fe) (and several other elements) in the collar material have high thermal neutron capture cross sections.

Figure 4-61. A plot of normalized neutron count rate versus energy as a function of collar wall thickness, indicating that the proportion of epithermal to thermal neutrons detected in the tool increases with increasing collar wall thickness.

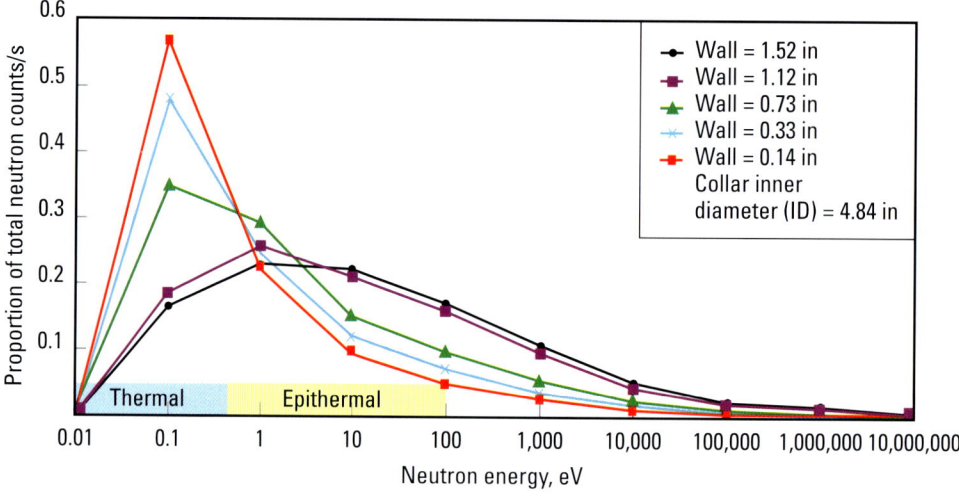

The thermal neutron capture effect on the measurement is further reduced by taking the ratio of the responses of two detectors, one placed near and the other far from the neutron source. Because both detectors exhibit a thermal neutron response, taking the ratio of their count rates significantly reduces the thermal capture response through cancellation. A much-reduced thermal response is achieved without sacrificing count rate, effectively reducing Equation 4-14 to

$$N(x) \approx N_0 e^{-\alpha.HI.x} e^{-\beta.\rho.x}$$ (Equation 4-15).

Figure 4-62 shows a comparison of the near/far ratio dynamic range of the adnVISION LWD tool and the EcoScope tool. The uncorrected ratio from the EcoScope tool has a slightly smaller dynamic range (relative change of the ratio from 0 to 100 pu) because of the longer slowing-down length and the increased density effect associated with the high-energy neutrons.

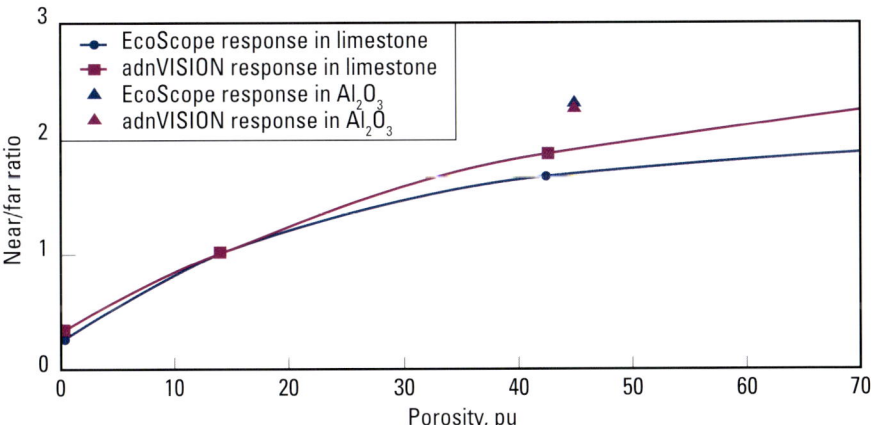

Figure 4-62. Near/far ratio of the EcoScope tool (blue), compared with that of the adnVISION tool (pink). The two triangles are measured in an aluminum oxide (Al_2O_3) block, simulating a shale response.

The density effect could be minimized by placing both neutron detectors close to the source, but this would limit the DOI of the measurement. In the EcoScope tool, the neutron detectors are positioned to give greater DOI than previous-generation LWD neutron measurements. An explicit density correction is applied using one of the two density measurements made by the EcoScope tool.

The dynamic range after correcting for the density effect is shown in **Figure 4-63**. The effect of the correction is indicated by the aluminum oxide (Al_2O_3) block points (representing shale with a grain density 3.97 g/cm³), which now lie on the same curve as the limestone points. The density correction has only a small effect at low porosities, but there is a large improvement in the effective dynamic range at high porosities. By taking into account the density effect, the HI-based thermal neutron porosity, BPHI, from the fully corrected ratio closely represents the HI of the formation.

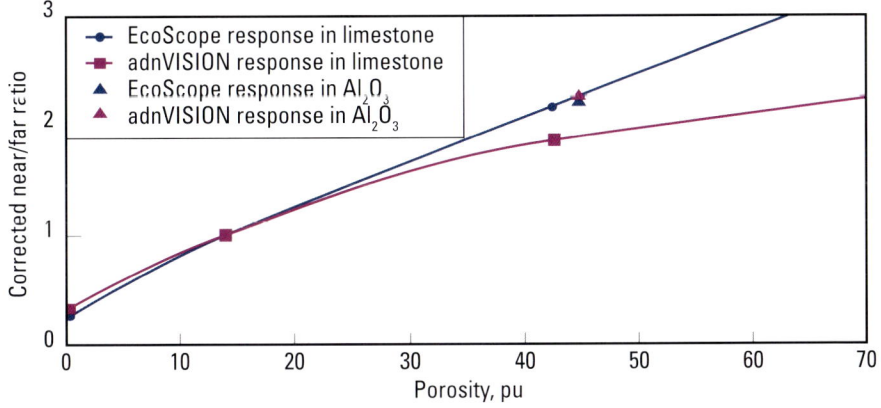

Figure 4-63. The density-corrected near/far ratio (blue), compared with the previous-generation LWD tool near/far ratio (pink). The density-corrected ratio is used for the BPHI measurement.

The effect of the density correction on the accuracy of the neutron measurement is small. The two curves in **Figure 4-64** show the effect on the porosity of an error on the density. Even with a large error on the density, the porosity remains within the specified accuracy.

Figure 4-64. Effect on the density-corrected limestone porosity of a 0.015-g/cm^3 (blue) and a 0.07-g/cm^3 (pink) density uncertainty. If the density measurement is within the specified accuracy, the bias of the porosity does not exceed 0.5 pu at 40 pu—even with a large density error, the resulting bias of the porosity answer is less than 1 pu below 20 pu.

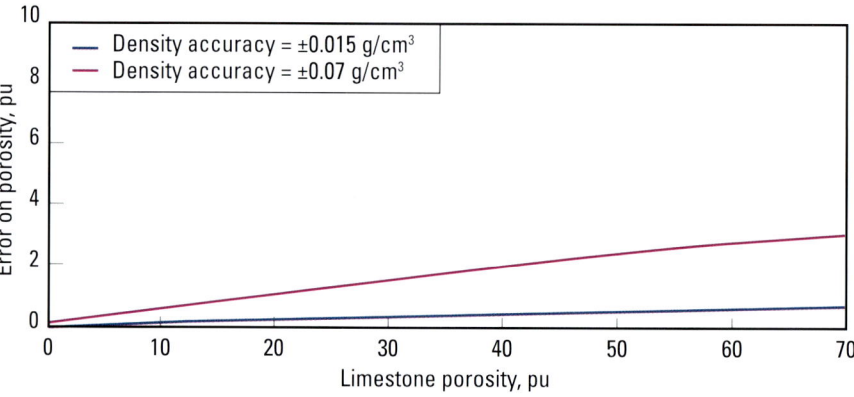

Isolating the ideal HI response,

$$N(x)_{\text{density-corrected}} = N(x)e^{\beta \cdot \rho \cdot x} \approx N_0 e^{-\alpha \cdot HI \cdot x}$$ (Equation 4-16).

where

$N(x)_{\text{density-corrected}}$ = measured neutron count rate corrected for the density effect.

This equation shows that the measured detector count rate, $N(x)$, corrected for the density effect, $e^{\beta \cdot \rho \cdot x}$, will respond only to the neutron source output, N_0, and the HI effect, $e^{-\alpha \cdot HI \cdot x}$. Because the number of neutrons emitted per second by the PNG may vary over time, a monitor detector is positioned close to the PNG source to provide a count rate that is proportional to the neutron source output. This count rate is used to normalize the near and far detector count rates to the source output.

The neutron response can be visually approximated as a vector in a three-dimensional (3D) space where the dimensions correspond to formation density, HI, and sigma. Because the majority of the sigma response is eliminated by thermal neutron absorption in the collar material and use of the near/far ratio, for simplicity, **Figure 4-65** shows only the density and HI dimensions.

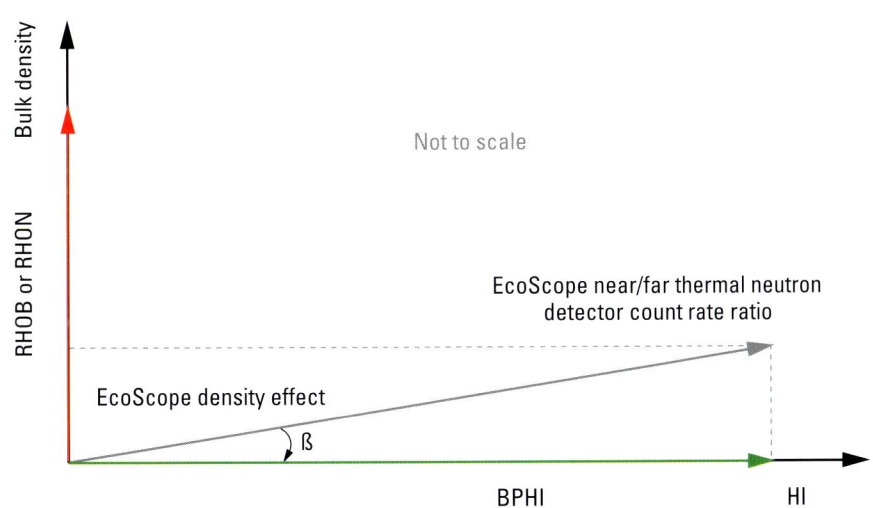

The density sensitivity, which is related to β, is a constant for a fixed detector spacing. The EcoScope near detector is positioned so as to have negligible density sensitivity. Density correction is applied to the far detector counts so that when the near/far detector count rate ratio is taken, an HI response is obtained. This is the basis of the BPHI measurement delivered by the EcoScope system.

4.7.2. Hydrogen index environmental corrections

After PNG neutron output normalization and density correction have been applied to the raw count rates, the environmental effects of the borehole, temperature, and formation salinity require correction.

Figure 4-66 compares the corrections for the adnVISION TNPH measurement with those for the EcoScope BPHI measurement under the same conditions. The significant reduction in the borehole size correction is because of the longer source-detector spacing and the high-energy neutrons from the EcoScope PNG traveling farther in the formation before slowing. Both these reasons result in the deeper DOI of the EcoScope BPHI measurement compared with the adnVISION TNPH measurement. This reduction in borehole size sensitivity is a significant advantage for a real-time neutron measurement where the size of the borehole may not be well known. The mud HI correction is also smaller for the same reasons.

Note that the formation salinity correction for the TNPH measurement reduces the apparent porosity at low salinities and increases the apparent porosity at high formation salinity and porosity. This change in the sign of the correction is because at low salinities, the capture effect of the chlorine in the water dominates, reducing the number of neutrons reaching the detectors and hence, increasing the apparent porosity. In high-porosity formations filled with high-salinity water, the reduction in the HI of the water due to displacement by large quantities of salt partially counteracts the capture effect, resulting in the bend in the correction curve. The EcoScope BPHI formation salinity correction is consistently positive because the measurement is almost immune to the capture effect, so the only correction with increasing fluid salinity is to account for the decrease in fluid HI.

The correction charts shown are simplifications of the corrections applied in the acquisition software where the coupling between the borehole size and mud property effects are respected by applying linked corrections.

Borehole correction is used to compensate for the size of the borehole, assuming a centered tool. Even though the neutron measurement is not azimuthally focused, an effect due to the tool position in the borehole remains. While taking the ratio of the near and far count rates compensates partially for the standoff effect, and the deeper DOI of the EcoScope neutron measurement further reduces the borehole and the eccentering effects, a standoff correction is still required. This correction is accomplished by computing the apparent porosity from the near detector count rate and comparing it with the ratio porosity. It is computed using an approach similar to the spine-and-ribs algorithm used for density standoff compensation (refer to Section 4.4.2.1) based on the difference between the two apparent porosities. The applied standoff correction is presented as DBPHI.

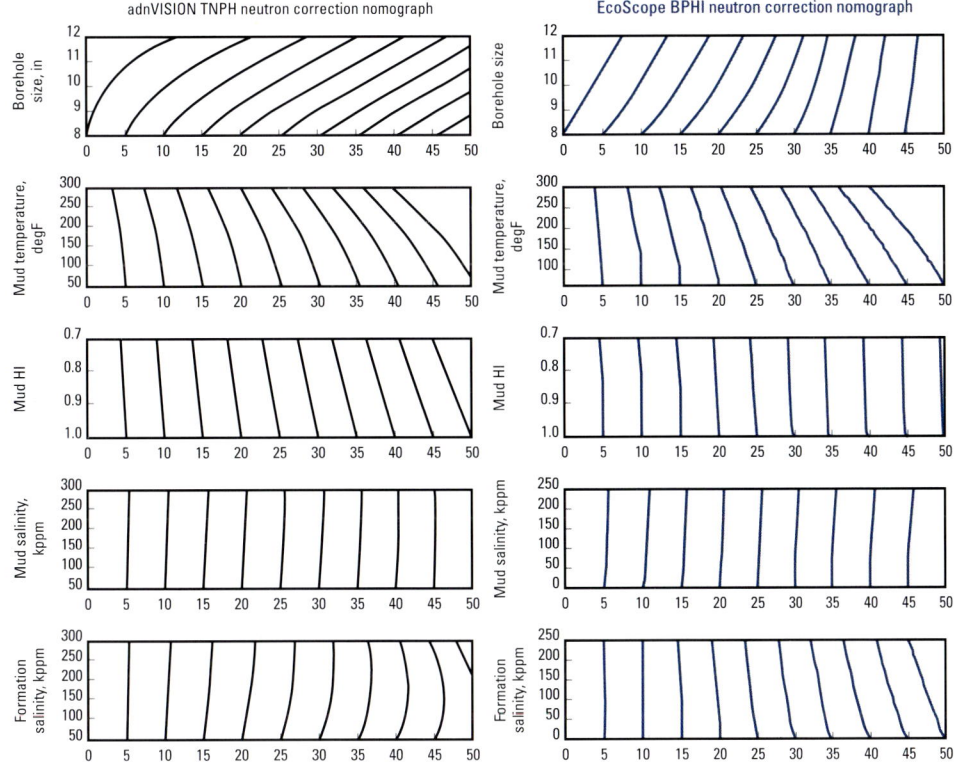

Figure 4-66. Environmental correction nomographs, showing the significantly smaller environmental corrections required to the EcoScope BPHI measurement compared to the adnVISION TNPH.

While the dominant effect on the neutron response is because of interactions with the hydrogen in the pore fluids, the neutrons also interact with the atoms in the solid part of the formation. Various solids (lithologies) are composed of different elements and so have slightly differing effects on the neutrons. The three most common clean lithologies are

- quartz (sandstone)—SiO_2

- calcite (limestone)—$CaCO_3$

- dolomite (dolostone)—$CaCO_3MgCO_3$.

The neutron detector count rates are calibrated such that the transform to porosity gives the correct value in a clean, freshwater-filled limestone. As quartz is not as effective at slowing neutrons as calcite in formations of the same porosity, more neutrons are counted at the detectors if the lithology is sandstone than if the lithology is limestone. Increased count rate implies a lower apparent porosity (the effect is equivalent to having less hydrogen in the pore space to slow the neutrons, allowing them to be detected).

Figure 4-67 shows this lithology effect on the EcoScope BPHI. For example, in a limestone formation of 18 pu filled with freshwater, BPHI calculated on a limestone matrix (green solid line) will read 18 pu, but if the lithology is sandstone, BPHI (limestone) will read 16 pu. In the same circumstances, the adnVISION TNPH measurement (green dashed line) will also read 18 pu in the limestone lithology, but 15 pu in the sandstone.

Dolomite is more effective at slowing neutrons than calcite, so the count rate at the detectors will be lower in dolomite, resulting in a higher apparent porosity. In 18-pu dolomite, BPHI (limestone) will read 20.5 pu while the adnVISION TNPH will read 22.5 pu.

The appropriate lithology should be used in calculating the neutron response. If the formation is primarily sandstone, and a neutron response calculated on the assumption of a limestone formation is used, then the neutron response will consistently read too low.

As seen in Figure 4-67, the lithology effect is significantly smaller on the BPHI measurement than on the adnVISION TNPH, resulting in less error if the formation lithology is not accurately defined.

Figure 4-67. Porosity equivalency chart showing the effect of the three major lithologies on the adnVISION TNPH and EcoScope BPHI measurements. The closer spacing of the solid lines indicates that the EcoScope BPHI provides a neutron porosity with less lithology dependence than the adnVISION TNPH.

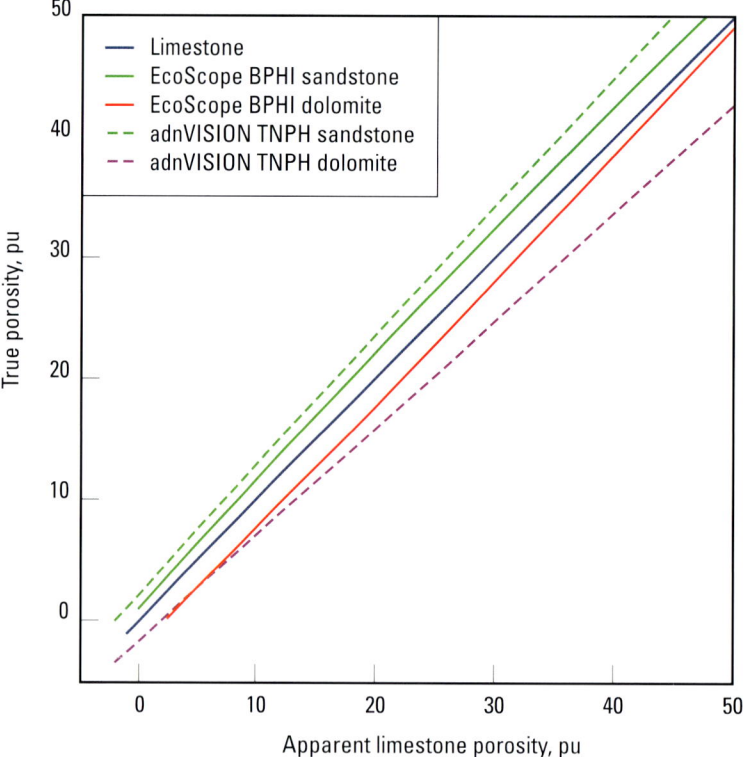

4.7.3. Hydrogen index applications

The density and neutron measurements are commonly used together. The density acts as the primary porosity measurement, but the fluid and formation densities must be known to derive an accurate porosity from the density according to the equation

$$\phi = \frac{\rho_{grain} - \rho_{bulk}}{\rho_{grain} - \rho_{fluid}}$$

(Equation 4-17).

Because the neutron responds primarily to the HI of the fluid, comparing the two measurements on a compatible scale helps diagnose whether the "standard conditions" of a freshwater-filled limestone exist in the formation.

For example, a limestone-compatible scale plots the density from 1.7 to 2.7 g/cm^3 against the limestone HI response plotted from 60 to 0 pu, as shown in **Figure 4-68**. For a 60-pu clean limestone formation of grain density 2.7 g/cm^3 filled with freshwater with a density of 1 g/cm^3, the density equation

$$\rho_{bulk} = \phi\, \rho_{fluid} + (1 - \phi)\, \rho_{grain}$$

(Equation 4-18)

gives a bulk density of 1.68 g/cm^3. For simplicity, the scale is generally rounded to 1.7 g/cm^3.

Note that the curves in Figure 4-68 represent typical field data (all environmental corrections other than formation salinity applied) in an uninvaded formation.

Figure 4-68. Separation between the density and HI measurements. When presented on a lithology-compatible scale (limestone-compatible in this example), these reveal whether the formation is a clean, freshwater-filled layer of the selected lithology.

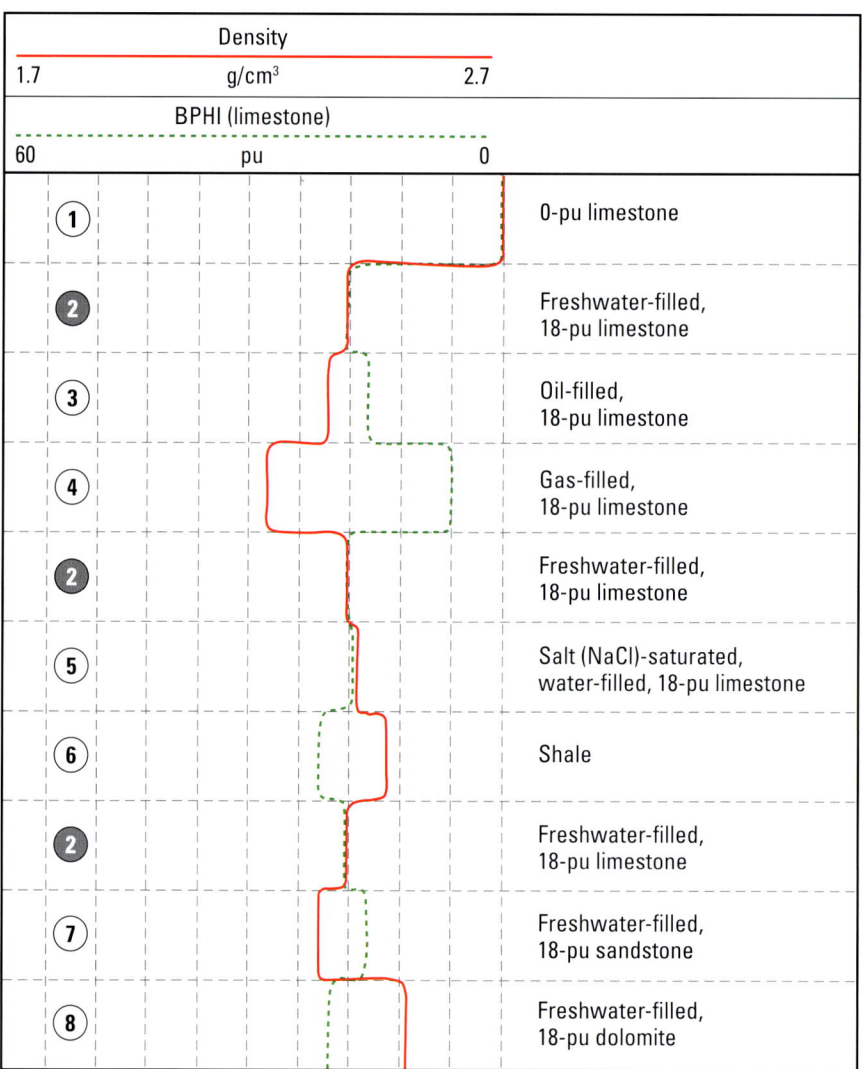

With both the density and HI measurements plotted on a limestone-compatible scale, the two curves will overlay when a clean, water-filled limestone of any porosity is encountered, as shown in Rows 1 and 2 of Figure 4-68.

If the freshwater is replaced by oil, the density measurement will decrease because the oil has a lower density than the water it replaces. The HI measurement is also likely to decrease slightly, as the HI of the oil may be slightly less than water. This creates the hydrocarbon separation, typically 6 pu or less, seen in Row 3.

If the pores are filled with gas, the separation becomes greater. The formation density is significantly lower because the density of gas is much lower than water. The HI response is also much lower, as the HI of gas is very low. Even though methane (CH_4) has more hydrogen per molecule than water or oil, the molecules in gas are much farther apart than in a liquid, so the hydrogen per unit volume is lower. These two effects result in the gas separation seen in Row 4.

The next row of Figure 4-68 returns to the base case of freshwater-filled 18-pu limestone where the curves overlay.

If the freshwater is replaced by salty water, a slight reverse separation is created. The density of water with dissolved salt is greater than freshwater, so the density measurement increases. The dissolved salt pushes the water molecules slightly apart, resulting in a small decrease in the HI, as seen in Row 5. The curve separation is caused by the proportional increase in fluid density due to the addition of salt being greater than the decrease in HI.

Row 6 shows a typical separation in shale. In general, shale has a higher matrix density than limestone because of the presence of heavy minerals, resulting in the density measurement reading higher. The HI response in shale is high because of significant quantities of adsorbed water and hydroxyl (OH^-) ions associated with the clay and silt material in the shale. A large separation caused by both the HI response and density measurement reading high is generally attributable to the presence of shale.

The next row of Figure 4-68 returns to the base case of freshwater-filled, 18-pu limestone where the curves overlay.

If the formation is filled with freshwater, but the matrix is sandstone rather than limestone, the separation seen in Row 7 of Figure 4-68 will be observed. The density will decrease because the matrix density of sandstone (2.65 g/cm^3) is lower than the matrix density of limestone (2.71 g/cm^3). In this example, the porosity remains 18 pu, and the formation is filled with freshwater, so the HI associated with the fluid does not change from the limestone case. The reduction in the measured HI response is because of the differing scattering, neutron reaction, and capture interactions that the neutrons experience in sandstone relative to limestone. Solid sandstone allows more neutrons through to the detectors than solid limestone. Increasing neutron count results in decreased apparent HI response. This lithology effect is shown in Figure 4-67.

If the matrix is dolomite rather than limestone, the effect is reversed, as shown in Row 8 of Figure 4-68. The dolomite has a higher matrix density (2.85 g/cm^3) than limestone, so the measured bulk density is increased. As with the sandstone case, the HI due to the fluid in the pores of the formation does not change, but the neutron interactions with the atoms in the dolomite matrix let through fewer neutrons to the detectors than limestone, so the neutron response is increased in dolomite compared with limestone.

An alternative method of visualizing this lithology interdependence between the density and neutron measurements is to crossplot the measurements, as shown in **Figure 4-69**.

Figure 4-69. Density versus BPHI crossplot for the EcoScope responses in freshwater-filled formations.

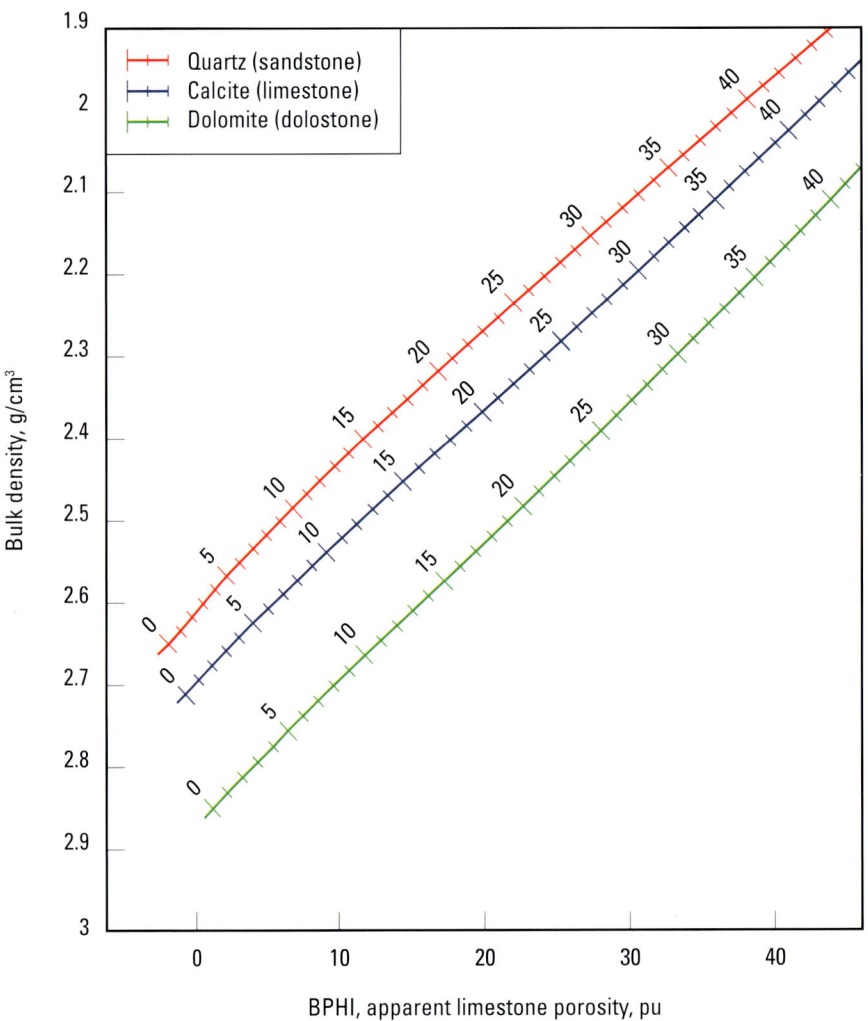

In the simple case of a freshwater-filled limestone, the HI measurement can be used to derive formation porosity directly. The primary application of the HI measurement, however, is as a complement to the density measurement to assist in identifying the formation lithology and fluids. This information is required for calculation of the true formation porosity.

4.7.4. Hydrogen index log quality control

The shop calibration must have been performed less than 1 month before the acquisition date and be in tolerance. Refer to the Schlumberger LWD Quality Control Manual for details.

QC should be applied to the density measurement used in the density correction to ensure that it represents true formation density. If azimuthal density information is to be used, then rotation of the tool over the interval of interest must be verified.

When plotted on a lithology-compatible scale, the density and neutron responses should overlay in a freshwater-filled formation after all environmental corrections have been applied.

The EcoScope hardware incorporates checks to verify correct performance of the acquisition detectors and electronics. Additional checks to ensure that the measurements fall within the range expected for downhole formation are outlined in the QC logic shown in **Table 4-37**.

Table 4-37. HI QC flag logic.

QC_BPHI	Hardware	Measurement
Green	Monitor high-voltage control is AUTO and	−3 pu ≤ BPHI ≤ 60 pu and
	Neutron monitor high-voltage control loop status is FINE and	−4 pu ≤ BPHI − BPHI_UNC ≤ 10 pu and
	Monitor counts in range 50 (2,200 counts/s) ≤ MON ≤ 200 (8,800 counts/s) and	Density (conventional of neutron-derived) QC flag green or yellow (depending on the selected density for the correction) and
	Helium-3 monitor high-voltage control loop status is FINE and	
	Helium-3 monitor high-voltage control is AUTO and	ROP ≤ 450 ft/h and
	Thermal near counts in range 1,000 counts/s ≤ NE2T ≤ 30,000 counts/s and	Bit size ≤ 9.875 in or
	Thermal far counts in range 10 counts/s ≤ FA1T ≤ 8,000 counts/s and	Selected caliper ≤ 12 in and related QC flag green
	10 counts/s ≤ FA2T ≤ 8,000 counts/s and	
	FAR1/FAR2 ratio in range 88% ≤ FAR1/FAR2 ≤ 120%	
Yellow	Neutron monitor high-voltage control loop status is COARSE or	60 pu < BPHI ≤ 100 pu or
	Helium-3 monitor high-voltage control loop status is COARSE or	−10 pu ≤ BPHI < −3 pu or
	Neutron board status word Bits 5, 6, 7, 13 = 1, indicating hardware warning or	10 pu < BPHI − BPHI_UNC ≤ 25 pu or
		−15 pu ≤ BPHI − BPHI_UNC < −4 pu
	Thermal near counts outside range 1,000 counts/s ≤ NE2T ≤ 30,000 counts/s or	ROP > 450 ft/h or
	Thermal far counts outside range 8,000 counts/s < FA1T ≤ 10,000 counts/s or 8,000 counts/s < FA2T ≤ 10,000 counts/s or	Bit size > 9.875 in or
		Selected caliper > 12 in or related QC flag yellow
	FAR1/FAR2 ratio outside range 88% ≤ FAR1/FAR2 ≤ 120%	QC flag red on the density channel used for the density correction.

QC_BPHI	Hardware	Measurement
Red	Neutron monitor high-voltage control loop status is SEARCH or	BPHI > 100 pu or
	Neutron monitor high-voltage control is MANUAL or HOLD or	BPHI < –10 pu or
	Neutron measurements configuration and calibration parameters not received or	BPHI – BPHI_UNC > 25 pu or
	Monitor counts out range MON < 50 (2,200 counts/s) or > 200 (8,800 counts/s) or	BPHI – BPHI_UNC < –15 pu or
	Thermal near counts out of range	
	NE2T < 1,000 counts/s or > 30,000 counts/s	
	Helium-3 monitor high-voltage control status is SEARCH or	
	Helium-3 monitor high-voltage control is MANUAL or HOLD or	
	Neutron board status word PNSW set for neutrons OFF or	
	Thermal far counts out of range	
	FA1T < 10 counts/s or	
	FA2T < 10 counts/s or	
	FA1T > 10,000 counts/s or	
	FA2T > 10,000 counts/s or	
	Memory board status = full	
White		No new BPHI tick available for more than 2 ft or
		BPHI is absent value or
		Near or far counts/s input is absent

4.7.5. Hydrogen index channels and parameters

4.7.5.1. Real time

Table 4-38. HI real-time parameters.

Transmitted Channel	Description	Unit
NEU_DCOR_OPT_RT	Density correction source for neutron processing, real time	(Bottom, average, neutron, none)
BS_RT	Bit size	in
GRAD	Real-time formation temperature gradient	degF/ft
RTST	Real-time surface temperature	degF
RW	Connate water resistivity	ohm.m
TWS	Water sample temperature	degF
OBMF	Oil-base mud	(Yes/no)
RT_BSAL	Mud salinity	ppk
MATRIX	Main matrix encountered	(Limestone, sandstone, dolomite)
T_IN	Mud temperature in	degF
MRIN	Mud resistivity in	ohm.m
MWIN	Mud weight in	lbm/galUS
SEABDEPTH	Water depth	ft

Table 4-39. HI real-time transmitted channels.

Transmitted Channel	Description
Azimuthal Average	
TNEA_DH_ECO_RT	Thermal neutron near count rates, average, real time, computed downhole
TFAR_DH_ECO_RT	Thermal neutron far count rates, average, real time, computed downhole
QC_TNPH_DH_ECO_RT	Thermal neutron quality indicator, real time, computed downhole
Quadrants	
TNEAB[†]_DH_ECO_RT	Thermal neutron near count rates, bottom, real time, computed downhole
TFARB[†]_DH_ECO_RT	Thermal neutron far count rates, bottom, real time, computed downhole
Density Required for Correction	
ROBB_DH_ECO_RT[‡]	Bulk density, bottom, real time, computed downhole
RHOB_DH_ECO_RT[‡]	Bulk density average, real time, computed downhole
RHON_DH_ECO_RT[†]	Bulk density from neutron, average, real time, computed downhole

[†] Corresponding near and far count rate pairs exist for each of the four quadrants—bottom (B), left (L), up (U), and right (R).

[‡] A near and far count rate pair plus one of the density channels must be transmitted to be able to deliver a neutron porosity measurement.

Table 4-40. HI real-time computed channels.

Computed Channel	Description
Azimuthal Average	
BPHI_UNC_ECO_RT	Uncorrected best thermal neutron porosity, average, real time
BPHI_ECO_RT	Best thermal neutron porosity, average, real time
Quadrants	
BPHB_UNC_ECO_RT	Uncorrected best thermal neutron porosity, bottom, real time
BPHB_ECO_RT	Best thermal neutron porosity, bottom, real time
BPHL_UNC_ECO_RT	Uncorrected best thermal neutron porosity, left, real time
BPHL_ECO_RT	Best thermal neutron porosity, left, real time
BPHU_UNC_ECO_RT	Uncorrected best thermal neutron porosity, up, real time
BPHU_ECO_RT	Best thermal neutron porosity, up, real time
BPHR_UNC_ECO_RT	Uncorrected best thermal neutron porosity, right, real time
BPHR_ECO_RT	Best thermal neutron porosity, right, real time

4.7.5.2. Recorded mode

Table 4-41. HI recorded-mode parameters.

Parameter Name	Description	Unit
OBMF_RM	Oil-base mud	(Yes/no)
MW_RM	Mud weight (recorded mode)	lbm/galUS
BS_RM	Bit size (recorded mode)	in
SEABDEPTH	Water depth	ft
BSAL_RM	Mud salinity (recorded mode)	ppk
MST_RM	Mud sample temperature (recorded mode)	degF
TWS_RM	Temperature of connate water (recorded mode)	degF
RMS_RM	Resistivity of mud sample (recorded mode)	ohm.m
RWS_RM	Resistivity of connate water	ohm.m
GCSE	Generalized caliper selection	(BS, UCAL, DENS, BCAL)
MATR	Rock matrix for neutron porosity corrections	(Limestone, sandstone, dolomite)
NEU_FTUBE_OPT	Far thermal tube selection	(Both, one, two)
NEU_DCOR_OPT	Density correction source for neutron processing	(Bottom, average, neutron, none)
NEU_TEMPCOR_OPT	Temperature correction source for neutron processing	(Tool_temp, annulus_temp, interpolated_temp)
NEU_PRESCOR_OPT	Pressure correction source for neutron processing	(Annulus_press, interpolated_press)

Table 4-42. HI recorded-mode channels.

Output Channel	Description
Azimuthal Average	
BPHI_UNC	Uncorrected best thermal neutron porosity, average
BPHI	Best thermal neutron porosity, average
DBPHI	Best thermal neutron porosity correction, average
QC_BPHI	Best neutron porosity quality indicator
TICK_NEU	Neutron samples
TAB_NEU	Neutron time after bit
Quadrants	
BPHB_UNC	Uncorrected best thermal neutron porosity, bottom
BPHB	Best thermal neutron porosity, bottom
DBPHB	Best thermal neutron porosity correction, bottom
BPHL_UNC	Uncorrected best thermal neutron porosity, left
BPHL	Best thermal neutron porosity, left
DBPHL	Best thermal neutron porosity correction, left
BPHU_UNC	Uncorrected best thermal neutron porosity, up
BPHU	Best thermal neutron porosity, up
DBPHU	Best thermal neutron porosity correction, up
BPHR_UNC	Uncorrected best thermal neutron porosity, right
BPHR	Best thermal neutron porosity, right
DBPHR	Best thermal neutron porosity correction, right

4.7.6. Hydrogen index measurement specifications

Table 4-43. BPHI porosity (with density correction) specifications at 200-ft/h ROP, assuming a three-level spatial average.

Item	Value	Remarks
Range	0 to 100 pu	
Axial resolution	15 in (11 in with enhanced resolution processing)	
Accuracy		
Below 10 pu	0.5 pu	
From 10 to 50 pu	5%	
Statistical precision		
Borehole size = 8 in	0.8 pu	At 30 pu and 200 ft/h
Borehole size = 8.5 in	0.9 pu	At 30 pu and 200 ft/h
Borehole size = 10 in	1.2 pu	At 30 pu and 200 ft/h

4.8. Thermal neutron response

A thermal neutron response emulating the TNPH measurement from the previous-generation tool, the adnVISION azimuthal density neutron tool, is also provided by the EcoScope tool.

4.8.1. Thermal neutron theory of measurement

Please refer to Section 4.6 for details on neutron generation, interactions, and detection.

As outlined in Section 4.7.1, after thermal neutron absorption in the walls of the LWD collar and the partial thermal neutron capture effect cancellation by taking the ratio of near and far neutron detectors, the resulting neutron response has little thermal neutron capture sensitivity. The EcoScope neutron detector responses can be approximated in terms of the formation HI and density by the equation

$$N(x) \approx N_0 e^{-\alpha.HI.x} e^{-\beta.\rho.x} \qquad \text{(Equation 4-19)}.$$

where

$e^{-\beta.\rho.x}$ = density sensitivity term
$e^{-\alpha.HI.x}$ = HI sensitivity term.

The HI response is extracted by applying $e^{\beta.\rho.x}$ to N(x), normalized for N_0.

$$\frac{N(x)}{N_0} e^{\beta.\rho.x} \approx e^{-\alpha.HI.x} \qquad \text{(Equation 4-20)}.$$

The high-energy (14-MeV) neutron PNG source and long spacing of the EcoScope far neutron detector from the source create a deeper neutron measurement than the adnVISION TNPH, but they also increase the measurement sensitivity to the formation density. In the EcoScope tool, this sensitivity is explicitly accounted for by the density correction applied to the far neutron count rate. The adnVISION neutron response is also sensitive to the formation density, though to a lesser degree because of the lower energy of the neutrons emitted by the adnVISION chemical source (4 MeV) and the closer detector to source spacing. However, this density sensitivity is not corrected for in the TNPH measurement delivered by the adnVISION tool.

By reapplying the appropriate density effect to the HI response according to the adnVISION TNPH density sensitivity factor, β_{adn}, it is possible to emulate the adnVISION TNPH measurement from the EcoScope data. **Figure 4-70** shows schematically how the raw EcoScope near/far thermal neutron ratio can be corrected for the density effect by factoring out the density sensitivity, β_{Eco}, to create the basis for the BPHI measurement. By factoring back in the smaller adnVISION density sensitivity, the EcoScope TNPH measurement is created, which emulates the adnVISION TNPH.

Figure 4-70. The EcoScope neutron ratio, transformed to BPHI through correction for the density sensitivity. Subsequent inclusion of a smaller density effect provides an emulation of the adnVISION TNPH measurement.

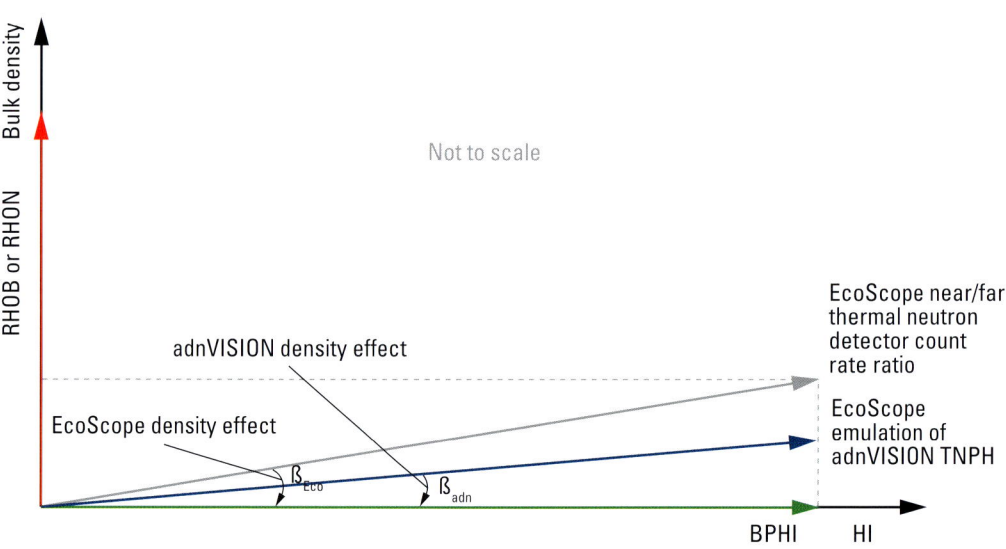

Figure 4-71 shows a partial density correction that emulates the adnVISION TNPH response. The EcoScope TNPH is derived in this manner. The aluminum oxide (Al_2O_3) points (which represent the shale response) overlay indicating that the EcoScope TNPH accurately emulates the adnVISION TNPH even in shale.

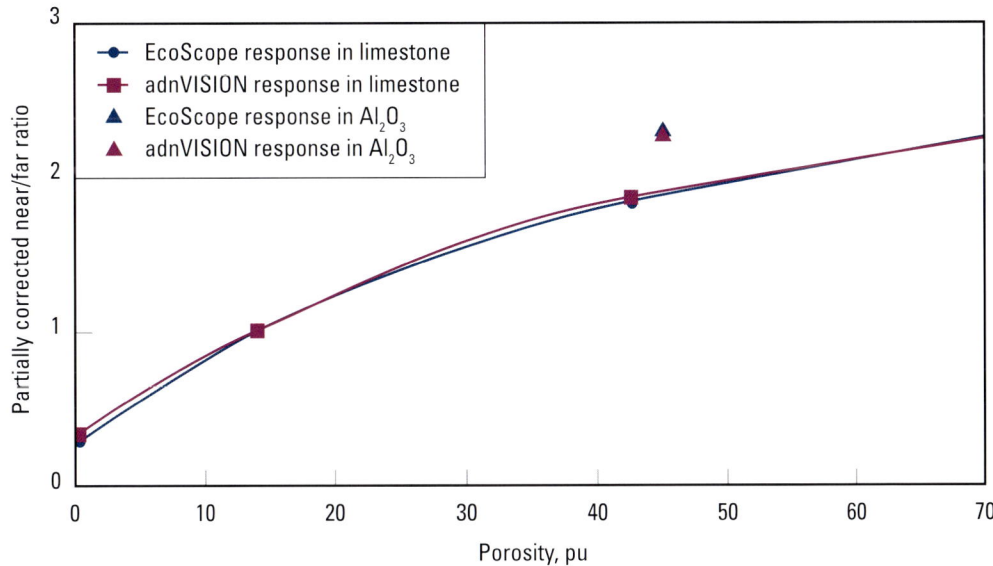

Figure 4-71. Comparison of the partially density-corrected EcoScope TNPH near/far neutron ratio with that from the adnVISION TNPH it is designed to emulate. The overlay indicates that the measurement responses are very similar.

Figure 4-72 shows good agreement between the TNPH responses from the two tools in shaly sand (high gamma ray intervals) and gas sand. In the shaly intervals, the EcoScope BPHI measurement shows a lower reading. In the interval between X250 and X320, which is a gas sand, the EcoScope TNPH reads slightly lower than the adnVISION TNPH because of the deeper DOI of the EcoScope TNPH and the location of the EcoScope tool in the BHA. The EcoScope measurements can be located closer to the bit, so they see less invasion effect than the corresponding measurements from the adnVISION tool. This effect is clearly seen on the density measurements (which have a shallower DOI than the neutron measurements); the EcoScope density reads lower than the adnVISION density, indicating less invasion at the earlier time.

Figure 4-72. Comparison of EcoScope and adnVISION densities and TNPH responses over a shaly sand and gas sand sequence. The separation in both the density and neutron curves in the zone below X250 is because of filtrate invasion occurring between the two logging passes.

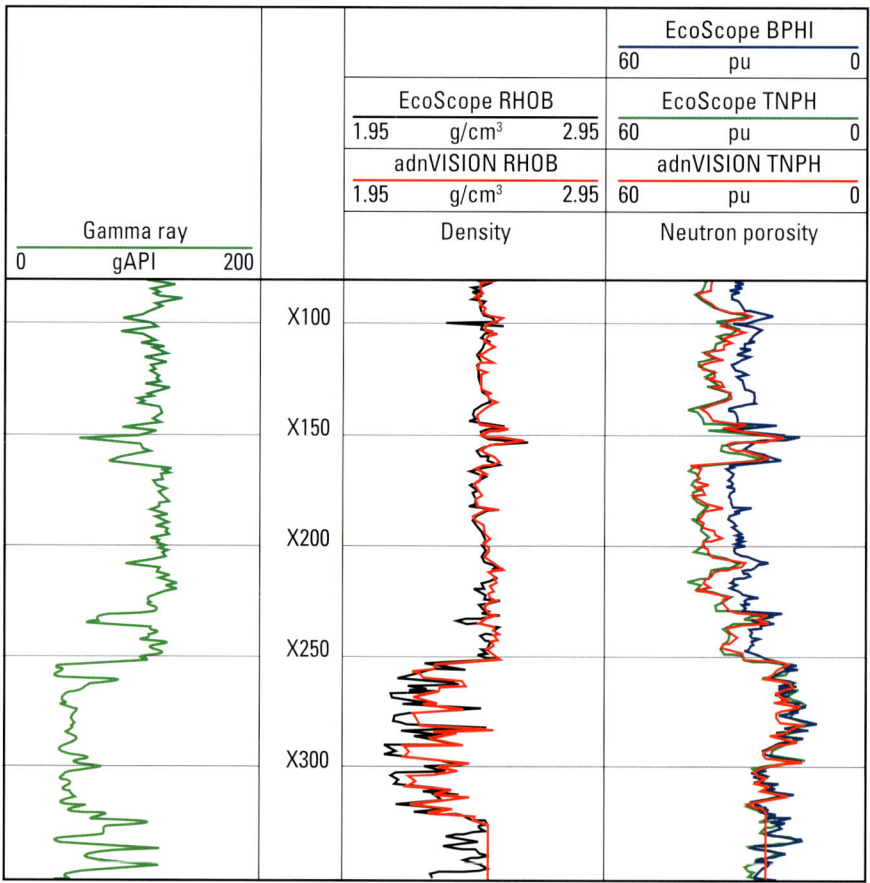

The EcoScope and adnVISION TNPH measurements generally match; however, the responses may not be identical in some circumstances because of the greater DOI of the EcoScope TNPH compared with the adnVISION TNPH. The combined effect of a higher-energy neutron source and greater source detector spacing results in the EcoScope TNPH measurement being slightly deeper than the adnVISION equivalent (**Figure 4-73**).

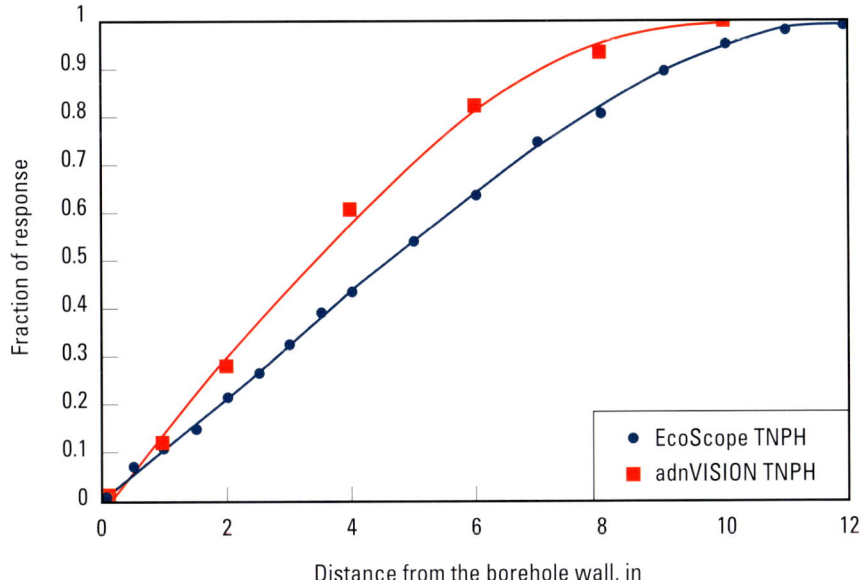

Figure 4-73. Comparison of adnVISION and EcoScope TNPH DOIs.

The lower thermal neutron capture sensitivity of the EcoScope TNPH may also result in a small difference in comparison with field logs of adnVISION TNPH. Field log data is generally corrected for all environmental corrections except formation salinity, as this depends on the formation porosity, water salinity, and water saturation, which are generally not computed simultaneously with log acquisition because they require some interpretation input. In low-salinity formations or after appropriate formation salinity correction, the results should be in close agreement.

4.8.2. Thermal neutron response environmental corrections

After PNG neutron output normalization and density corrections have been applied to the raw count rates, the environmental effects of the borehole, temperature, and formation salinity require correction.

Figure 4-74 compares the corrections for the adnVISION TNPH measurement with those for the EcoScope TNPH measurement under the same conditions. The significant reduction in the borehole size correction is because of the deeper DOI of the EcoScope TNPH measurement compared with the adnVISION TNPH measurement, as outlined previously. This greater depth is a significant advantage for a real-time neutron measurement where the size of the borehole may not be well known. The mud HI correction is also smaller for the same reasons.

As with the BPHI measurement, the formation salinity correction for the EcoScope TNPH measurement is less complex than that for the adnVISION TNPH because the EcoScope measurements have significantly less thermal neutron capture sensitivity.

The correction charts shown are simplifications of the corrections applied in the acquisition software where the coupling between the borehole size and mud property effects are taken into account by applying linked corrections.

Figure 4-74. Comparison of adnVISION TNPH and EcoScope TNPH environmental correction nomographs.

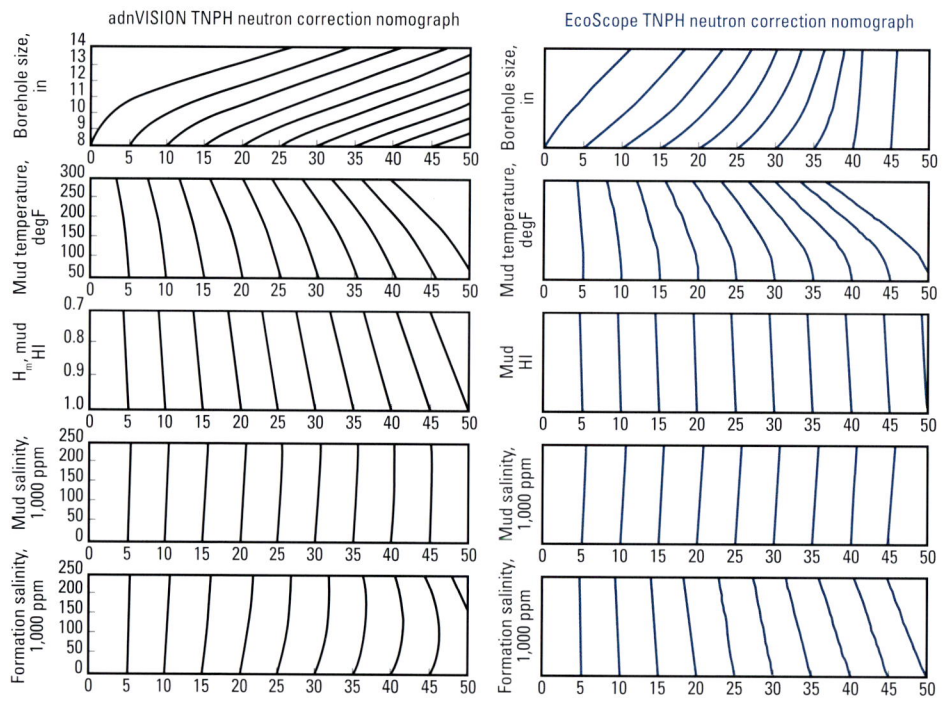

While the dominant effect on the neutron response is because of interactions with the hydrogen in the pore fluids, the neutrons also interact with the atoms in the solid part of the formation. Various solids (lithologies) are composed of different elements and so have slightly differing effects on the neutrons. The three most common clean lithologies are

- quartz (sandstone)—SiO_2
- calcite (limestone)—$CaCO_3$
- dolomite (dolostone)—$CaCO_3MgCO_3$.

The neutron detector count rates are calibrated such that the transform to porosity gives the correct value in a clean, freshwater-filled limestone. As quartz is not as effective at slowing neutrons as calcite, in formations of the same porosity, more neutrons are counted at the detectors if the lithology is sandstone than if the lithology is limestone. Increased count rate implies a lower apparent porosity (the effect is equivalent to having less hydrogen in the pore space to slow the neutrons, allowing them to be detected).

Figure 4-75 shows the lithology effect on the EcoScope TNPH. The effect on the adnVISION TNPH measurement is almost identical, showing the validity of the EcoScope TNPH as an emulation of the adnVISION TNPH measurement.

For both the EcoScope and the adnVISION TNPHs, a limestone formation of 18 pu filled with freshwater will read 18 pu, but a sandstone of the same porosity will read 15 pu on both. An 18-pu dolomite will read approximately 22.5 pu on both the adnVISION and EcoScope TNPH measurements.

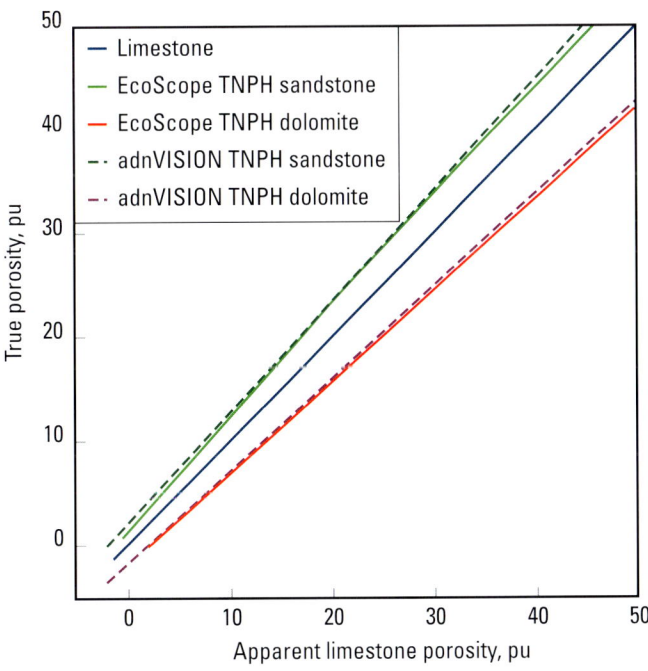

Figure 4-75. Comparison of adnVISION TNPH and EcoScope TNPH lithology sensitivities.

4.8.3. Thermal neutron response applications

TNPH from the EcoScope tool is an emulation of the adnVISION TNPH measurement and is intended primarily for correlation with previously acquired TNPH data.

The TNPH measurement can be used in conjunction with the density measurement similar to the method outlined for the HI measurement. However, the thermal neutron measurement response is more complex, making interpretation more difficult.

The technique for plotting the thermal neutron response against the formation density measurement uses the same scale logic as outlined previously. For example, a limestone-compatible scale plots the density from 1.7 to 2.7 g/cm³ against the limestone thermal neutron response plotted from 60 to 0 pu, as shown in **Figure 4-76**. For a 60-pu clean limestone formation with a 2.7-g/cm³ grain density filled with freshwater with a density of 1 g/cm³, Equation 4-18 gives a bulk density of 1.68 g/cm³. For simplicity, the scale is generally rounded to 1.7 g/cm³.

For comparison, both the TNPH (blue dashed line) and BPHI (green dotted line) are presented in Figure 4-76. Note that the curves represent typical field data (all environmental corrections other than formation salinity applied) in an uninvaded formation.

With both the density and thermal neutron measurements plotted on a limestone-compatible scale, the two curves will overlay when a clean, water-filled limestone of any porosity is encountered, as shown in Rows 1 and 2.

If the freshwater is replaced by oil, the density measurement will decrease because the oil has a lower density than the water it replaces. The thermal neutron measurement is also likely to decrease slightly because the HI of the oil may be slightly less than water. This decrease creates the hydrocarbon separation, typically 6 pu or less, seen in Row 3. The thermal neutron response reads slightly lower than the HI response, as the small drop in the formation density due to oil replacing water allows more neutrons through to the detector. This additional density effect is generally small in oil-bearing formations.

If the pores are filled with gas, the separation becomes greater, as shown in Row 4. The formation density is significantly lower because the density of gas is much lower than water. The thermal neutron response is also significantly lower because the HI of gas is very low. In addition, the reduced formation density allows more neutrons through to the detectors, pushing the thermal neutron response even lower. This is referred to as the excavation effect in some literature. It is actually a density effect. So while the BPHI reads the correct formation HI, the TNPH reads too low because this density effect is not corrected. The reduced density and fluid HI effects result in the gas separation. Note that the responses shown in Figure 4-76 are for an uninvaded formation. If invasion occurs, the density neutron separation is generally reduced. When comparing EcoScope to adnVISION TNPH in gas formations, the EcoScope TNPH often reads lower because it is acquired in the shallower invasion conditions closer to the bit.

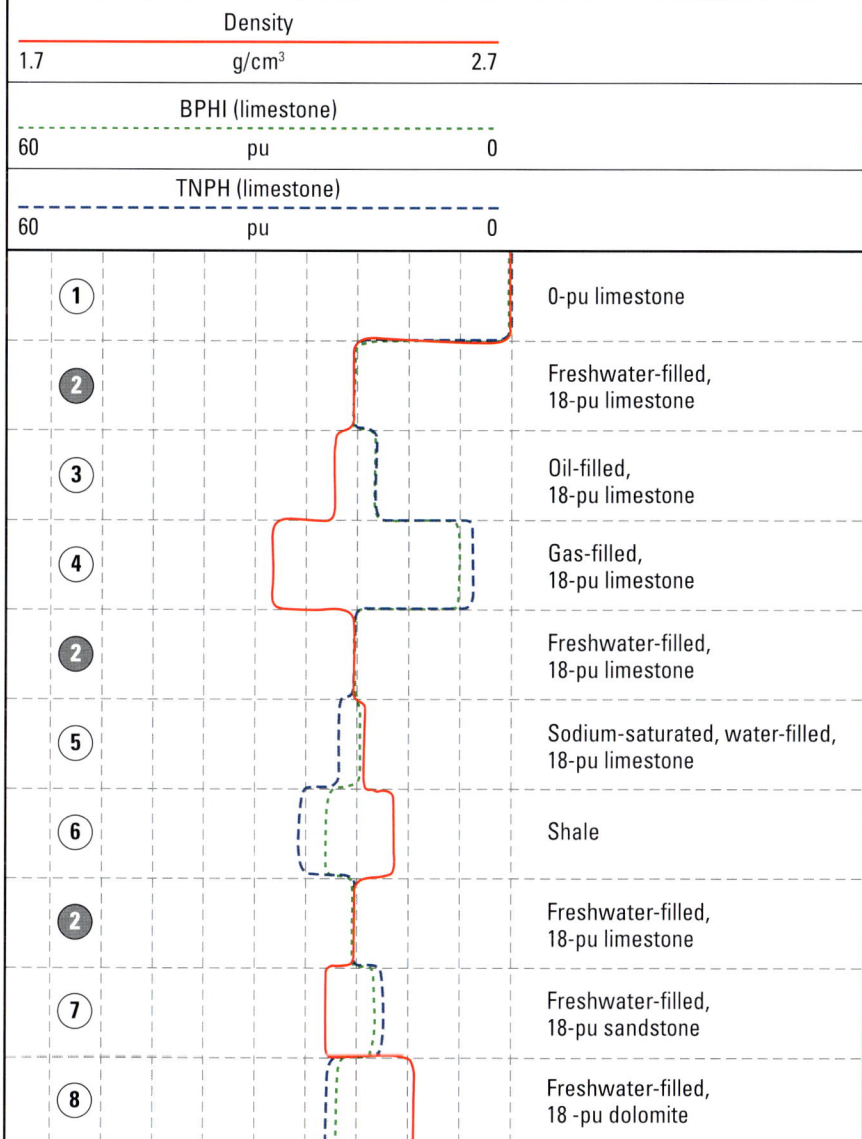

Density		
1.7	g/cm³	2.7
BPHI (limestone)		
60	pu	0
TNPH (limestone)		
60	pu	0

1. 0-pu limestone
2. Freshwater-filled, 18-pu limestone
3. Oil-filled, 18-pu limestone
4. Gas-filled, 18-pu limestone
2. Freshwater-filled, 18-pu limestone
5. Sodium-saturated, water-filled, 18-pu limestone
6. Shale
2. Freshwater-filled, 18-pu limestone
7. Freshwater-filled, 18-pu sandstone
8. Freshwater-filled, 18-pu dolomite

Figure 4-76. Separation between the density and neutron measurements. When presented on a lithology-compatible scale (limestone-compatible in this example), the separation reveals whether the formation is a clean, freshwater-filled layer of the selected lithology.

The next row of Figure 4-76 returns to the base case of freshwater-filled limestone where the curves overlay.

If the freshwater is replaced by salty water, a slight reverse separation is created. The density of water with dissolved salt is greater than freshwater, so the density measurement increases. The dissolved salt pushes the water molecules apart slightly, resulting in a small decrease in the HI, as seen in Row 5. The curve separation between the density and HI measurements is caused by the proportional increase in fluid density due to the addition of salt being greater than the decrease in HI. The effect is amplified on the thermal neutron response because the higher density allows fewer neutrons through to the detectors, resulting in an exaggerated thermal neutron response. In addition, for the adnVISION TNPH prior to formation thermal neutron capture correction, the capture of thermal neutrons by the chlorine in the saltwater further reduces the neutron counts in the detectors, further increasing the separation. This example shows the three formation properties of HI, density, and thermal neutron capture cross section involved in understanding the TNPH response. BPHI simplifies the interpretation by having a dominant sensitivity to the HI only.

Row 6 shows a typical separation in shale. In general, shale has a higher matrix density than limestone because of the presence of heavy minerals, resulting in the density measurement reading higher. The HI response in shale is high because of significant quantities of adsorbed water and hydroxyl ions associated with the clay and silt material in the shale. The effect is exaggerated on the thermal neutron response because of the increased density effect. In addition, for the adnVISION TNPH prior to formation thermal neutron capture correction, the capture of thermal neutrons in the shale further reduces the neutron counts in the detectors, further increasing the separation.

Row 7 of Figure 4-76 returns to the base case of freshwater-filled 18-pu limestone where the curves overlay.

If the formation is filled with freshwater, but the matrix is sandstone rather than limestone, the separation seen in Row 7 is observed. The density will decrease as the matrix density of sandstone (2.65 g/cm^3) is lower than the matrix density of limestone (2.71 g/cm^3). In this example, the porosity remains 18 pu and is filled with freshwater, so the HI associated with the fluid does not change from the limestone case. The reduction in the measured thermal neutron and HI responses is because of the differing scattering, neutron reaction, and capture interactions that the neutrons experience in sandstone relative to limestone. Solid sandstone allows more neutrons through to the detectors than solid limestone. Increasing neutron counts result in decreased apparent thermal neutron and HI responses. This lithology effect is shown in Figure 4-75. As TNPH has more lithology effect than BPHI, the separation is slightly increased between the density and TNPH measurements.

If the matrix is dolomite rather than limestone, the effect is reversed, as shown in Row 8. The dolomite has a higher matrix density (2.85 g/cm^3) than limestone, so the measured bulk density is increased. As with the sandstone case, the HI due to the fluid in the pores of the formation does not change, but the neutron interactions with the atoms in the dolomite matrix let through fewer neutrons to the detectors than limestone, so the neutron response is increased in dolomite compared with limestone. As in the sandstone case, the lithology effect on TNPH is greater than on BPHI, so the separation is slightly increased between the density and TNPH measurements.

An alternative method of visualizing this lithology interdependence between the density and neutron measurements is to crossplot the measurements, as shown in **Figure 4-77**.

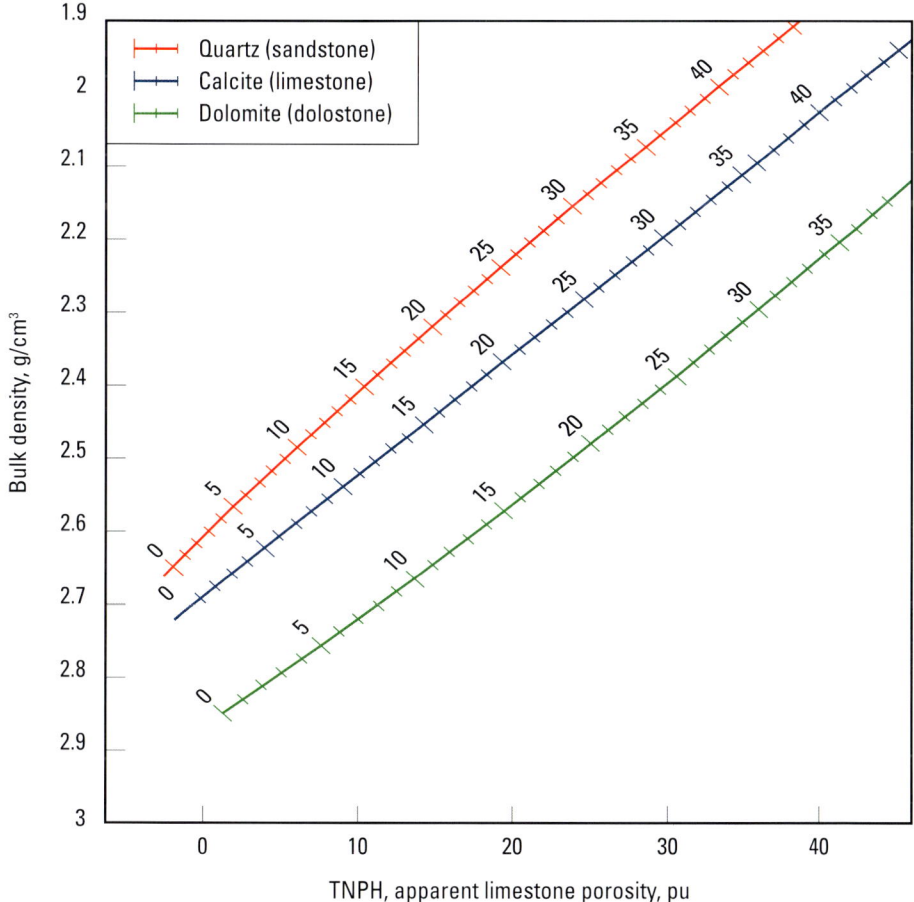

Figure 4-77. Density versus thermal neutron crossplot for the EcoScope responses in freshwater-filled formations.

In the simple case of a freshwater-filled limestone, the thermal neutron and HI measurements can be used to derive formation porosity directly. However, in more complex formations, the superposition of density and thermal capture effects on the TNPH response makes interpretation increasingly difficult. The BPHI measurement was developed to address these limitations to provide a complement to the density measurement to assist in identifying the formation lithology and fluids so that the true formation porosity can be determined.

4.8.4. Thermal neutron response log quality control

The shop calibration must have been performed less than 1 month before the acquisition date and must be in tolerance. Refer to the Schlumberger LWD Quality Control Manual for details.

QC should be applied to the density measurement used in the density correction to ensure that it represents true formation density. If azimuthal density information is to be used, then rotation of the tool over the interval of interest must be verified.

When plotted on a lithology-compatible scale, the density and neutron responses should overlay in a freshwater-filled formation after all environmental corrections have been applied.

The EcoScope hardware incorporates checks to verify correct performance of the acquisition detectors and electronics. Additional checks to ensure that the measurements fall within the range expected for downhole formations are outlined in the QC logic shown in **Table 4-44**.

Table 4-44. Thermal neutron response QC flag logic.

QC_TNPH	Hardware	Measurement
Green	Monitor high-voltage control is AUTO and	−3 pu ≤ TNPH ≤ 60 pu and
	Neutron monitor high-voltage control loop status is FINE and	−4 pu ≤ TNPH − TNPH_UNC ≤ 15 pu and
	Monitor counts in range 50 (2,200 counts/s) ≤ MON ≤ 200 (8,800 counts/s) and	Density (conventional or neutron-derived) QC flag green or yellow (depending on the selected density for the correction) and
	Helium-3 monitor high-voltage control loop status is FINE and	
	Helium-3 monitor high-voltage control is AUTO and	
	Thermal near counts in range 1,000 counts/s ≤ NE2T ≤ 30,000 counts/s and	ROP ≤ 450 ft/h and
		Bit size ≤ 9.875 in or
	Thermal far counts in range 10 counts/s ≤ FA1T ≤ 8,000 counts/s and 10 counts/s ≤ FA2T ≤ 8,000 counts/s and	Selected caliper ≤ 12 in and related QC flag green
	FAR1/FAR2 ratio in range 88% ≤ FAR1/FAR2 ≤ 120%	
Yellow	Neutron monitor high-voltage control loop status is COARSE or	60 pu < TNPH ≤ 100 pu or
		−10 pu ≤ TNPH < −3 pu or
	Helium-3 monitor high-voltage control loop status is COARSE or	15 pu < TNPH − TNPH_UNC ≤ 25 pu or
	Neutron board status word Bits 5, 6, 7, 13 = 1 indicating hardware warning or	−15 pu ≤ TNPH − TNPH_UNC < −4 pu or
	Thermal near counts outside range 1,000 counts/s ≤ NE2T ≤ 30,000 counts/s or	ROP > 450 ft/h or
	Thermal far counts outside range 8,000 counts/s < FA1T ≤ 10,000 counts/s o 8,000 counts/s < FA2T ≤ 10,000 counts/s or	Bit size > 9.875 in or
		Selected caliper > 12 in or related QC flag yellow or
	FAR1/FAR2 ratio outside range 88% ≤ FAR1/FAR2 ≤ 120%	QC flag red on the density channel used for the density correction.

QC_TNPH	Hardware	Measurement
Red	Neutron monitor high-voltage control loop status is SEARCH or	TNPH > 100 pu or
	Neutron monitor high-voltage control is MANUAL or HOLD or	TNPH < −10 pu or
	Neutron measurements configuration and calibration parameters not received or	TNPH − TNPH_UNC > 25 pu or
	Monitor counts out range: MON < 50 (2,200 counts/s) or > 200 (8,800 counts/s) or	TNPH − TNPH_UNC < −15 pu or
	Thermal near counts out of range NE2T < 1,000 counts/s or > 30,000 counts/s	
	Helium-3 monitor high-voltage control loop status is SEARCH or	
	Helium-3 monitor high-voltage control is MANUAL or HOLD or	
	Neutron board status word PNSW set for neutrons OFF or	
	Thermal far counts out of range FA1T < 10 counts/s or FA2T < 10 counts/s or FA1T > 10,000 counts/s or FA2T > 10,000 counts/s or	
	Memory board status = full	
White		No new TNPH tick available for more than 2 ft or
		TNPH is absent value or
		Near or far counts/s input is absent

4.8.5. Thermal neutron response channels and parameters

4.8.5.1. Real time

Table 4-45. Thermal neutron response real-time parameters.

Parameter Name	Description	Unit
BS_RT	Bit size	in
GRAD	Real-time formation temperature gradient	degF/ft
RTST	Real-time surface temperature	degF
RW	Connate water resistivity	ohm.m
TWS	Water sample temperature	degF
OBMF	Oil-base mud	(Yes/no)
RT_BSAL	Mud salinity	ppk
MATRIX	Main matrix encountered	(Limestone, sandstone, dolomite)
T_IN	Mud temperature in	degF
MRIN	Mud resistivity in	ohm.m
MWIN	Mud weight in	lbm/galUS
SEABDEPTH	Water depth	ft
NEU_DCOR_OPT_RT	Density correction source for neutron processing, real time	(Bottom, average, neutron, none)

Table 4-46. Thermal neutron response real-time transmitted channels.

Transmitted Channel	Description
Azimuthal Average	
TNEA_DH_ECO_RT	Thermal neutron near count rates, average, real time, computed downhole
TFAR_ DH_ECO _RT	Thermal neutron far count rates, average, real time, computed downhole
QC_TNPH_DH_ECO_RT	Thermal neutron quality indicator, real time, computed downhole
Quadrants	
TNEAB[†]_ DH_ECO _RT	Thermal neutron near count rates, bottom, real time, computed downhole
TFARB[†]_ DH_ECO _RT	Thermal neutron far count rates, bottom, real time, computed downhole
Density Required for Correction	
ROBB_ DH_ECO _RT[‡]	Bulk density, bottom, real time, computed downhole
RHOB_ DH_ECO _RT[‡]	Bulk density average, real time, computed downhole
RHON_ DH_ECO _RT[‡]	Bulk density from neutron, average, real time, computed downhole

[†] Corresponding near and far count rate pairs exist for each of the four quadrants—bottom (B), left (L), up (U), and right (R).

[‡] A near and far count rate pair, plus one of the density channels, must be transmitted to be able to deliver a neutron porosity measurement.

Table 4-47. Thermal neutron response real-time computed channels.

Computed Channel	Description
Azimuthal Average	
TNPH_UNC_ECO_RT	Uncorrected thermal neutron porosity, average, real time
TNPH_ECO_RT	Thermal neutron porosity, average, real time
Quadrants	
TNPB_UNC_ECO_RT	Uncorrected thermal neutron porosity, bottom, real time
TNPB_ECO_RT	Thermal neutron porosity, bottom, real time
TNPL_UNC_ECO_RT	Uncorrected thermal neutron porosity, left, real time
TNPL_ECO_RT	Thermal neutron porosity, left, real time
TNPU_UNC_ECO_RT	Uncorrected thermal neutron porosity, up, real time
TNPU_ECO_RT	Thermal neutron porosity, up, real time
TNPR_UNC_ECO_RT	Uncorrected thermal neutron porosity, right, real time
TNPR_ECO_RT	Thermal neutron porosity, right, real time

4.8.5.2. Recorded mode

Table 4-48. Thermal neutron response recorded-mode parameters.

Parameter Name	Description	Unit
OBMF_RM	Oil-base mud	(Yes/no)
MW_RM	Mud weight (recorded mode)	lbm/galUS
BSAL_RM	Mud salinity (recorded mode)	ppk
MST_RM	Mud sample temperature (recorded mode)	degF
TWS_RM	Temperature of connate water (recorded mode)	degF
RMS_RM	Resistivity of mud sample (recorded mode)	ohm.m
RWS_RM	Resistivity of connate water	ohm.m
GCSE	Generalized caliper selection	(BS—bit size, UCAL—ultrasonic caliper, DCAV—density caliper)
MATR	Rock matrix for neutron porosity corrections	(Limestone, sandstone, dolomite)
NEU_FTUBE_OPT	Far thermal tube selection	(Both, one, two)
NEU_DCOR_OPT	Density correction source for neutron processing	(Bottom, average, neutron, none)
NEU_TEMPCOR_OPT	Temperature correction source for neutron processing	(Tool_Temp, Annulus_Temp, Interpolated_Temp)
NEU_PRESCOR_OPT	Pressure correction source for neutron processing	(Annulus_Press, Interpolated_Press)

Table 4-49. Thermal neutron response recorded-mode channels.

Output Channel	Description
Azimuthal Average	
TNPH	Thermal neutron porosity, average
TNPH_UNC	Uncorrected thermal neutron porosity, average
QC_TNPH	Thermal neutron porosity quality indicator
TICK_NEU	Neutron samples
TAB_NEU	Neutron time after bit
Quadrants	
TNPB_UNC	Uncorrected thermal neutron porosity, bottom
TNPB	Thermal neutron porosity, bottom
TNPL_UNC	Uncorrected thermal neutron porosity, left
TNPL	Thermal neutron porosity, left
TNPU_UNC	Uncorrected thermal neutron porosity, up
TNPU	Thermal neutron porosity, up
TNPR_UNC	Uncorrected thermal neutron porosity, right
TNPR	Thermal neutron porosity, right

4.8.6. Thermal neutron response measurement specifications

Table 4-50. TNPH porosity (with density correction) specifications at 200-ft/h ROP with a three-level spatial average.

Item	Description	Unit
Range	0 to 100 pu	
Axial resolution	15 in (11 in with enhanced resolution processing)	
Accuracy[‡]		
Below 10 pu	0.5 pu	
Above 10 pu	5%	
Statistical precision		
Borehole size = 8 in	0.9 pu	At 30 pu and 200 ft/h
Borehole size = 8.5 in	1.0 pu	At 30 pu and 200 ft/h
Borehole size = 10 in	1.4 pu	At 30 pu and 200 ft/h

4.9. Neutron-gamma density

The EcoScope tool utilizes the PNG and a suite of detectors to determine formation density from gamma rays induced by the interaction of high-energy neutrons with the formation. This density is referred to as the sourceless, or neutron-gamma, density. The NGD can be used to replace the traditional gamma-gamma density measurement, allowing conventional formation evaluation without the use of chemical nuclear sources. However, the sourceless density cannot be focused azimuthally to form an image, nor can the photoelectric factor be derived with the NGD as it is with the azimuthal density. However, the EcoScope spectroscopy measurement provides improved lithology information to that usually provided by the PEF measurement. The sourceless density measurement is also deeper and less dependent on good contact with the formation than the azimuthal measurement.

4.9.1. Neutron-gamma density theory of measurement

The NGD measurement (RHON) is based on the detection of neutron-induced gamma rays at detectors placed far from the neutron source. High-energy neutrons from the PNG excite the nuclei of atoms in the formation during inelastic collisions. The nuclei emit gamma rays as they deexcite. The gamma ray flux at the detectors is therefore influenced by the neutron transport to the point of the gamma-ray-producing neutron interaction and the subsequent transport of the gamma rays from their origin to the gamma ray detector, as shown in **Figure 4-78**.

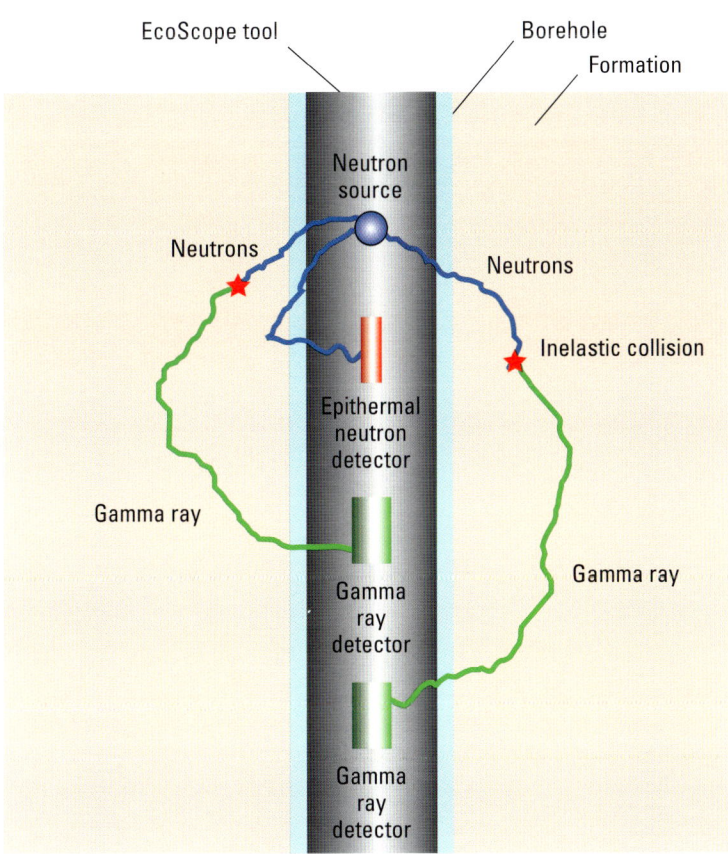

Figure 4-78. The NGD measurement, which uses an epithermal neutron detector to characterise the size of the neutron cloud around the tool, and two gamma ray detectors to provide a compensated density measurement based on the neutron-induced gamma rays.

To eliminate the influence of thermal neutron effects on the measurement, only gamma rays produced by inelastic collisions between nuclei and high-energy neutrons are measured. The gamma rays from neutron capture are subtracted out.

The technique is not as straightforward as the gamma-gamma density measurement. The gamma ray source is not a point source. It is an extended source, the size of which depends on the HI of the formation and, to a lesser extent, on the formation lithology. An epithermal neutron detector positioned close to the source is used to assess the size of the high-energy neutron cloud around the PNG and thus, the volume from which the inelastic gamma rays are emitted.

Figure 4-79 shows a simplified diagram of the high-energy neutron interaction cloud (light blue, representing the volume within which neutrons have enough energy to create inelastic gamma rays) around the PNG and the corresponding gamma ray interaction volume around the gamma ray detector.

In a high-porosity, low-density formation (left) the abundance of hydrogen in the pores of the formation rapidly slows the neutrons, keeping the neutron cloud relatively small. This slowing reduces the size of the effective gamma ray source and increases the average distance gamma rays have to travel to reach the detector.

In a lower-porosity, higher-density formation (right), the amount of hydrogen in the formation is reduced, and the slowing-down length increases, resulting in a larger induced gamma ray source. This also decreases the average distance gamma rays have to travel to reach the detector.

Figure 4-79. Simplified diagram of the neutron cloud and gamma ray attenuation volume for high-porosity, low-density (left) and low-porosity, high-density (right) liquid-filled formations.

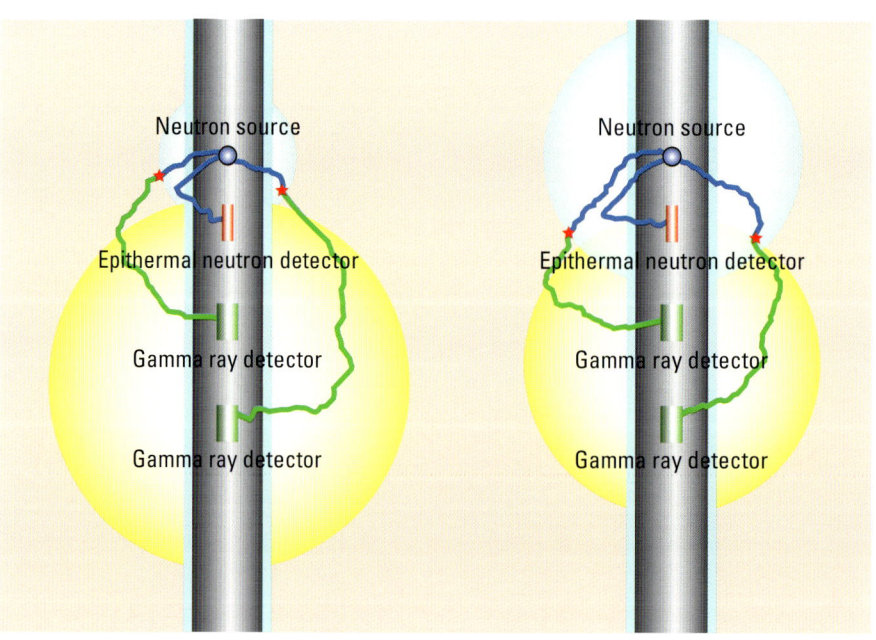

As the size of the high-energy neutron cloud and the related induced gamma ray source is dependent on the HI of the formation, the measured inelastic gamma ray counts can be schematically represented in terms of the formation density and HI, as shown in **Figure 4-80**. The cloud size is determined using an epithermal neutron detector.

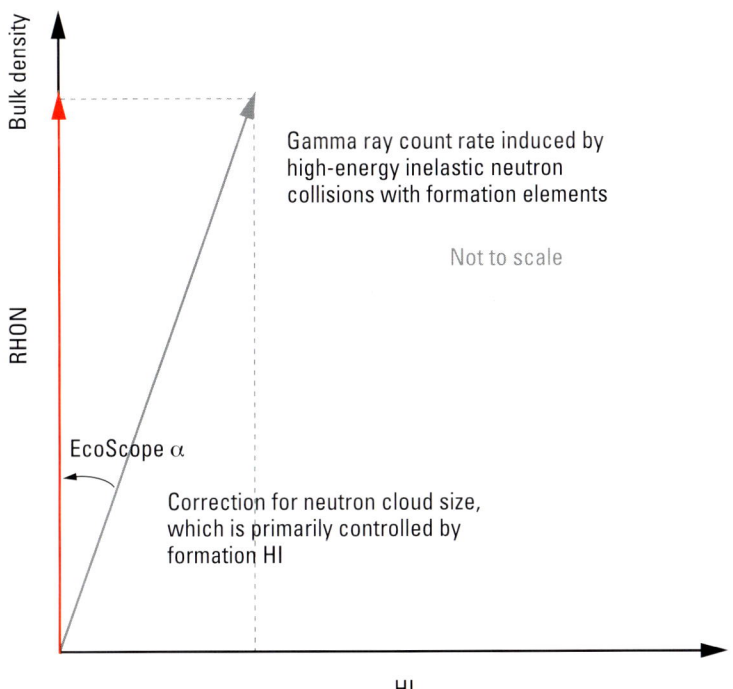

Gamma ray count rate induced by high-energy inelastic neutron collisions with formation elements

Not to scale

EcoScope α

Correction for neutron cloud size, which is primarily controlled by formation HI

Bulk density

RHON

HI

Figure 4-80. The EcoScope inelastic gamma ray count rate (thick gray line), sensitive to the formation properties of density and HI.

In comparison with the gamma-gamma density, the NGD measurement has several key advantages and a few disadvantages.

The NGD measurement involves high-energy neutrons and gamma rays, both of which have significant penetrating capabilities, resulting in a deep measurement in comparison with the conventional density.

In addition to being substantially deeper, the NGD measurement is also significantly less sensitive to tool-formation contact. There are no density windows in the collar as with the conventional density measurement, so there is no need to align the windows with the borehole wall while drilling in sliding mode with a directional motor.

One key benefit is that the measurement requires no chemical radioactive source, eliminating personnel exposure and the risks associated with shipment, storage, and possible abandonment of a chemical nuclear source in the reservoir.

The use of high-energy neutrons and gamma rays makes the measurement difficult to focus. Consequently, the RHON measurement is an azimuthal average, and no density image or azimuthal density caliper is available with the sourceless density. In addition, the axial resolution is not as tightly focused as the gamma-gamma density measurement.

The use of high energy gamma rays also eliminates the possibility of a photoelectric measurement (which requires very low energy gamma rays) being available with the sourceless density as it is with the gamma-gamma density. Formation lithology information, traditionally inferred from the PEF measurement is provided by the EcoScope capture spectroscopy measurement.

The two step process from neutron to gamma ray and the subsequent scattering of the gamma rays results in reduced accuracy (0.02 g/cm^3) for the sourceless density in comparison with the conventional density (0.015 g/cm^3).

A comparison of the sourceless, NGD with the conventional gamma-gamma density in a Middle East carbonate is shown in **Figure 4-81**. Note that the bottom quadrant PEF log, PEB, is only available with the gamma-gamma density.

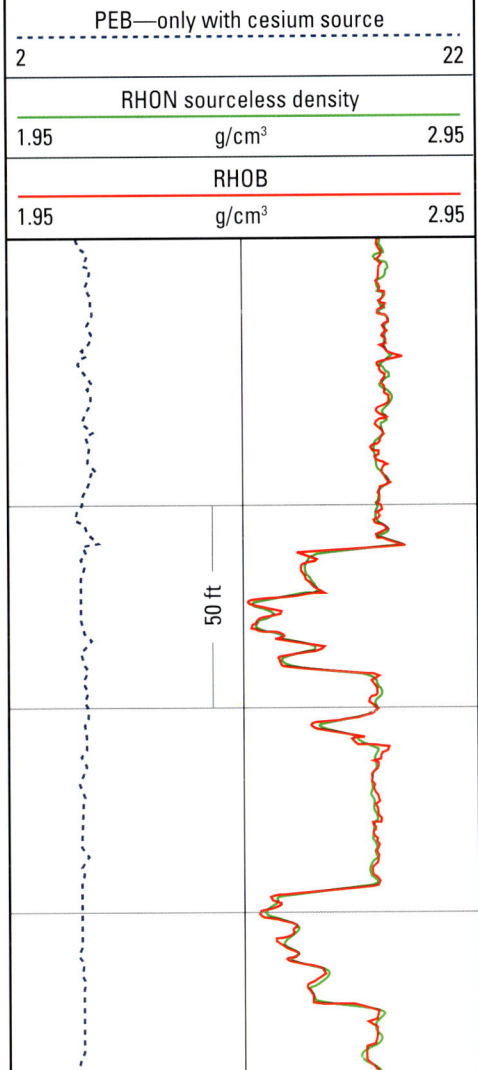

PEB—only with cesium source		
2		22
RHON sourceless density		
1.95	g/cm³	2.95
RHOB		
1.95	g/cm³	2.95

50 ft

Figure 4-81. Neutron-gamma (RHON) and conventional gamma-gamma (RHOB) density logs in a Middle East carbonate reservoir are compared.

4.9.2. Neutron-gamma density environmental corrections

The use of a second inelastic gamma ray detector in the tool permits borehole compensation similar to the spine-and-ribs technique used for the gamma-gamma density. The density correction applied is labeled DRHN.

The standoff-compensated density is subsequently corrected for borehole size and mud properties.

4.9.3. Neutron-gamma density applications

NGD applications are identical to the gamma-gamma density applications other than where images are involved. Comparison with the neutron response for fluid and lithology determination is performed in the same manner. The significantly deeper measurement depth of investigation should be considered when comparing the NGD measurement to the shallower gamma-gamma density.

Unlike the gamma-gamma density measurement, the NGD has very little sensitivity to the contact between the tool and the borehole wall. This makes the NGD measurement ideal in situations where oriented tool-to-borehole wall contact cannot be guaranteed, resulting in poor gamma-gamma density measurements, such as when sliding while drilling with a motor. It also means that processing used to overcome borehole contact issues with the gamma-gamma density, such as IDD and spiraled borehole processing, are not required with the NGD.

4.9.4. Neutron-gamma density log quality control

The shop calibration must have been performed less than 1 month before the acquisition date and must be in tolerance. Refer to the Schlumberger LWD Quality Control Manual for details.

When plotted on a lithology-compatible scale, the density and neutron responses should overlay in a clean, freshwater-filled formation of the appropriate lithology after all environmental corrections have been applied. Refer to Section 4.8.3 for details.

The EcoScope hardware incorporates checks to verify correct performance of the acquisition detectors and electronics. Additional checks to ensure that the measurements fall within the range expected for downhole formation are outlined in the QC logic shown in **Table 4-51**.

Table 4-51. NGD QC flag logic.

QC_RHON	Hardware	Measurement
Green	All ok	1.7 g/cm^3 < RHON < 3.05 g/cm^3 and -0.1 g/cm^3 ≤ DRHN ≤ 0.15 g/cm^3
Yellow	High voltage out of range or Azimuthal system problem	1 g/cm^3 < RHON < 1.7 g/cm^3 or 3.05 g/cm^3 < RHON < 3.1 g/cm^3 or -0.2 g/cm^3 < DRHN < -0.1 g/cm^3 or 0.15 g/cm^3 < DRHN < 0.3 g/cm^3
Red	PNG or Monitor or LSn or SSn hardware flag set	RHON ≤ 1 g/cm^3 or RHON ≥ 3.1 g/cm^3 or DRHN ≤ -0.2 g/cm^3 or DRHN ≥ 0.3 g/cm^3
White		No new RHON data available for more than 2 ft.

4.9.5. Neutron-gamma density channels and parameters

4.9.5.1. Real time

Table 4-52. NGD real-time parameters.

Parameter Name	Description	Unit
BS_RT	Bit size	in
GRAD	Real-time formation temperature gradient	degF/ft
RTST	Real-time surface temperature	degF
RW	Connate water resistivity	ohm.m
TWS	Water sample temperature	degF
OBMF	Oil-base mud	Yes/no
RT_BSAL	Mud salinity	ppk
T_IN	Mud temperature in	degF
MRIN	Mud resistivity in	ohm.m
MWIN	Mud weight in	lbm/galUS
SEABDEPTH	Water depth	ft
NEU_DCOR_OPT_RT	Density correction source for neutron processing, real time	(Bottom, average, neutron, none)

Table 4-53. NGD real-time transmitted channels.

Transmitted Channel	Description
Azimuthal Average	
RHON_DH_ECO_RT	Bulk density from neutron, average, real time, computed downhole
DRHN_DH_ECO_RT	Bulk density from neutron correction, average, real time, computed downhole
QC_RHON_DH_ECO_RT	Pulsed neutron density quality indicator, real time, computed downhole
Quadrants	
RHNB_DH_ECO_RT	Bulk density from neutron, bottom, real time, computed downhole
RHNL_DH_ECO_RT	Bulk density from neutron, left, real time, computed downhole
RHNR_DH_ECO_RT	Bulk density from neutron, right, real time, computed downhole
RHNU_DH_ECO_RT	Bulk density from neutron, up, real time, computed downhole
DRHNB_DH_ECO_RT	Bulk density from neutron correction, bottom, real time, computed downhole
DRHNL_DH_ECO_RT	Bulk density from neutron correction, left, real time, computed downhole
DRHNR_DH_ECO_RT	Bulk density from neutron correction, right, real time, computed downhole
DRHNU_DH_ECO_RT	Bulk density from neutron correction, up, real time, computed downhole

4.9.5.2. Recorded mode

Table 4-54. NGD recorded-mode parameters.

Parameter Name	Description	Unit
MW_RM	Mud weight (recorded mode)	lbm/galUS
BS_RM	Bit size (recorded mode)	in
GRAD	Formation temperature gradient	degF/ft
RTST	Surface temperature	degF
RW	Connate water resistivity	ohm.m
TWS	Water sample temperature	degF
OBMF_RM	Oil-base mud	Yes/no
BSAL_RM	Mud salinity (recorded mode)	ppk
RT_BSAL	Mud salinity	ppk
T_IN	Mud temperature in	degF
MRIN	Mud resistivity in	ohm.m
MWIN	Mud weight in	lbm/galUS
SEABDEPTH	Water depth	ft
NEU_DCOR_OPT	Density correction source for neutron processing, real time	(Bottom, average, neutron, none)

Table 4-55. NGD recorded-mode channels.

Output Channel	Description
Azimuthal Average	
RHON	Bulk density from neutron, average
DRHN	Bulk density from neutron correction, average
QC_RHON	Pulsed neutron density quality indicator
Quadrants	
RHNB	Bulk density from neutron, bottom
RHNL	Bulk density from neutron, left
RHNR	Bulk density from neutron, right
RHNU	Bulk density from neutron, up
DRHNB	Bulk density from neutron correction, bottom
DRHNL	Bulk density from neutron correction, left
DRHNR	Bulk density from neutron correction, right
DRHNU	Bulk density from neutron correction, up
RLNB	Long-spacing bulk density from neutron, bottom
RLNL	Long-spacing bulk density from neutron, left
RLNR	Long-spacing bulk density from neutron, right
RLNU	Long-spacing bulk density from neutron, up
ROLN	Long-spacing bulk density from neutron, average
ROSN	Short-spacing bulk density from neutron, average
RSNB	Short-spacing bulk density from neutron, bottom
RSNL	Short-spacing bulk density from neutron, left
RSNR	Short-spacing bulk density from neutron, right
RSNU	Short -pacing bulk density from neutron, up

4.9.6. Neutron-gamma density measurement specifications

Table 4-56. NGD precision and accuracy estimates.

Item	Value
Range	1 to 3.05 g/cm^3
Axial resolution	18 in
Accuracy[†] at 2.5 g/cm^3 and 200 ft/h	0.025 g/cm^3
Statistical precision at 2.5 g/cm^3 and 200 ft/h	0.02 g/cm^3

[†] These accuracy estimates assume an 8.5-in borehole in excellent condition and do not include other effects of the borehole environment such as temperature or tool rotation. These estimates are subject to change as the measurement is developed further.

4.10. Spectroscopy

The short-spaced gamma ray detector associated with the PNG is used to perform spectral analysis of the gamma rays resulting from neutron capture in the formation. The spectroscopy computations are performed downhole, allowing real-time identification of the elemental composition of the formation, formation grain nuclear response computation, and lithology determination. Volumetric lithology quantification is important for accurate porosity and saturation determination.

4.10.1. Spectroscopy theory of measurement

Once a neutron has lost the majority of its original energy through interactions with the atoms in a formation, it reaches thermal energy level. Eventually, the neutron is captured by an atom, which becomes a different isotope of the same element through the addition of the neutron to its nucleus. Some of the energy from this process is released as a set of gamma rays; their energies are characteristic of the element from which they were released. By measuring the gamma ray spectra emitted by a formation after bombardment with high-energy neutrons, the elemental composition of the formation can be determined.

The capture spectroscopy measurement consists of a sequence of events including

- emission of fast neutrons from a neutron source

- slowing down of fast neutrons to thermal energies due to collisions with nuclei in the formation and borehole

- capture of thermal neutrons by atoms in the formation and borehole and the subsequent emission of one or more prompt high-energy gamma rays

- detection of the prompt capture gamma rays

- spectral stripping of the measured pulse height spectrum to obtain the relative contributions of the gamma rays from the various elements to the total spectrum (elemental relative yields)

- oxide closure processing to account for the presence of elements which are not measured by capture spectroscopy (such as oxygen and carbon) and to convert the relative yields to the dry weight proportions of the elements in the formation

- conversion of the dry weight elemental proportions to formation grain nuclear response properties; for example, conversion from the dry weight proportions of silicon, calcium, iron, and sulfur (S) to the grain density of the formation

- conversion from the dry weight elemental proportions to dry weight mineral proportions and subsequently to the volumetric mineral proportions.

The production of gamma rays from neutrons interactions involves either

- high-energy inelastic reactions

- thermal capture reactions.

In the first case, a high-energy (several-MeV) neutron interacts with a nucleus in the borehole or the formation. As a result, the nucleus emits one or more gamma rays that are characteristic of the particular element (isotope). Inelastic gamma rays are mainly important for logging (carbon/oxygen), but this measurement is not yet incorporated in the EcoScope tool.

In the second case, a slow (thermal) neutron is absorbed by a nucleus in the formation, the borehole or the tool. While neutron absorption is unlikely at high-neutron energies, it occurs quite readily at thermal energies, especially in the presence of thermal neutron absorbers such as chlorine, boron (B), and gadolinium (Gd). When neutron capture occurs, the nucleus is left in an excited state. When the nucleus returns to its ground state, high-energy gamma rays are emitted. These gamma rays are referred to as prompt capture gamma rays because they are emitted immediately after neutron capture. These gamma rays are characteristic of the nucleus that captured the neutron. Many of the gamma rays emitted are very energetic and scatter throughout the formation and wellbore, resulting in a cascade of associated lower energy gamma rays.

When struck by an incoming gamma ray, the detector material scintillates—that is, it generates light, which is extracted from the crystal. The scintillation light is absorbed by the photocathode material in the photomultiplier tube. The photoelectric emission of an electron from the surface of the photocathode is followed by electron multiplication to increase the magnitude of the signal. The result is a composite gamma ray energy spectrum, which provides a measurement of the contribution of various elements in the tool, borehole, and formation **(Figure 4-82)**.

The measured spectrum is the linear summation of gamma ray signals from elements in and around the tool. The fraction of the spectral area due to gamma rays from a particular element is called the relative elemental yield. In most well logging situations, the main contributions to the measured capture gamma ray spectra come from the elements hydrogen (H), chlorine (Cl), silicon (Si), calcium (Ca), iron (Fe), sulfur (S), potassium (K), titanium (Ti), gadolinium (Gd), magnesium (Mg), chromium (Cr), nickel (Ni), and barium (Ba).

Figure 4-82. Neutron capture spectroscopy, allowing the elemental composition of a formation to be determined by examining the prompt gamma rays emitted after neutron bombardment.

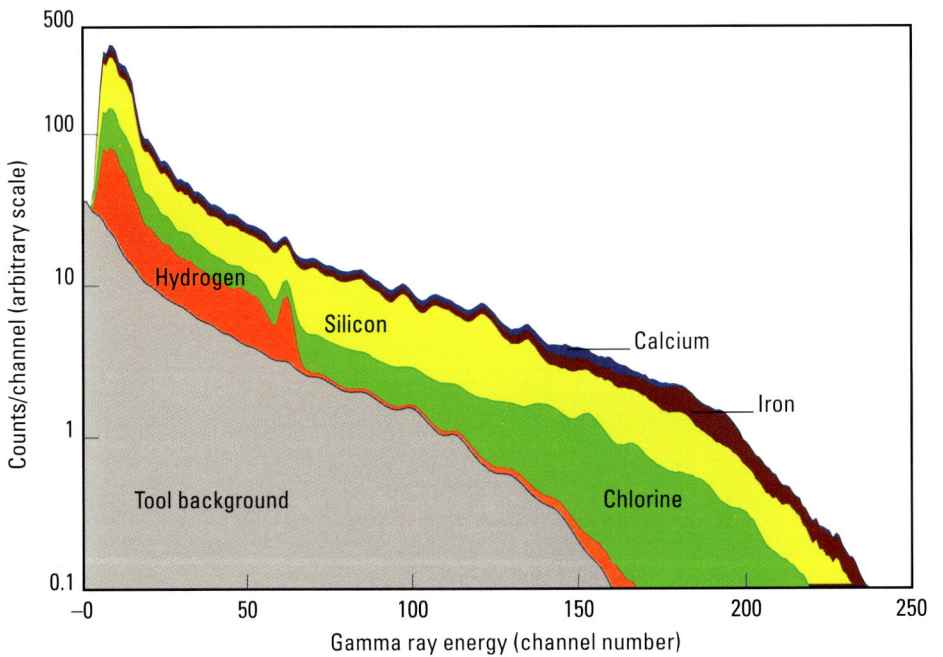

For a given tool and detector configuration, each element has a characteristic distribution of gamma rays called an elemental standard spectrum.

Spectral stripping is a weighted least-squares regression used to decompose the measured spectrum into the relative proportions of the various elemental standard spectra. Spectral stripping processing delivers the relative yields of the various elements in the tool, borehole, and formation. The elemental composition of the tool is known so the tool background contribution can be subtracted (gray shaded region in Figure 4-82) leaving the relative yields of the elements in the formation.

The relative elemental yields can be used to determine the ratio of one element to another in the formation (e.g., calcium/silicon) but do not directly deliver the absolute concentration or weight fraction of any element.

Oxides closure processing converts the relative elemental yields to elemental concentrations that can be used for quantitative interpretation. This is accomplished using an oxides closure model, which is derived from the knowledge that in sedimentary rocks, most elements exist in their oxide forms and that the sum of the rock forming oxides is 1. In terms of the elements, this is expressed as

$$SiO_2 + TiO_2 + Al_2O_3 + Fe_2O_3 + MgO + CaO + Na_2O + K_2O + CO_2 + P_2O_5 + H_2O + SO_2 = 1 \qquad \text{(Equation 4-21)}.$$

Although only a subset of these rock forming elements is available in the form of relative yields from the capture spectroscopy measurements, it is possible to adapt this equation to an oxides closure model that permits conversion from yields to concentrations. Several factors must be taken into consideration. First, unmeasured elements must be accounted for by relying on natural correlations between elements in sedimentary rocks. For example, aluminum is not measured, but computed from the elements silicon, calcium, and iron. The complementary relationship between these elements was discovered by examining a large database of sedimentary rock chemistry and mineralogy.

$$Al_{derived} = 0.39(100 - 2.139Si_{measured} - 2.497Ca_{measured} - 1.99Fe_{measured}) \qquad \text{(Equation 4-22)}.$$

Second, as the input measurements are relative yields, the measurement sensitivity of each element must be taken into account. Some elements are more easily detected than others. Each tool has sensitivity coefficients quantifying how easily each element is detected. Calcium has a sensitivity coefficient around 1.6 compared with 1 for silicon. This means that calcium atoms can be detected 60% more easily in the spectra than silicon atoms.

Third, not all of the measured yields come from the rock. For example, the analyzed yields include tool background, as well as hydrogen and chlorine from both borehole and formation fluids. In order to derive only the rock elemental concentrations, the undesired yields are left out of the oxides closure model, and the remaining yields are normalized to unity.

The oxides closure model can be expressed as

$$F \left| X_{Si} \frac{Y_{Si}}{S_{Si}} + X_{Ca} \frac{Y_{Ca}}{S_{Ca}} + X_S \frac{Y_S}{S_S} + X_{Ti} \frac{Y_{Ti}}{S_{Ti}} + ... + X_{FeAl} \frac{Y_{Fe}}{S_{Fe}} \right| = 1 \qquad \text{(Equation 4-23)},$$

where

Y_i = the relative yield of element i

S_i = the sensitivity of element i to the prompt neutron capture measurement

X_i = the oxide association factor used to convert the element to its appropriate oxide or oxide and related elements (for silicon and titanium, this term simply converts silicon to SiO_2 and titanium to TiO_2; for calcium, the model assumes that calcium primarily resides as $CaCO_3$, thus accounting for both the CaO and CO_2 terms in Equation 4-21)

F = the normalization factor that compensates for the fact that chlorine, hydrogen, and the tool background are eliminated from the analysis and that the yields are relative and divided by their sensitivities.

Using this oxides closure model, it is possible to compute the weight fraction of each element in the formation using the relationship

$$W_i = F \frac{Y_i}{S_i} \qquad \text{(Equation 4-24)},$$

where

W_i = the weight fraction of element i.

These dry weight elemental concentrations can be used to compute grain properties such as the grain density, sigma, and neutron responses. For example, formation grain density can be directly approximated as a linear combination of the elements silicon, calcium, iron, and sulfur with a standard error of only 0.015 g/cm^3 according to the relationship

$$\rho_{grain} = 2.620 + 0.0490Si + 0.2274Ca + 1.993Fe + 1.193S \qquad \text{(Equation 4-25)},$$

where Si, Ca, Fe, and S are weight fractions of the elements silicon, calcium, iron, and sulfur as derived from the oxide closure processing. Pure quartz has a dry weight silicon fraction of 0.47. Substituting this value into the equation produces a grain density of 2.65 g/cm^3, corresponding to the grain density of quartz.

The dry weight elemental concentrations can also be converted into the dry weight and volumetric proportions of the major mineral groups. Standard spectral conversion processing delivers the proportions of

pyrite   = pyrite
anhydrite = anhydrite + gypsum
clay   = kaolinite + illite + smectite + chlorite + glauconite
carbonate = calcite + dolomite
siderite   = siderite
QFM   = quartz + feldspar + mica

based on the measured silicon, calcium, iron, and sulfur concentrations, along with the derived aluminum concentration. The interpretation is based on an extensive core database comprising chemical concentrations and mineralogy measurements on hundreds of sedimentary rocks. A sequential logic is generally used. For example, the sequence

1. pyrite—FeS_2
2. anhydrite—$CaSO_4$
3. clay—$Al_wSi_xO_y(OH)_z$
4. dolomite—$CaMg(CO_3)_2$
5. calcite—$CaCO_3$
6. siderite—$FeCO_3$
7. quartz feldspar mica—QFM

may be used such that the amount of pyrite is determined first. The appropriate proportions of sulfur and iron are then subtracted from their respective dry weights. The remaining sulfur is allocated to anhydrite. The appropriate proportion of calcium is then subtracted from the dry weight of calcium, the remainder of which is allocated to carbonate (dolomite and calcite). The processing continues allocating elements to the minerals and calculating the residual until it has deduced the formation mineralogy.

Note that through the use of elemental capture spectroscopy, the formation mineralogy has been determined without reference to any other measurement and that porosity is not required for the solution. **Figure 4-83** shows neutron capture spectroscopy processing, involving spectral stipping to derive the elemental yields and dry weight concentrations.

Quantitative lithology volumes from capture spectroscopy significantly simplify porosity and subsequent saturation evaluation, particularly in complex lithologies.

Figure 4-83. Neutron capture spectroscopy processing, which involves spectral stripping to derive the elemental relative yields followed by oxides closure to determine the dry weight concentrations of each of the elements present. Formation lithology and grain properties such as grain density can be computed from the dry weight concentrations.

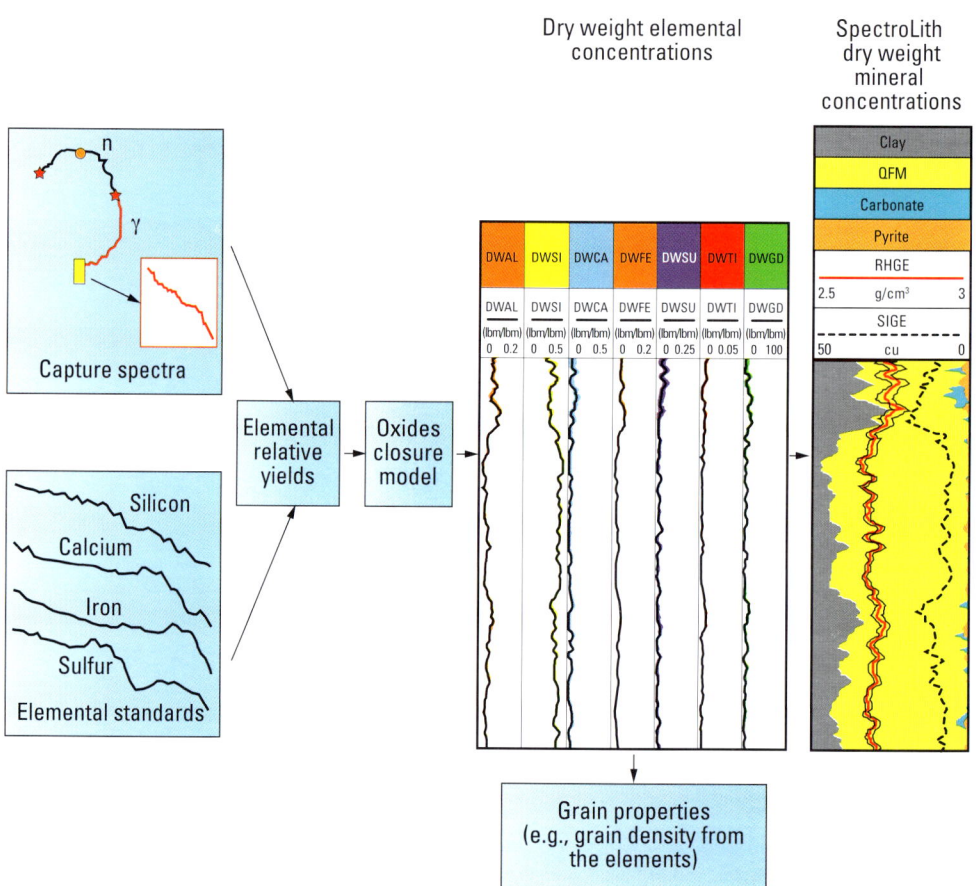

4.10.2. Spectroscopy environmental corrections

The elemental capture spectroscopy measurement responds to the elements in the proximity of the detector. Consequently, the spectral response from the elements in the tool must be removed. Using the known composition and spectral response of the collar material, the tool background signature is subtracted from the measured spectra.

The following elements in the borehole must be accounted for.

- The chlorine- and hydrogen-relative yields are not used in the oxide closure model as these elements are often associated with fluids in the borehole and formation pores. Elemental spectroscopy is targeted at evaluating the composition of the solid formation grain.

- The presence of barite ($BaSO_4$) in the mud requires that the barium yield be determined and a corresponding proportion of the sulfur yield removed to leave only the sulfur yield associated with the sulfur in the grain of the formation.

- Potassium is not used in the processing beyond oxide closure due to the possibility of potassium salts, such as potassium chloride (KCl), being used in the mud system.

4.10.3. Spectroscopy applications

Spectroscopy is used to determine the elemental composition of the formation grain, its mineralogy, and response to nuclear measurements such as density, sigma, thermal, and epithermal neutron. Independent determination of these responses allows the corresponding measurements of bulk density, sigma, and neutron response to be used to assess the true formation porosity and fluid content with greater accuracy.

The response of the formation solid (grain or matrix) to nuclear measurements, such as density, sigma, and neutron response, can either be computed from the elemental composition or derived from the volumetric proportions of the minerals and their known responses. Knowledge of the grain density, for example, improves the accuracy of the calculated density porosity by reducing the uncertainty on this key input in the density to porosity transform equation

(Equation 4-26),

$$\phi = \frac{\rho_{grain} - \rho_{bulk}}{\rho_{grain} - \rho_{fluid}}$$

where

ρ_{grain} = density of the formation solid.

Studies have shown that the common assumption of simple formation mineralogy generally results in underestimation of the formation porosity. This underestimation is because small quantities of heavy minerals are present in many formations, resulting in a higher grain density than if only the major mineralogy is present. Spectroscopy allows accurate grain density to be calculated by quantifying the proportions of the elements in the formation, including those in heavy minerals that increase the formation grain density.

When the spectroscopy-derived grain density is used to calculate the porosity from density, the resulting porosity is often higher, shown in **Figure 4-84** by the percentage increase in porosity observed on several hundred cores.

Figure 4-84. Formation porosity, which tends to be systematically underestimated when the mineralogy is assumed to be simple. Spectroscopy allows quantitative evaluation of the mineralogy, including the small proportions of heavy minerals which are often overlooked, resulting in more accurate evaluation and generally increased formation porosity.

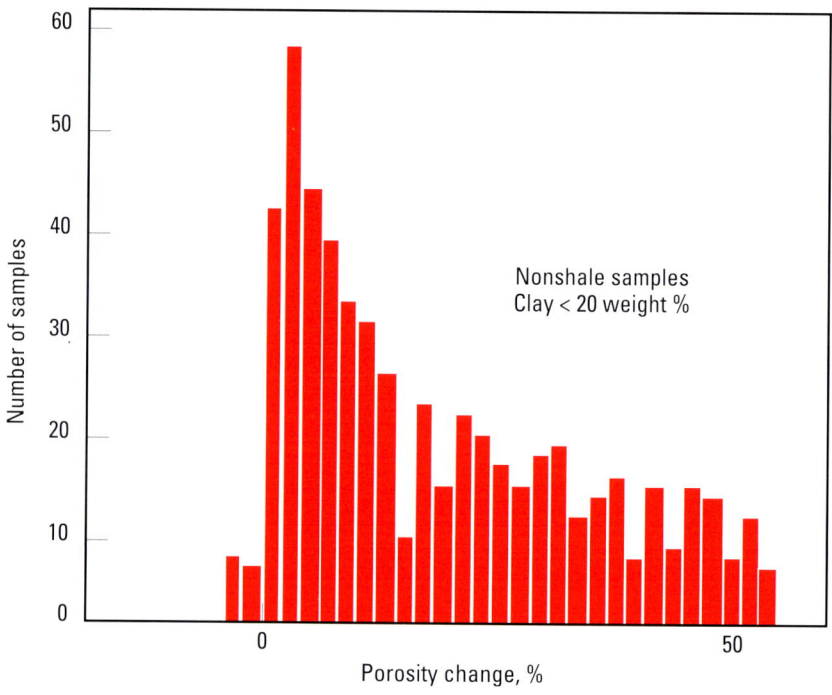

The sigma response of the formation grain can also be derived from the elements measured by capture spectroscopy. Grain sigma is a key parameter in calculating water saturation from sigma as seen in the equation

$$S_w = \frac{(\Sigma_{bulk} - \Sigma_{grain}) + \phi \times (\Sigma_{grain} - \Sigma_{hc})}{\phi \times (\Sigma_{water} - \Sigma_{hc})}$$

(Equation 4-27),

Where

Σ_{bulk} = measured bulk sigma

Σ_{grain} = grain sigma

Σ_{water} = sigma of the water

Σ_{hc} = sigma of the hydrocarbon.

The use of a continuous spectroscopy log to derive the sigma response of the grain allows accurate saturation determination by accounting for formation mineralogy changes on a level-by-level basis.

The response of thermal and epithermal neutron measurements to the formation grain can also be derived from knowledge of the grain elemental composition. The neutron measurements respond primarily to hydrogen, which is predominantly found in the fluids in the pore space. However, neutrons also interact with elements in the formation grains resulting in differing neutron responses depending on the mineralogy. The neutron measurements generally respond according to the volumetric proportions of the formation components, as summarized in the equation

$$N_{bulk} = (1 - \phi) \times N_{grain} + \phi \times N_{fluid} \qquad \text{(Equation 4-28)},$$

where

N_{bulk} = measured bulk neutron response
N_{grain} = neutron response to the formation grain
N_{fluid} = neutron response to the fluid.

As the HI of the fluid is the dominant effect on the neutron response, this equation is often simplified to

$$HI_{bulk} = \phi \times HI_{fluid} \qquad \text{(Equation 4-29)}$$

so that the formation porosity is approximated by

$$\phi = \frac{HI_{bulk}}{HI_{fluid}} \qquad \text{(Equation 4-30)}.$$

To take the formation effect on the neutron response into account the equation should actually be

$$\phi = \frac{N_{grain} - N_{bulk}}{N_{grain} - N_{fluid}} \qquad \text{(Equation 4-31)},$$

where

$$N_{fluid} \approx HI_{fluid}.$$

By explicitly taking into account the neutron response of the grain, the accuracy of the porosity derived from neutron measurements is improved.

Spectroscopy delivers the elemental composition of the formation grain, allowing the response of the nuclear measurements to the solid in the formation to be calculated. The bulk nuclear measurements respond to both the solid and the fluids in the formation. By accurately quantifying the nuclear measurement response to the formation grain, the proportion of the bulk measurement due to the solids can be identified, allowing more accurate determination of the proportion of the formation that is composed of fluid (the porosity) and the nature of the fluid in the pores.

The mineralogical composition of the formation is derived from the elemental information extracted from the capture spectroscopy measurement. SpectroLith processing transforms the dry weight elemental concentrations into quantitative, dry weight proportions of the minerals in the formation, simplifying formation evaluation in complex lithologies (**Figure 4-85**).

Quantitative dry weight proportion of clay is one of the key outputs from SpectroLith processing. In a traditional interpretation without spectroscopy, clay and fine-grained silt are lumped together as shale. High and low thresholds are often applied to the gamma ray measurement to approximate the shale volume from a linear correlation with the gamma ray response. Not only are there significant uncertainties associated with the application of thresholds to the gamma ray measurement due to variations in clay type or radioactive mineral content for example, but the gamma ray measurement itself may be responding to differing proportions of uranium salts in the formation. In extreme cases, this response can cause a clean sand to be interpreted as shale because of the high gamma ray associated with the uranium salts deposited in the pores.

By analyzing the elemental composition of the formation, the clay content can be distinguished from the silt, and the volumetric proportion of clay in the formation can be quantified.

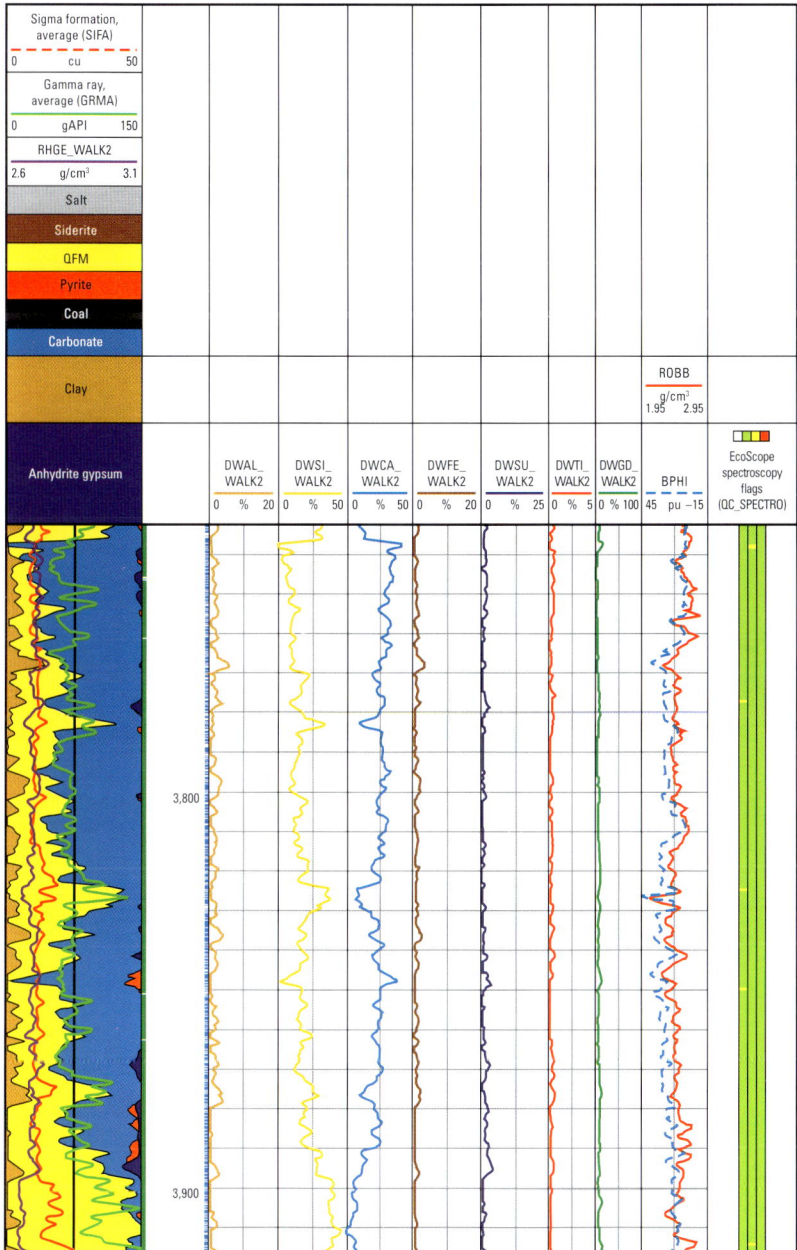

Figure 4-85. The elemental composition of the formation (right tracks), transformed into a dry weight proportion mineralogy of the formation (left track).

Figure **4-86** is an example where the gamma ray underestimates the volume of clay above 1,400 ft, but agrees with the core-validated, spectroscopy-derived clay volume below 1,400 ft. This difference demonstrates that while the gamma ray approximation may be applicable in some formations, it may be misleading in others.

Figure 4-86. Elemental spectroscopy, which determines the proportion of clay in the formation by its chemical signature, rather than depending on a loose correlation between clay content and formation gamma ray response, as is often assumed. In this example, the gamma ray underestimates the volume of clay above 1,400 ft but agrees with the core-validated, spectroscopy-derived clay volume below 1,400 ft.

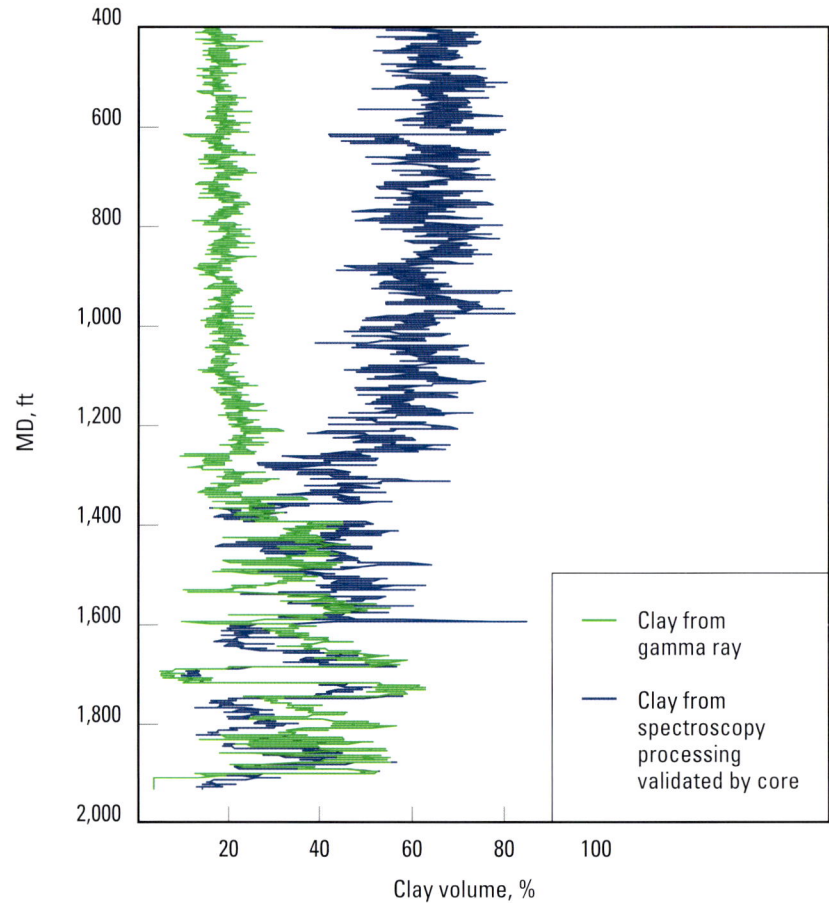

In addition to the volumetric proportion of clay, spectroscopy analysis also delivers the volumetric proportion of the other minerals in the formation, which aids in interpretation of measurements such as the density-neutron separation and sigma response, allowing better identification of the formation fluids and porosity.

Knowledge of formation lithology can also be used for permeability transforms, such as the k-Lambda equation, which use porosity and mineralogy information to estimate formation permeability.

Formation water salinity can be estimated from the thermal neutron capture spectroscopy. As shown in Figure 4-82, both chlorine and hydrogen are visible in the spectrum. Their contribution to the spectrum is calculated during spectral stripping. However, as both chlorine and hydrogen are associated with formation fluids they are not included in the oxide closure processing which is designed to determine the elemental composition of the formation matrix. The apparent salinity channel, ASAL, is computed from the ratio of the relative yields of chlorine and hydrogen derived from the spectral stripping and gives an indication of the salinity of the fluids surrounding the tool. In an 8.5-in hole, the EcoScope collar occupies much of the borehole volume, so the majority of the signal is contributed by the chlorine and hydrogen in the formation.

Figure 4-87 shows an example, discussed in more detail in Section 5.5.10, in which the ASAL (dashed blue line in Tracks 3 and 5), provides strong supporting evidence for a formation water salinity of 150 ppk rather than the anticipated value of 250 ppk.

Track 1 shows no separation in the resistivities, suggesting that there is minimal invasion, so the sigma measurement can be used to derive the uninvaded formation water saturation. Selecting a water salinity of 250 ppk (red vertical line in Track 3) does not agree with the ASAL value and results in a discrepancy between the resistivity- and sigma-derived water saturations (blue and red lines, respectively, in Track 4). When the salinity is changed to 150 ppk (vertical red line in Track 5) such that it overlays the ASAL curve, the independent resistivity- and sigma-derived saturations agree (blue and red lines, respectively, in Track 6). The independent water saturations crossconfirm each other, and in conjunction with the spectroscopy-derived water salinity, provide strong evidence that the in situ water salinity is 150 ppk rather than the expected value of 250 ppk. The spectroscopy-derived water salinity measurement provides valuable information in zones of uncertain water salinity.

Figure 4-87. ASAL (dashed blue line in Tracks 3 and 5), computed from the relative yields of chlorine and H. The channel provides a direct estimate of fluid salinity.

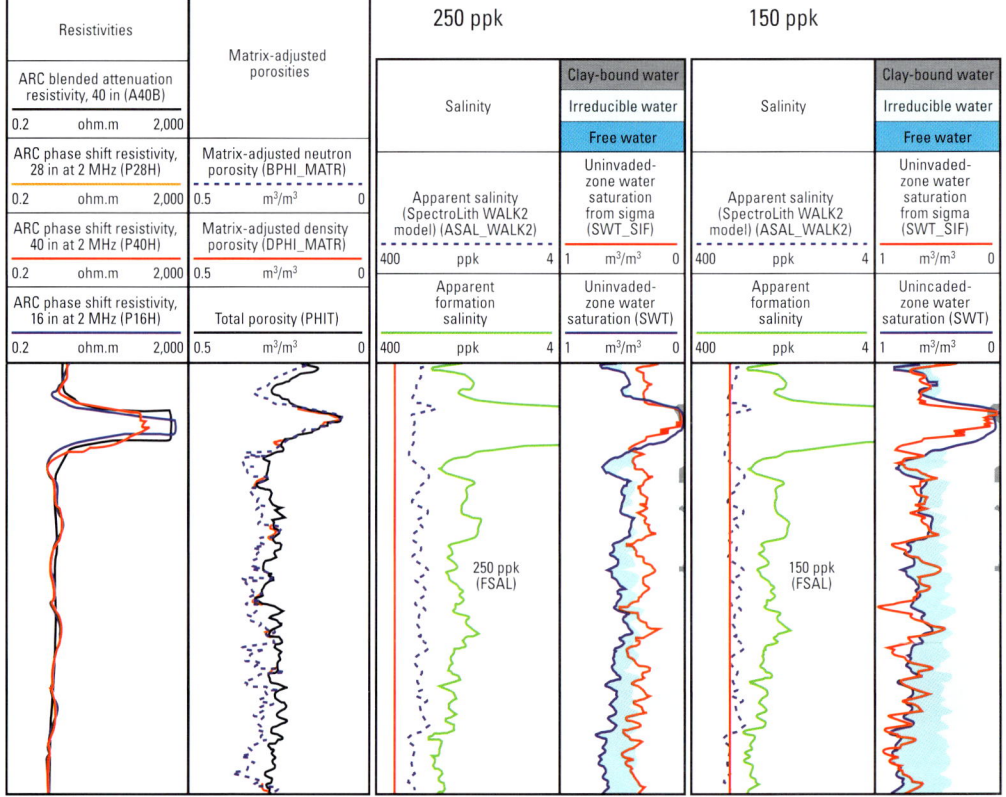

Chemostratigraphic correlation by identification of marker beds within a complex sequence is made possible by the availability of the elemental composition of the formation. **Figure 4-88** shows an example where correlations are made based on the dry weight percent of silicon in the formation (blue lines). Calcium is also used to make correlations (red lines), which verify the silicon correlations.

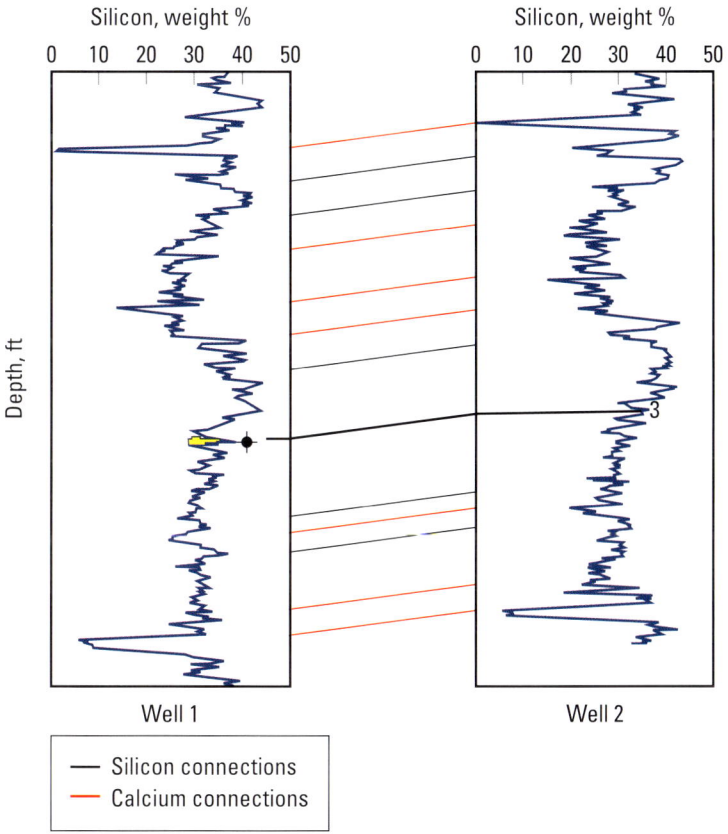

Figure 4-88. Chemostratigraphic correlation, using marker elements to link layers.

4.10.4. Spectroscopy log quality control

Drilling ROP is the single biggest factor affecting the quality of the spectroscopy measurement. Lower ROP improves the measurement statistics.

An acquisition job planner is available to predict the required ROP for accurate spectroscopy data (**Figure 4-89**). The quality of the spectroscopy data is largely driven by statistical effects. This planner is designed to provide a quantitative estimate of the precision of the measurement with the current EcoScope hardware in differing environments. Under poor environmental conditions, it is strongly recommended to slow down the ROP as much as possible to recover some of the performance lost due to the nature of the environment. This planner computes the required ROP to achieve acceptable performance.

Figure 4-89. A spectroscopy acquisition planner, provided to predict data quality.

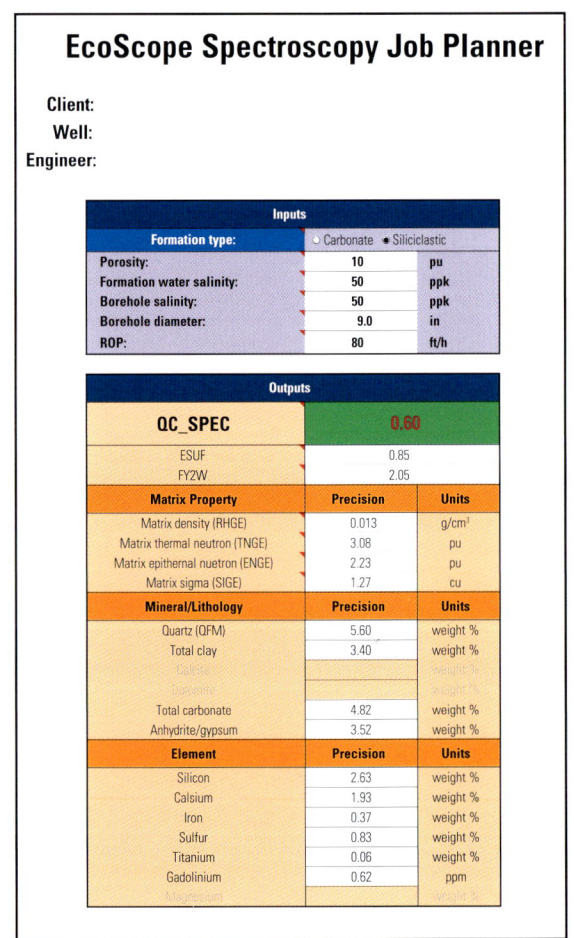

QC_SPEC

This QC flag is a composite of the ESUF and FY2W QC indicators. The value and coding is the same as the other QC indicators for the EcoScope.

- Below 1 is **GREEN (data quality should be acceptable)**.

- Between 1 and 2 is **YELLOW (data quality may be questionable)**.

- Above 2 is **RED (data quality is probably not acceptable)**.

The values of QC_SPEC also correspond to the real-time transmitted C_SPEC channel. Note that the real-time QC_SPEC channel will only contain discrete values of 0, 1, or 2.

ESUF

The elemental capture spectroscopy statistical uncertainty factor is a derived number that includes factors like count rate statistics, environmental conditions, and detector performance and is the best indicator of overall spectroscopy precision. All of the EcoScope statistical uncertainties are directly proportional to ESUF. ESUF can be improved by reducing ROP.

FY2W

The yield-to-weight (from yields to weights—FY2W) factor is a measure of the impact of the nonformation gamma rays on the computation of the elemental concentrations and the derived quantities. It is independent of the ROP and detector effects and can therefore be used to judge the impact of the environment itself on the measurement quality. A high FY2W indicates that a below-average ROP is required to recover spectroscopy precision (ESUF).

The EcoScope hardware incorporates checks to verify correct performance of the acquisition detectors and electronics. Additional checks to ensure that the measurements fall within the range expected for downhole formation are outlined in the QC logic shown in **Table 4-57**.

Table 4-57. Spectroscopy QC flag logic.

QC_SPEC	Hardware	Measurement
Green	All ok	$0.97 < MQGN^† < 1.03$ and $-1.5 < MAOF^‡ < 1.5$ and $0 < MARD^§ < 6$ and $0 < FY2W < 7$ and $-5 <$ absolute value of the sum of negative relative yields < 0
Yellow	High voltage out of range or Gain loop not in AUTO_REMOTE or Insufficient count rate	$0.95 > MQGN > 0.97$ or $1.03 > MQGN > 1.05$ $-3 > MAOF > -1.5$ or $1.5 > MAOF > 3$ $6 > MARD > 10$ or $7 > FY2W > 10$ or $-7 <$ absolute value of sum of negative relative yields < -5
Red	SSn failure or PNG failure	$0.95 > MQGN$ or $MQGN > 1.05$ $-3 > MAOF$ or $MAOF > 3$ $6 > MARD$ or $MARD > 10$ $FY2W > 10$ or Absolute value of sum of negative relative yields < -7
White		No new spectroscopy data available for more than 2 ft.

† Marquardt level-by-level optimization gain factor (a fit parameter in the level-by-level marquardt optimization that adjusts for gain in the spectral energy calibration).
‡ Marquardt optimization offset (a fit parameter in the marquardt optimization cycle that adjusts for offset in the spectral energy calibration).
§ Marquardt optimization resolution degradation factor (a fit parameter in the marquardt optimization cycle that adjusts for degradation in the spectral resolution with temperature).

4.10.5. Spectroscopy channels and parameters

4.10.5.1. Real time

Table 4-58. Spectroscopy real-time parameters.

Parameter Name	Description	Unit
SPL_SULFUR_MIN	SpectroLith sulfur mineral option	Anhydrite_pyrite (hard-coded)
SPL_CLAY_MODEL	SpectroLith clay model	Arenite (hard-coded)
SPL_MG_OPT†	SpectroLith magnesium processing option	On, off (default is off)

† The magnesium processing option determines whether magnesium is stripped from the capture spectra and used in the subsequent processing to distinguish dolomite from calcite. This option should only be used in wells where dolomite is expected.

Table 4-59. Spectroscopy real-time transmitted channels.

Transmitted Channel	Description
CCA_DH_ECO_RT	Capture calcium relative yield, real time, computed downhole
CFE_DH_ECO_RT	Capture iron relative yield, real time, computed downhole
CSI_DH_ECO_RT	Capture silicon relative yield, real time, computed downhole
CSUL_DH_ECO_RT	Capture sulfur relative yield, real time, computed downhole
CGD_DH_ECO_RT	Capture gadolinium relative yield, real time, computed downhole
CMG_DH_ECO_RT	Capture magnesium relative yield, real time, computed downhole
FY2W_DH_ECO_RT	Factor from yield to weight, real time, computed downhole
QC_SPEC_DH_ECO_RT	Spectroscopy quality indicator, real time, computed downhole

Table 4-60. Spectroscopy real-time computed channels.

Computed Channel	Description
CTI_ECO_RT	Capture titanium relative yield, real time
DWSI_ECO_RT	Dry weight fraction silicon, real time
DWCA_ECO_RT	Dry weight fraction ca, real time
DWFE_ECO_RT	Dry weight fraction iron + 0.14Al, real time
DWSU_ECO_RT	Dry weight fraction sulfur, real time
DWGD_ECO_RT	Dry weight fraction gadolinium, real time
DWAL_ECO_RT	Dry weight fraction pseudoaluminum, real time
DWTI_ECO_RT	Dry weight fraction titanium, real time
DWMG_ECO_RT	Dry weight fraction magnesium, real time
Mineral Dry Weight Fractions	
WANH_ECO_RT	Dry weight fraction anhydrite/gypsum, real time
WCLA_ECO_RT	Dry weight fraction clay, real time
WCAR_ECO_RT	Dry weight fraction carbonate, real time
WCLC_ECO_RT	Dry weight fraction calcite, real time
WDOL_ECO_RT	Dry weight fraction dolomite, real time
WPYR_ECO_RT	Dry weight fraction pyrite, real time
WQFM_ECO_RT	Dry weight fraction QFM, real time
WSID_ECO_RT	Dry weight fraction siderite, real time
Grain Responses Computed from the Elements	
PEGE_ECO_RT	Grain PEF response from elements, real time
RHGE_ECO_RT	Grain density from elemental concentrations, real time
SIGE_ECO_RT	Grain sigma from elemental concentrations, real time
TNGE _ECO_RT	Grain thermal neutron response from elemental concentrations, real time

4.10.5.2. Recorded mode

Table 4-61. Spectroscopy recorded-mode parameters.

Parameter Name	Description	Unit
SPL_SULFUR_MIN[†]	SpectroLith sulfur mineral option	Anhydrite, pyrite, anhydrite_pyrite (default), none
SPL_CLAY_MODEL[‡]	SpectroLith clay model	Arenite (default), arkose, subarkose, none
SPL_MG_OPT[§]	SpectroLith magnesium processing option	On, off (default is off)

[†] Use pyrite for sandstone formations and anhydrite for carbonate formations. In complex formations, using both is an option, but results in reduced precision.
[‡] Use arenite or subarkose for sandstone formations and arenite for carbonate formations. Using subarkose changes the relation between the estimated aluminum content and the clay weight fraction. The result is that shaly sections appear even shalier, and the clean sections become even cleaner. The transition point is just short of 25% clay.
[§] This option determines whether magnesium is stripped from the capture spectra and used in the subsequent processing to distinguish dolomite from calcite. Only use in wells where dolomite is expected.

Table 4-62 . Spectroscopy recorded-mode channels.

Output Channel	Description
CK_WALK2[†]	Capture potassium relative yield (WALK2 model)
CBA_WALK2[†]	Capture barium relative yield (WALK2 model)
CCA_WALK2[†]	Capture calcium relative yield (WALK2 model)
CCHL_WALK2[†]	Capture chlorine relative yield (WALK2 model)
CFEU_WALK2[†]	Capture iron relative yield (Uncorrected, WALK2 model)
CGD_WALK2[†]	Capture gadolinium relative yield (WALK2 model)
CHY_WALK2[†]	Capture hydrogen relative yield (WALK2 model)
CK_WALK2[†]	Capture potassium relative yield (WALK2 model)
CNC_WALK2[†]	Capture nickel-chromium relative yield (WALK2 model)
CSI_WALK2[†]	Capture silicon relative yield (WALK2 model)
CSUU_WALK2[†]	Capture sulfur relative yield (uncorrected, WALK2 model)
CTI_WALK2[†]	Capture titanium relative yield (WALK2 model)
CMG_MGWALK	Capture magnesium relative yield (MGWALK model)
DWAL_WALK2[†]	Dry weight fraction pseudoaluminum (WALK2 model)
DWAL_SIG_WALK2[†]	Statistical uncertainty in aluminum (WALK2 model)
DWCA_WALK2[†]	Dry weight fraction calcium (WALK2 model)
DWCA_SIG_WALK2[†]	Statistical uncertainty in calcium (WALK2 model)
DWFE_WALK2[†]	Dry weight fraction iron + aluminum-0.14 (WALK2 model)
DWFE_SIG_WALK2[†]	Statistical uncertainty in dry weight iron (WALK2 model)
DWGD_WALK2[†]	Dry weight fraction gadolinium (WALK2 model)
DWGD_SIG_WALK2[†]	Statistical uncertainty in gadolinium (WALK2 model)
DWK_WALK2[†]	Dry weight fraction potassium (WALK2 model)
DWK_SIG_WALK2[†]	Statistical uncertainty in potassium (WALK2 model)
DWSI_WALK2[†]	Dry weight fraction silicon (WALK2 model)
DWSI_SIG_WALK2[†]	Statistical uncertainty in silicon (WALK2 model)

Output Channel	Description
DWSU_WALK2[†]	Dry weight fraction sulfur (WALK2 model)
DWSU_SIG_WALK2[†]	Statistical uncertainty in dry weight sulfur (WALK2 model)
DWTI_WALK2[†]	Dry weight fraction titanium (WALK2 model)
DWTI_SIG_WALK2[†]	Statistical uncertainty in dry weight titanium (WALK2 model)
DWMG_MGWALK	Dry weight fraction magnesium (MGWALK model)
DWMG_SIG_MGWALK	Statistical uncertainty in dry weight magnesium (MGWALK model)
FY2W	Factor from yield to weight
QC_SPEC	Spectroscopy quality indicator
TAB_SPC	Spectroscopy time after bit
TICK_SPC	Spectroscopy samples
WANH_WALK2[†]	Dry weight fraction anhydrite/gypsum (WALK2 model)
WCLA_WALK2[†]	Dry weight fraction clay (WALK2 model)
WCAR_WALK2[†]	Dry weight fraction carbonate (WALK2 model)
WQFM_WALK2[†]	Dry weight fraction QFM (WALK2 model)
WPYR_WALK2[†]	Dry weight fraction pyrite (WALK2 model)
WSID_WALK2[†]	Dry weight fraction siderite (WALK2 model)
WCOA_WALK2[†]	Dry weight fraction coal (WALK2 model)
WEVA_WALK2[†]	Dry weight fraction salt (WALK2 model)
WCLC_MGWALK	Dry weight fraction calcite (MGWALK model)
WDOL_MGWALK	Dry weight fraction dolomite (MGWALK model)
RHGE_WALK2[†]	Grain density from elemental concentrations (WALK2 model)
PEGE_WALK2[†]	Grain PEF from elemental concentrations (WALK2 model)
UGE_WALK2[†]	Grain volumetric PEF from elemental concentrations (WALK2 model)
ENGE_WALK2[†]	Grain epithermal neutron porosity from elemental concentrations (WALK2 model)
TNGE_WALK2[†]	Grain thermal neutron porosity from elemental concentrations (WALK2 model)
SIGE_WALK2[†]	Grain sigma from elemental concentrations, real time (WALK2 model)
ASAL_WALK2[†]	Apparent salinity (WALK2 model)

[†] These channels may also be derived from the MGWALK model (includes magnesium), in which case the channel name has the _MGWALK modifier rather than the _WALK2 modifier.

4.10.6. Spectroscopy measurement specifications

Table 4-65. Thermal capture gamma ray spectroscopy specifications at 100-ft/h ROP in a 10-pu water-filled carbonate formation.

Item	Description	Remarks
Axial Resolution		
Intrinsic	15 in	Based on tool hardware
Effective	1.7ft at ROP of 100ft/h	Based on data averaging
Elements		
Yields	Silicon, calcium, iron, sulfur, titanium, gadolinium, potassium, magnesium, barium, hydrogen, chlorine	
Dry weight	Silicon, calcium, iron, sulfur, titanium, gadolinium, magnesium, (pseudoaluminum)	Oxide closure processing (derived aluminum)
Lithology (weight fractions)	QFM Calcite Dolomite (if magnesium stripping is performed) Anhydrite[†] Pyrite[†] Clay Siderite	SpectroLith processing
Matrix Properties	Grain density Grain thermal neutron Grain epithermal neutron Grain PEF and grain U Grain sigma	
Precision	Absolute weight errors (%) at ROP of 100 ft/h in an 8.5-in borehole filled with 50-ppk salinity water drilled thorugh a water-saturated, 10-pu carbonate formation	
Elemental Dry Weights		
Silicon	2.26 weight %	
Calcium	1.74 weight %	
Iron	0.39 weight %	
Sulfur	0.72 weight %	
Titanium	0.05 weight %	
Gadolinium	0.53 ppm	
Magnesium	1.61 weight %	
Lithology		
QFM	4.8 weight %	
Calcite	8.22 weight %	
Dolomite	12.37 weight %	
Clay	3.27 weight %	
Anhydrite/gypsum	3.05 weight %	
Accuracy[‡]		

[†] User has the option.

[‡] Precision and accuracy are a non-trivial function of formation and tool properties. The given values are calculated for water filled formations at 50-ppk salinity, consisting of the given mineral group, with a formation closure fraction FY2W=2 and resolution degradation factor MARD=2.5.

4.11. Thermal neutron capture cross section (sigma)

The sigma measurement involves determining the time decay constant of the neutron capture gamma ray spectrum. Sigma is primarily sensitive to the presence of chlorine. The measurement has a variety of applications, which include distinguishing the fluids (water, oil, or gas) in the formation and identification of low-resistivity pay (LRP) zones. In a saline water environment, it can be used to calculate the formation oil saturation independent of resistivity measurements.

4.11.1. Sigma theory of measurement

Sigma is the macroscopic thermal neutron capture cross section of the formation. The measurement makes use of the fact that after slowing down to thermal energy, neutrons linger in the formation and the borehole for several hundred microseconds, undergoing multiple collisions with nuclei in the surrounding material. Their capture by formation (and borehole) nuclei results in the emission of one or more gamma rays from the resulting excited nucleus.

As the neutron population near the tool declines because of capture and drift of the neutrons farther and farther away from the tool (diffusion), the neutron and gamma ray count rates observed in the tool detectors decrease. The primary cause of the decrease lies in the decline of the neutron population because of neutron capture. The measurement is therefore sensitive to the presence of thermal absorbers in the formation surrounding the tool.

Thermal neutron capture cross sections for typical elements found in geologic formations are given in **Table 4-66**. While gadolinium has an extremely high thermal neutron capture cross section, it is generally only found in tiny quantities (a few ppm) and so does not have a large impact. Chlorine is the most common element with a high thermal neutron capture cross section found in geological formations. Sigma, therefore, is a very good indicator of the chlorine concentration around the tool, and thus of the formation fluid salinity.

Table 4-66. Thermal neutron capture cross sections for typical elements found in geologic formations.

Element	Symbol	Thermal Neutron Capture Cross Section, b
Gadolinium	Gd	49,000
Chlorine	Cl	35.5
Iron	Fe	2.56
Sulfur	S	0.53
Calcium	Ca	0.43
Hydrogen	H	0.3326
Aluminum	Al	0.232
Silicon	Si	0.171
Magnesium	Mg	0.063
Carbon	C	0.0035
Oxygen	O	0.00019

The primary EcoScope sigma measurement is acquired by a single gamma ray detector. The gamma ray count rate is determined as a function of time, as shown in **Figure 4-90**. Initially, the decrease in the count rate is because of the effect of the tool and borehole in the immediate proximity of the detector. The borehole effect is relatively small for sigma acquired using the EcoScope tool because the large LWD collar displaces the majority of the mud in the borehole.

Figure 4-90. Sigma, the macroscopic thermal neutron capture cross section of the formation, which is closely related to the formation salinity.

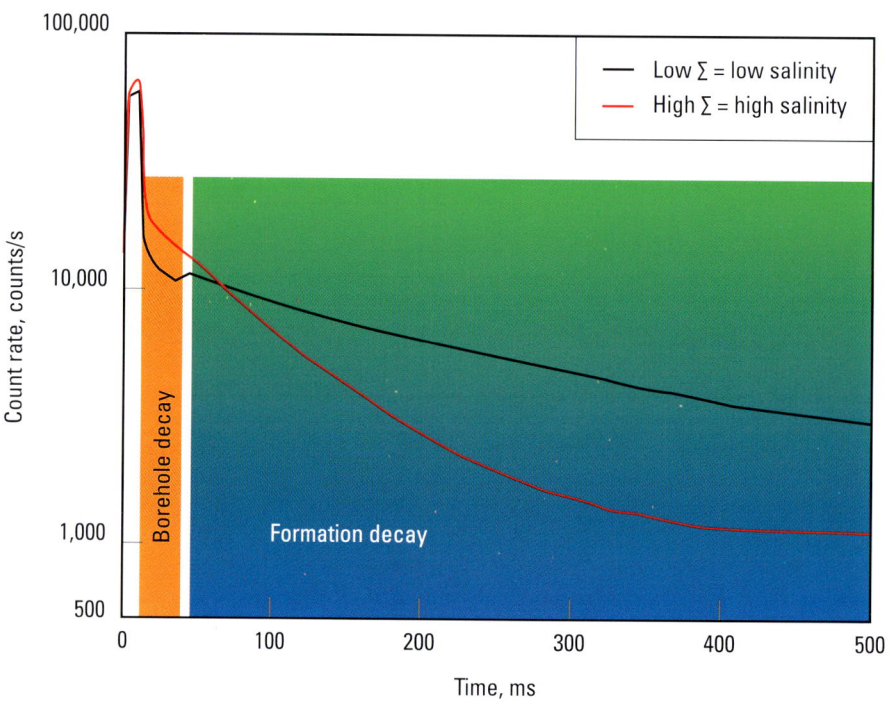

Assuming a single exponential decay, the decay rate of the gamma rays is given by

$$N_{(t)} = N_i e^{-t/\tau}$$ (Equation 4-32),

where
$N_{(t)}$ = count rate at time, t
N_i – initial count rate.

Sigma is related to the decay rate of the counts in the detector by Equation 4-11. Sigma is a volumetric measurement related primarily to the chlorine content of the formation. Because chlorine generally occurs dissolved in the formation water, sigma can be used to derive resistivity-independent water saturation as long as the formation water is sufficiently saline to produce a usable sigma contrast between water and hydrocarbons. This is particularly useful in formations where the traditional resistivity-based methods of estimating water saturation fail to provide reliable results (such as in some LRP zones).

Interpretation of the sigma log for uninvaded formation water saturation determination requires that the measurement be acquired before significant invasion. The EcoScope propagation resistivity array, which is collocated with the sigma measurement, can be used to assess whether invasion has occurred before the sigma data is acquired. However, the propagation resistivity measurements are not very sensitive to very shallow or resistive invasion. To confirm that the sigma measurement is acquired before significant invasion, multiple DOI sigma measurements are acquired using both gamma ray and neutron count rates in the various detectors associated with the PNG. Shallow, medium, and deep sigma measurements are obtained. Where these measurements overlay, the formation can be considered uninvaded for sigma interpretation. Where they separate, any subsequent sigma interpretation needs to account for the effect of invasion.

4.11.2. Sigma environmental corrections

The apparent formation sigma is primarily sensitive to the neutron capture cross section of the formation, but diffusion of neutrons away from the tool and detectors after the PNG neutron burst results in an accelerated decay. In high-HI formations, the neutrons are rapidly slowed and remain in the proximity of the tool. In low-HI formations, the neutrons bounce off the large nuclei of the heavier elements in the formation without losing much energy and are able to diffuse away from the tool. When a neutron is captured, the emitted gamma ray is less likely to make its way back to the detector, so the signal from these distant neutrons, which diffused away from the tool, is lost. The diffusion effect is more significant in low-HI formations.

The borehole size and salinity have a small effect on the sigma response. The relatively small influence of the borehole size and salinity compared with wireline sigma measurements is because of the LWD collar occupying the majority of the borehole volume, displacing the mud and therefore minimizing the borehole effect. These effects are corrected based on mud salinity and hole size inputs.

4.11.3. Sigma applications

Sigma provides formation water saturation without the use of resistivity measurements in saline water environments. The traditional resistivity-based Archie formula for formation water saturation becomes inaccurate when the rock grains themselves conduct some of the measure current, because the Archie equation assumes that the only current path is through the fluids. Sigma provides a method for calculating water saturation that avoids this pitfall. In addition, Archie's parameters a, m, and n and even the resistivity itself are avoided, which makes sigma particularly useful for calculating water saturation in LRP environments. The traditional Archie equation is

$$S_w = \sqrt[n]{\frac{a}{\phi^m} \times \frac{R_w}{R_t}}$$

(Equation 4-33),

where
a = empirically derived constant
m = cementation exponent
n = saturation exponent
R_w = in situ water resistivity.

The transform from sigma to water saturation can be derived from the fact that Σ_{bulk} is equal to the volumetric contribution of grain and fluid sigmas, Σ_{grain} and Σ_{fluid}, respectively.

$$\Sigma_{bulk} = (1 - \phi) \times \Sigma_{grain} + \phi \times \Sigma_{fluid}$$

(Equation 4-34).

Σ_{fluid} is equal to the volumetric contribution of Σ_{water} and Σ_{hc}.

$$\Sigma_{bulk} = (1 - \phi) \times \Sigma_{grain} + \phi \times [\Sigma_{water} \times S_w + \Sigma_{hc} \times (1 - S_w)]$$

(Equation 4-35).

This can be transposed into a linear equation for water saturation.

$$S_w = \frac{(\Sigma_{bulk} - \Sigma_{grain}) + \phi \times (\Sigma_{grain} - \Sigma_{hc})}{\phi \times (\Sigma_{water} - \Sigma_{hc})}$$

(Equation 4-36).

Typical sigma values for common formation components are shown graphically in **Figure 4-91**. Note the large difference between the sigma value of hydrocarbons and salty water.

Figure 4-91. Typical sigma values for common formation components.

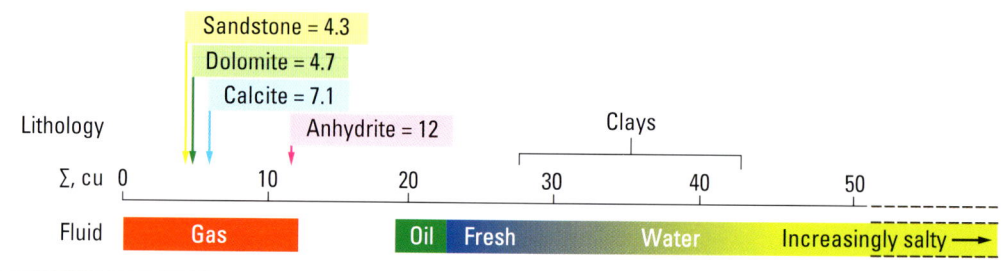

In shaly formations, sigma tends to correlate with the gamma ray measurement as a result of the high capture cross section of specific elements in some of the clay minerals. In clean formations, sigma tends to anticorrelate with resistivity, as shown in **Figure 4-92**. Note the resistivity scale increasing to the right and the sigma scale increasing to the left. In a formation filled with salty water, such as that shown in the lower part of Figure 4-92, sigma reads high because of the presence of chlorine in the water, and the resistivity reads low because of the conductivity of the salty water.

Where hydrocarbons displace salty water in the pores, the lack of chlorine in the hydrocarbon results in a decrease in the total chlorine content of the formation and hence, in the sigma measurement. The formation resistivity increases because the conductive water has been displaced by nonconductive hydrocarbon (middle interval in Figure 4-92). The upper interval shows an LRP zone composed of alternating thin beds of high and low resistivities. Because the beds are seen in parallel by the resistivity measurement, the measure current prefers to follow the path of least resistance through the low-resistivity beds. If these thin-bed effects are not taken in to account, water saturation computed from the resistivity measurements will indicate more water than is actually present in the formation. The sigma measurement is volumetric, which means that it is unperturbed by the path-of-least-resistance effect. The low sigma value suggests the presence of more hydrocarbon in the formation than indicated by the resistivity measurements.

Sigma is useful for saturation evaluation in LRP zones (such as where resistivity anisotropy complicates conventional saturation determination) because saturation can be determined without the use of resistivity measurements. Sigma is also useful where shoulder bed and proximity effects complicate the resistivity response and hence, conventional saturation calculations. Examples of these effects can be found in Chapter 5.

Figure 4-92. Sigma and resistivity, which anticorrelate in clean formations where hydrocarbons have displaced salty water. In the water interval, sigma is high, and resistivity is low. In the oil interval, sigma is low, and resistivity is high. The upper interval (orange) shows an LRP zone in which the volumetric sigma measurement gives a better indication of the formation water saturation because it is not influenced by the path-of-least-resistance effects that complicate the resistivity interpretation.

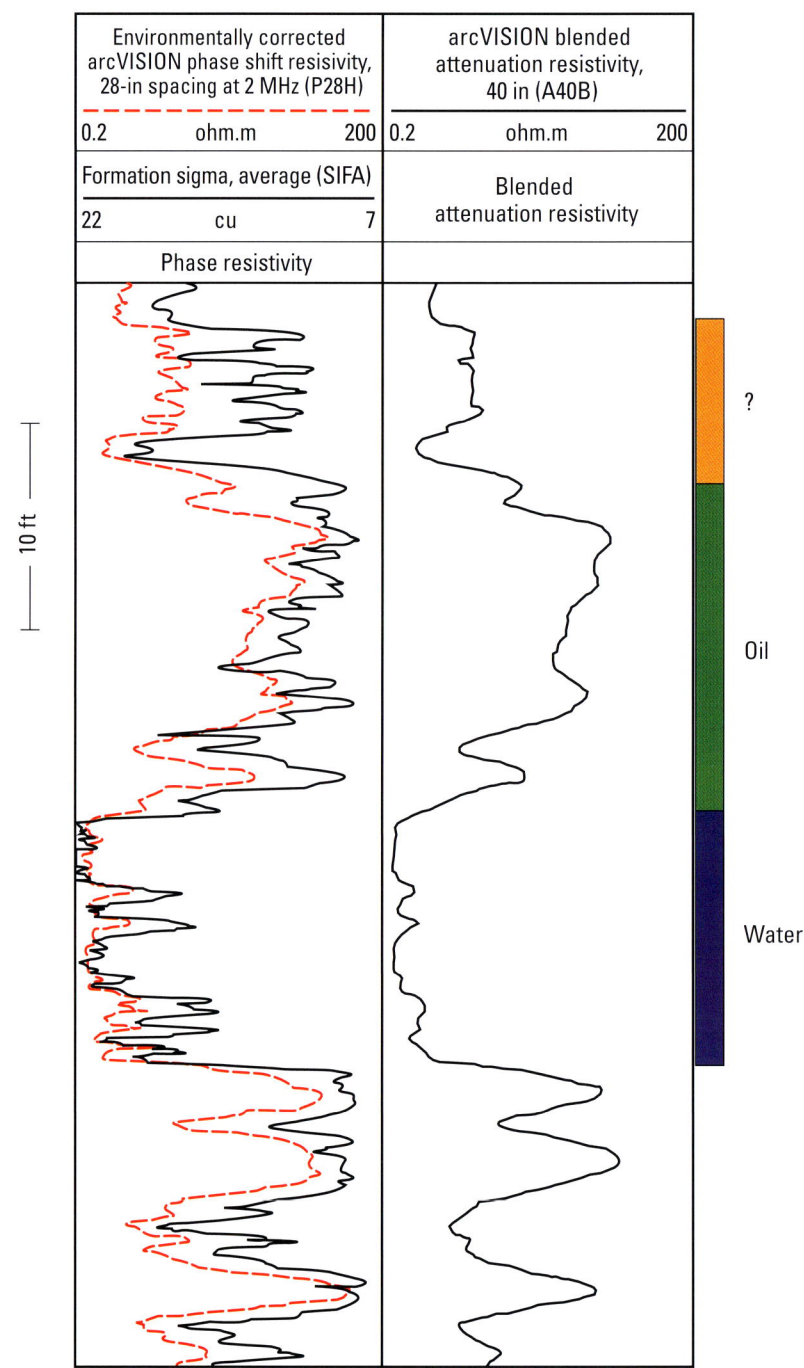

Interpretation of the sigma log for uninvaded formation water saturation determination requires that the measurement be acquired before significant invasion. The EcoScope propagation resistivity array, which is collocated with the sigma measurement, can be used to assess whether invasion has occurred before the sigma data is acquired. However, the propagation resistivity measurements are not very sensitive to very shallow or resistive invasions. To confirm that the sigma measurement is acquired before significant invasion, multiple DOI sigma measurements are acquired using both gamma ray and neutron count rates in the various detectors associated with the PNG. Shallow, medium, and deep sigma measurements are obtained. Where these measurements overlay, the formation can be considered uninvaded for sigma interpretation. Where they separate, any subsequent sigma interpretation needs to account for the effect of invasion.

Figure 4-93 shows a multiple-DOI EcoScope sigma log acquired in an 8.5-in borehole drilled with oil-base mud. In the left track, the shallow (blue), medium (green), and deep (red) sigma measurements acquired while drilling are compared with the medium measurement acquired while tripping out of the hole (solid black) after the formation has been open to invasion for several days. The theoretical sigma value for a water-filled formation (black dashed line) is also shown. In the upper oil-filled interval, either there is no invasion, or the oil-base mud filtrate displaces only the formation oil, so the oil saturation remains the same. In either case, the array of measurements overlay, indicating that they can be used to derive the uninvaded zone hydrocarbon saturation (right track). In the lower interval, the separation of the array sigma measurements indicates formation invasion, verified by the wipe log reading even lower several days later. This indicates oil-base mud invasion into a water zone. In this case, either an invasion-corrected resistivity or invasion-corrected sigma should be used to calculate the hydrocarbon saturation.

Figure 4-93. Sigma measurements, providing resistivity-independent water saturation evaluation in formations with sufficiently saline water. Multiple DOI sigma measurements help identify zones where mud filtrate invasion needs to be taken into account during interpretation of the data.

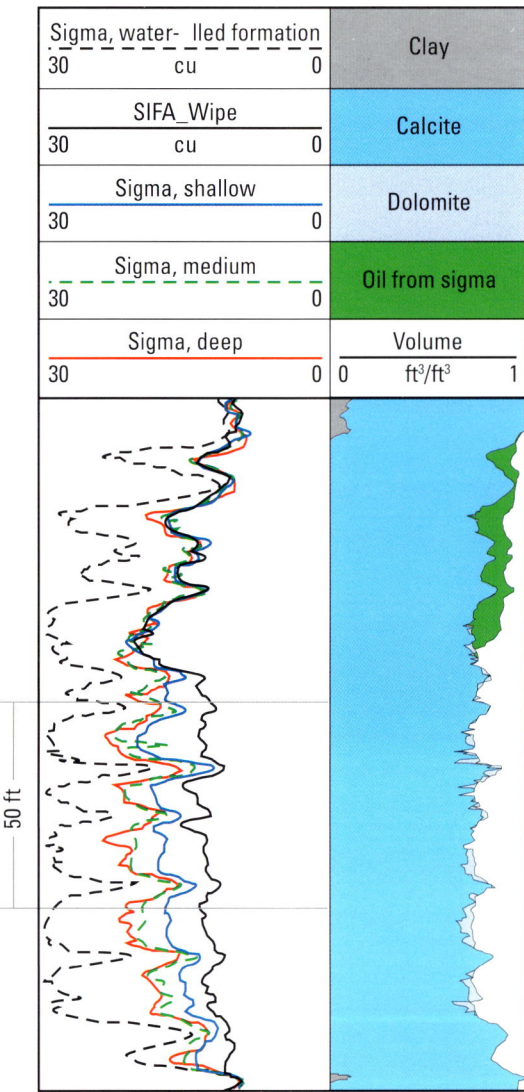

The accuracy of the water saturation derived from sigma improves with increasing water salinity and increasing porosity. **Figure 4-94** shows the uncertainty on the sigma-derived water saturation in an oil-bearing clean limestone as a function of the water salinity and porosity. For example, in a 20-pu limestone filled with 50-ppk-salinity water, the uncertainty on the calculated water saturation is approximately 27 su. If the water has a salinity of 140 ppk, the uncertainty is less than 9 su.

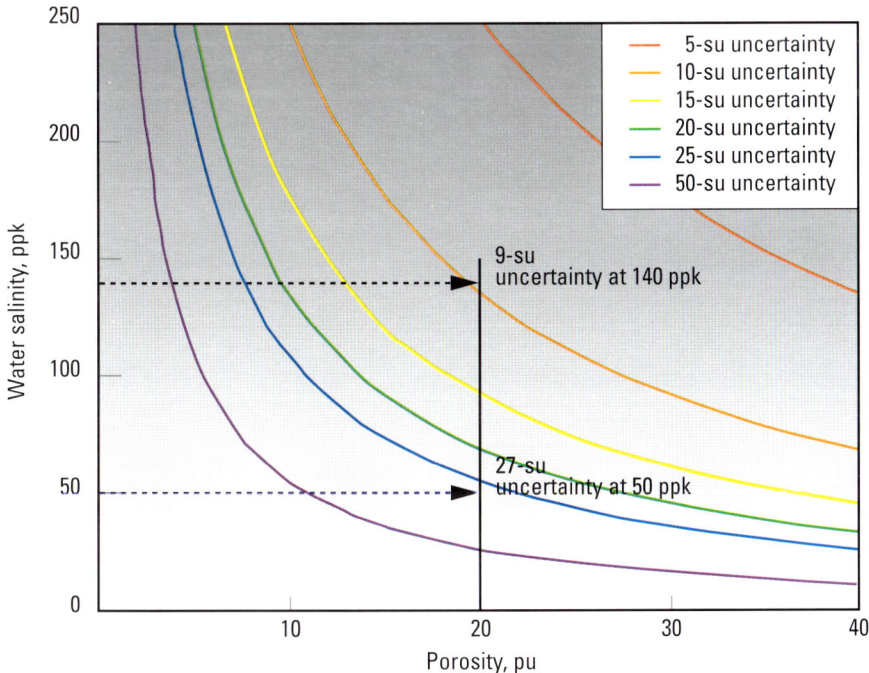

Figure 4-94 Sigma-derived water saturation uncertainty, which is a function of formation porosity and water salinity. For this clean limestone example, at 20 pu the uncertainty on the fluid saturations decreases by a factor of three as the salinity increases from to 140 ppk from 50.

The resistivity-independent saturation derived from the sigma measurement can be used in conjunction with resistivity measurements to back calculate in situ values for Archie's a, m, and n parameters. An example of this application is outlined in Section 5.5.8.

An alternative application of simultaneous sigma and resistivity data is for fluid typing. As shown in **Figure 4-91** gas and oil have different sigma values. If the hydrocarbon saturation is deduced using a resistivity-based equation, the sigma value of the hydrocarbon can then be calculated. A hydrocarbon sigma value of approximately 20 c.u. indicates oil. A lower value suggests the presence of gas.

Formation porosity and resistivity measurements are generally used in conjunction with an assumed formation water resistivity to derive a fluid saturation estimate through the Archie equation or equivalent. This estimate requires that the Archie parameters are known. Similarly, formation porosity and sigma measurements are generally used in conjunction with an assumed water sigma value to derive a fluid saturation estimate. These independent saturation techniques are shown in the left two columns of **Table 4-65**.

Both the formation water resistivity and sigma are primarily controlled by the water salinity. Unlike the water sigma, the water resistivity also has significant sensitivity to temperature.

Resistivity- and sigma-derived saturations can be calculated independently or used together to solve for an additional unknown, such as the water salinity, as shown in the right column of Table 4-65. The availability of all the required measurements in the EcoScope tool permits in situ evaluation of the formation water saturation and salinity, which is of particular importance in reservoirs with unknown water salinity, such as where injection water has mixed with the original formation water.

Table 4-65. Inputs and outputs from the Archie and sigma equations.

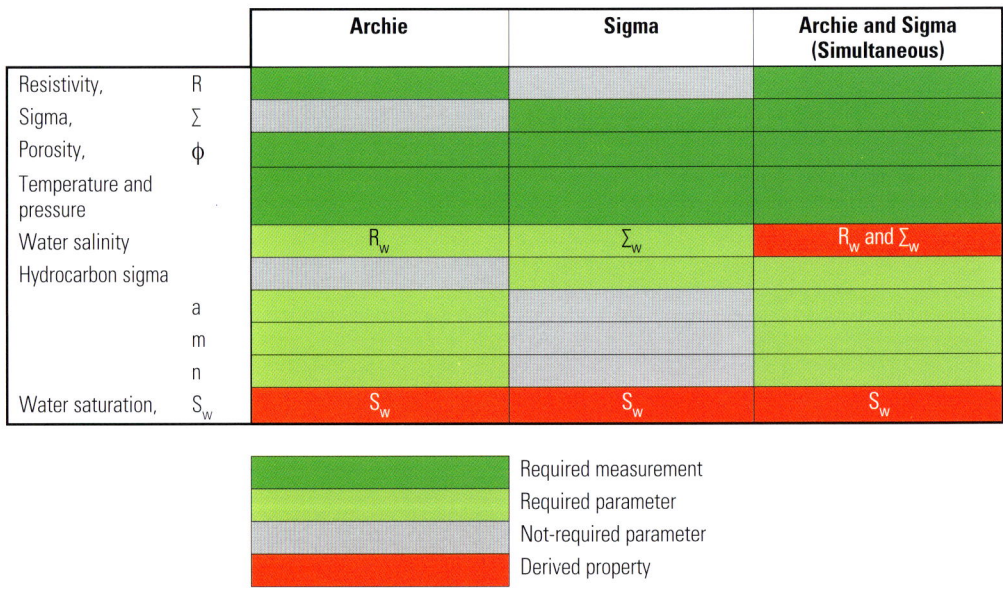

		Archie	Sigma	Archie and Sigma (Simultaneous)
Resistivity,	R			
Sigma,	Σ			
Porosity,	ϕ			
Temperature and pressure				
Water salinity		R_w	Σ_w	R_w and Σ_w
Hydrocarbon sigma				
	a			
	m			
	n			
Water saturation,	S_w	S_w	S_w	S_w

Required measurement
Required parameter
Not-required parameter
Derived property

An example of this simultaneous solution of the Archie and sigma equations to yield formation water saturation and salinity is outlined in Section 5.5.10.

4.11.4. Sigma log quality control

Sigma is a simple measurement based on the time decay of count rates. The measurement does not require calibration because it is a relative decay rate measurement.

The diffusion effect requires the largest correction to the raw data, so an appropriate formation HI measurement is required. BPHI provides the most direct HI measurement for this purpose. BPHI should be quality controlled before application to the sigma for diffusion correction.

The use of sigma for formation saturation evaluation assumes that this relatively shallow measurement is unaffected by invasion. The collocated propagation resistivity measurements can be used to identify invasion. Where the phase resistivity measurements separate, care should be taken to determine whether the sigma measurement is affected. Where multiple sigma measurements are available with different DOIs, separation between the shallower and deeper sigma measurements can be used as an invasion indicator. An invasion-corrected formation sigma can be derived from multiple DOI sigma measurements in a manner similar to the invasion-corrected resistivity derived from multiple resistivity measurements.

The EcoScope hardware incorporates checks to verify correct performance of the detectors and electronics. Additional checks to ensure that the measurements fall within the range expected for downhole formation are outlined in the QC logic shown in **Table 4-66**.

Table 4-66. Sigma QC flag logic.

QC_SIGM	Hardware	Measurement
Green	All ok	$5 < SIFA < 60$
Yellow	High voltage out of range or Azimuthal system problem or Gain loop in SEARCH (AUTO_LOCAL)	$60 < SIFA \leq 90$
Red	Detector failure or PNG failure	$SIFA > 90$ or $SIFA \leq 5$
White		No new Sigma data available for more than 2 ft.

4.11.5. Sigma channels and parameters

4.11.5.1. Real time

Table 4-67. Sigma real-time parameters.

Parameter Name	Description	Unit
RMS_RT	Resistivity of mud sample	ohm.m
RMT_RT	Mud sample temperature	degF
OBMF	Oil-base mud flag	NA[†]
RT_BSAL	Mud salinity	ppk
SIG_PCOR_OPT_RT	Porosity correction source for sigma processing, real time	BPHI, TNPH, none

[†] Not applicable.

Table 4-68. Sigma real-time transmitted channels.

Transmitted Channel	Description
Azimuthal Average	
RFSA_DH_ECO_RT[†]	Raw formation sigma, average, real time, computed downhole
QC_SIGM_DH_ECO_RT	Formation sigma quality indicator, real time, computed downhole
Quadrants	
RFSB_DH_ECO_RT	Raw formation sigma, bottom, real time, computed downhole
RFSL_DH_ECO_RT	Raw formation sigma, left, real time, computed downhole
RFSU_DH_ECO_RT	Raw formation sigma, up, real time, computed downhole
RFSR_DH_ECO_RT	Raw formation sigma, right, real time, computed downhole

[†] In addition to a raw formation sigma, a neutron porosity, preferably BPHI, is required to perform diffusion correction. See Section 4.7.5 for details of the channels required for real-time BPHI.

Table 4-69. Sigma real-time computed channels.

Computed Channel	Description
Azimuthal Average	
SIFA_ECO_RT	Sigma formation, average, real time
Quadrants	
SIFB_ECO_RT	Sigma formation, bottom, real time
SIFL_ECO_RT	Sigma formation, left, real time
SIFU_ECO_RT	Sigma formation, up, real time
SIFR_ECO_RT	Sigma formation, right, real time

4.11.5.2. Recorded mode

Table 4-70. Sigma recorded-mode parameters.

Parameter Name	Description	Unit
BS_RM	Bit size (recorded mode)	in
OBMF_RM	Oil-base mud flag	NA[†]
WPPV[‡]	Water phase as percent of total volume in oil-base mud	%
WPSL[‡]	Salinity of the water phase emulsified within the oil-base mud	ppk
BSAL_RM	Mud salinity (recorded mode)	ppk
GCSE	Generalized caliper selection	NA
MST_RM	Mud sample temperature (recorded mode)	degF
RMS_RM	Resistivity of mud sample (recorded mode)	ohm.m
SIG_PCOR_OPT	Porosity correction source for sigma processing	BPHI (default), TNPH, none

[†] Not applicable.
[‡] WPPV and WPSL are used to compute BSAL in oil-base mud systems.

Table 4-71. Sigma recorded-mode transmitted channels.

Output Channel	Description
Azimuthal Average	
RFSA	Raw formation sigma, average
RFSA_SH	Raw formation sigma, shallow
RFSA_MD	Raw formation sigma, medium
RFSA_DP	Raw formation sigma, deep
SIFA	Sigma formation, average
SIFA_SH	Sigma formation, shallow
SIFA_MD	Sigma formation, medium
SIFA_DP	Sigma formation, deep
QC_SIGM	Formation sigma quality indicator, average
QC_SIGM_SH	Formation sigma quality indicator, shallow
QC_SIGM_MD	Formation sigma quality indicator, medium
QC_SIGM_DP	Formation sigma quality indicator, deep
Quadrants	
RFSB	Raw formation sigma, bottom
RFSL	Raw formation sigma, left
RFSR	Raw formation sigma, right
RFSU	Raw formation sigma, up
SIFB	Sigma formation, bottom
SIFL	Sigma formation, left
SIFU	Sigma formation, up
SIFR	Sigma formation, right

4.11.6. Sigma measurement specifications

Table 4-72. Sigma measurement specifications at 200-ft/h ROP.

Item	Value
Range	5 to 100 cu
Axial resolution	15 in
Accuracy	1 cu ($\Sigma < 20$ cu) 5% ($\Sigma > 20$ cu)
Precision at 200 ft/h	0.3 cu ($\Sigma < 20$ cu) 1.5% ($\Sigma > 20$ cu)

4.12. Azimuthal density caliper

The magnitude of the standoff compensation implicit in the azimuthal density measurement is used to determine the size of the borehole around the density stabilizer. The EcoScope density caliper also incorporates information from the shallow PEF response for a more robust caliper measurement under a wide range of conditions.

4.12.1. Azimuthal density caliper theory of measurement

The density caliper is a measurement of standoff between the formation and the tool. The density caliper uses the two azimuthally focused density detectors to derive 16-radii measurements around the borehole. Rather than solve for formation density and density standoff correction, as is done for the formation density measurement, the data from the two detectors is used to solve for the geometrical standoff distance, given the mud density by the user. The density standoff, is derived using an equation of the form

$$\text{standoff}_{\text{density}} = k_d \times |(\rho_{\text{bulk}} - \rho_{\text{long-spacing}}) \; / \; (\rho_{\text{bulk}} - \rho_{\text{mud}})| \qquad \text{(Equation 4-37)},$$

where

$\text{standoff}_{\text{density}}$ = density standoff

k_d　　　　　 = empirically derived density caliper coefficient.

If the tool is in good contact with the borehole wall, then the density from the long-spacing detector should match the true formation density, so the standoff value is 0. As the standoff increases, the decrease in the density derived from the long-spacing detector is linearly related to the difference between the formation and mud densities.

This method generally works well in most typical mud systems. However, in the case of barite or cesium formate muds and in situations where there is a low density contrast between the mud and formation densities, a complementary approach is needed.

In addition to measuring density, the short- and long-spacing gamma ray detectors associated with the cesium source also measure the volumetric PEF. This measurement varies linearly with standoff according to the equation

$$\text{standoff}_U = k_U \times |(U - 1.3 \,|(\rho_{\text{long-spacing}})^2)| \qquad \text{(Equation 4-38)},$$

where

standoff$_U$ = volumetric photoelectric standoff

k_U = empirically derived volumetric PEF caliper coefficient.

$1.3 \,(\rho_{\text{long-spacing}})^2$ = true formation volumetric PEF based on a correlation with the deeper long-spacing density measurement. This provides the reference for volumetric PEF at 0 standoff.

The density and volumetric photoelectric standoff measurements are combined using a weighted average.

$$\text{Standoff} = W \times \text{standoff}_{\text{density}} + (1 - W) \times \text{standoff}_U \qquad \text{(Equation 4-39)},$$

where

W = standoff weighting factor.

W is computed from X, which compares U with $\rho - \rho_{\text{long-spacing}}$, as shown in **Figure 4-95**. Where the density difference is large, X becomes small, and the density standoff is used. Where the density difference is small, X becomes large (either positive or negative), and Standoff$_U$ is used. When $\rho > \rho_{\text{mud}}$, the transition between the two standoffs is controlled by the density difference, $\rho - \rho_{\text{mud}}$, using a moving threshold.

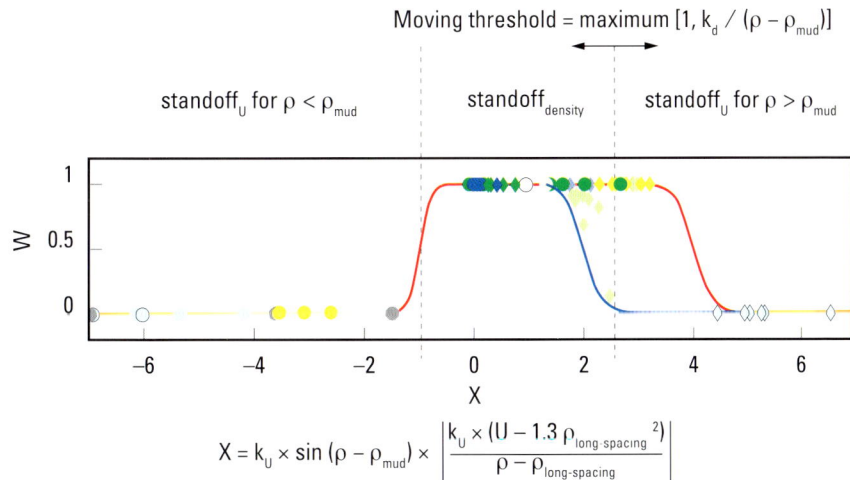

Figure 4-95. Thresholding logic of the standoff weighting factor, combining the density and PEF standoff measurements.

The moving threshold technique is used to accommodate various types of mud. Experimental data from nine different mud types in three different formation densities are plotted. Two cases of particular interest are a light cesium formate mud (green diamonds) with a high PEF and a very heavy mud (light blue diamonds and circles).

The blue line corresponds to the light cesium formate mud (1.445 g/cm^3, 100 b/cm^3). The red line corresponds to a heavier mud and has a higher transition threshold because the mud contrast is reduced (mud weight = 1.91 g/cm^3 → light blue). The high-density mud has little contrast with the formation, so the density standoff becomes unreliable. The volumetric PEF standoff is used in this situation.

Opposing pairs of weighted average standoffs from the 16 sectors around the hole are added to the diameter of the stabilizer where the measurements are taken to provide eight hole diameter measurements, as shown in **Figure 4-96**.

Figure 4-96. Adding opposing pairs of standoffs to the stabilizer diameter to compute the borehole diameter.

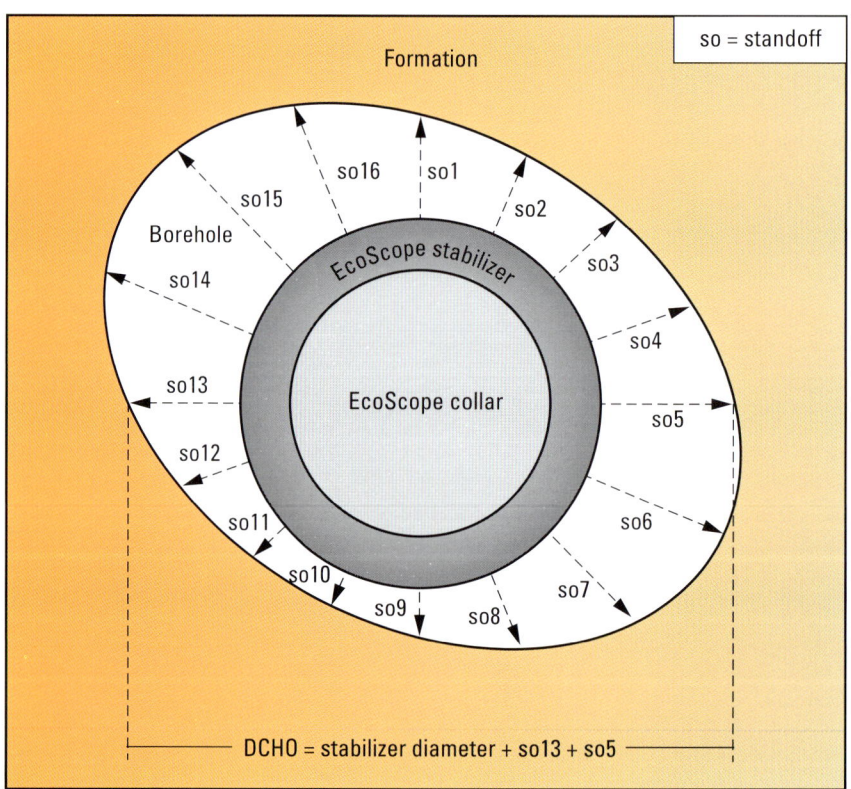

4.12.2. Azimuthal density caliper applications

Borehole size and shape information has numerous applications, including

- environmental correction of other logs for hole size and shape effects

- evaluation of borehole stability

- identification of stress directions on the borehole through determination of preferential ovalization or breakout orientation

- calculation of the hole volume

- calculation of the volume of cement required between the casing and borehole wall

- time-lapse evaluation of hole enlargement through multiple logging passes over an interval of interest.

4.12.3. Azimuthal density caliper log quality control

The following factors should be considered in quality controlling a density caliper log.

- **Excessively large borehole**
  If the distance between the detectors and the side of the borehole is too large, the standoff measurement will not be valid. This is closely related to the mud weight and volumetric PEF. The lighter the mud weight, the larger the borehole that can be measured.

- **Reference density**
  The standoff quality depends upon the reliability of the reference formation density. Density can be affected by factors such as erratic tool path, severe dogleg, and spiral hole. Where these conditions exist, the reference density should be quality controlled before computing standoff.

- **Tool rotation**
  The tool must be rotating to have valid diameter measurements.

The density caliper should be in agreement with the ultrasonic caliper under conditions where both are valid.

4.12.4. Azimuthal density caliper channels and parameters

4.12.4.1. Real time

Table 4-73. Azimuthal density caliper real-time parameters.

Parameter Name	Description	Unit
MW_RT	Mud weight, real time	lbm/galUS

Table 4-74. Azimuthal density caliper real-time transmitted channels

Transmitted Channel	Description
Azimuthal Average	
DCAV_DH_ECO_RT	Density caliper, average, real time, computed downhole
Horizontal and Vertical	
DCHO_DH_ECO_RT	Density caliper, horizontal, real time, computed downhole
DCVE_DH_ECO_RT	Density caliper, vertical, real time, computed downhole

4.12.4.2. Recorded mode

Table 4-75. Azimuthal density caliper recorded-mode parameters.

Parameter Name	Description	Unit
MW_RM	Mud weight (recorded mode)	lbm/galUS
STOH	Top of hole sector	NA[†]
IMAGE_MAX_DCRA	Image density caliper maximum scale value	NA
IMAGE_MIN_DCRA	Image density caliper minimum scale value	NA

[†] Not applicable.

Table 4-76. Azimuthal density caliper recorded-mode channels

Output Channel	Description
Azimuthal Average	
DCAV	Density caliper, average
Azimuthal	
DCHO	Density caliper, horizontal
DCVE	Density caliper, vertical
DCRA	Radii image from density caliper, 16-sector
DSOB	Density standoff, bottom quadrant
DSOL	Density standoff, left quadrant
DSO	Density standoff, upper quadrant
DSOR	Density standoff, right quadrant

4.12.5. Azimuthal density caliper measurement specifications

The following specifications are established for the standoff measurement (**Table 4-77**) and further translated into caliper specifications (**Table 4-78**). They were derived from laboratory measurements in uniform homogeneous formations in circular boreholes. They do not apply to cases where the drilling fluid has high photoelectric absorption (barite muds, cesium formates, or other fluids containing high atomic mass materials). The density caliper is not available in sliding mode.

Table 4-77. Azimuthal density standoff measurement specifications.

Standoff	Value
Range	
Mud weight < 10 lbm/galUS[†]	0 to 1 in
Mud weight < 14 lbm/galUS[†]	0 to 0.5 in
Axial resolution	6 in
Accuracy[‡]	0.2 in

[†] | Formation density − mud weight | > 0.6 g/cm^3 applies without distinction to oil- and water-base mud.

[‡] Uncertainty on mud weight not taken into account.

Table 4-78. Azimuthal density caliper specification for centered and eccentered tools.

Caliper Specification	Unit	Value	
Range		Eccentered	Centered
Mud weight , 10 lbm/galUS[†]	in	Stabilizer diameter (SSIZ) to SSIZ + 1	SSIZ to SSIZ + 2
Mud weight , 14 lbm/galUS[†]	in	SSIZ to SSIZ + 0.5	SSIZ to SSIZ + 1
Axial resolution	in	6	
Number of sectors		16	
Azimuthal resolution	sector	1	
Accuracy[‡]		Measured borehole caliper (BH) − SSIZ < 0.5 in.	0.5 < BH − SSIZ < 1 in
Density caliper, vertical (DCVE)	in	0.2 + 20% × (BH − SSIZ)	0.2 + 40% x (BH − SSIZ)[§]
Density caliper, horizontal (DCHO)	in	0.2 + 20% × (BH − SSIZ)	
Density caliper, average (DCAV)	in	0.2 + 20% × (BH − SSIZ)	

[†] | Formation density − mud weight | > 0.6 g/cm^3 applies without distinction to oil- and water-base mud.

[‡] Uncertainty on mud weight not taken into account.

[§] True caliper value lies between measured caliper value minus 40% of (BH − SSIZ) and measured caliper value plus 0.2 in.

4.13. Azimuthal ultrasonic caliper

The ultrasonic caliper measurement uses two ultrasonic transducers opposed at 180° and located just above the stabilizer to measure the borehole diameter, even while sliding. While rotating, a 16-sector (8-diameter) image of the borehole shape is acquired.

4.13.1. Azimuthal ultrasonic caliper theory of measurement

Ultrasonic waves are sound waves with frequencies above the range of human hearing. The frequency range of human hearing is between 20 and 20,000 Hz, while ultrasonic waves have frequencies between 195,000 and 2,000,000 Hz. The shorter wavelength of ultrasonic waves results in stronger reflections than the longer waves of audible sound.

Ultrasonic transducers are piezoelectric sensors that emit short ultrasonic pulses and then receive the reflected signal from the formation. In this pulse-echo mode, the same device is both the generator and the receiver of the ultrasonic signal. By measuring the time for the signal to travel from the transducer through the mud to the borehole wall, reflect off the formation, and return to the transducer through the mud, the two-way travel time for the ultrasonic signal is determined, as shown in **Figure 4-97**.

Figure 4-97. The ultrasonic pulse-echo measurement, which measures the two-way travel time of an ultrasonic pulse across the borehole and back to the transducer.

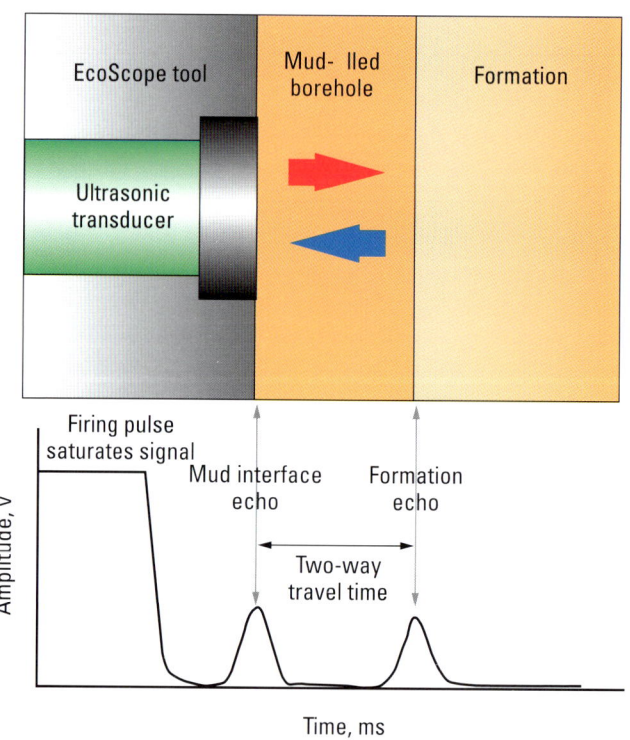

This travel time, divided by two to give the one-way travel time, is multiplied by the velocity of the ultrasonic wave through the mud to determine the standoff distance between the transducer face and borehole wall.

The EcoScope tool has two ultrasonic transducers, oriented 90° from the other detectors in the tool and 180° opposed to each other, as shown in **Figure 4-98**. This configuration makes a diameter caliper measurement possible, even when sliding. While rotating, the ultrasonic data is binned into 16 sectors, which are used to provide a borehole shape analysis of eight diameters.

The diameters are computed by adding the standoff computed from the two opposing ultrasonic transducers to the diameter of the tool at the location of the sensors.

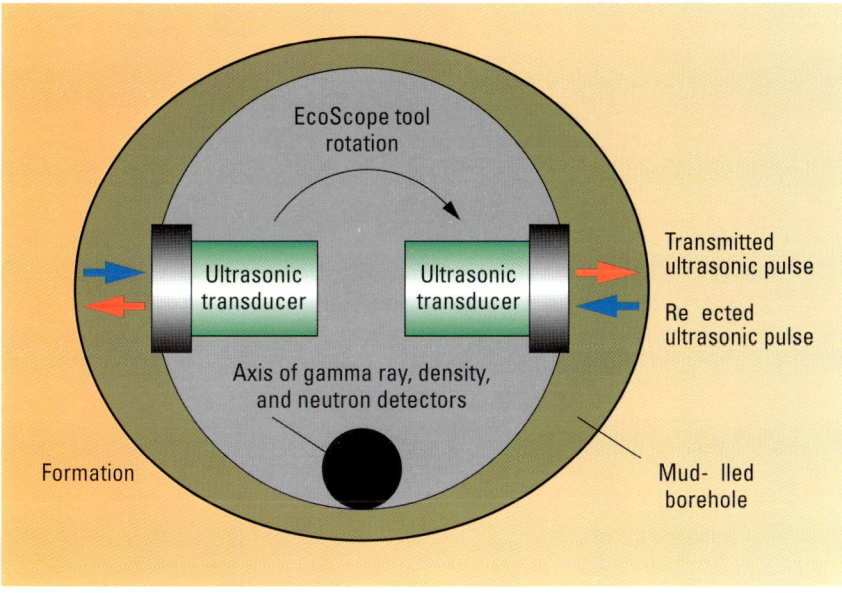

Figure 4-98. The ultrasonic transducers in the EcoScope tool, oriented to deliver a borehole diameter while sliding and an eight-diameter borehole shape analysis when rotating.

The ultrasonic measurements of the EcoScope tool provide

- 16-sector standoff image
- average borehole diameter
- 8-diameter borehole shape caliper
- vertical diameter
- horizontal diameter
- tool eccentricity.

4.13.2. Azimuthal ultrasonic caliper environmental corrections

The main uncertainty in the conversion of ultrasonic transit time to borehole diameter is the velocity of the ultrasonic pulse through the mud. This velocity is a function of the mud type, temperature, and pressure. Computed values are provided, but downhole conditions may differ from those assumed in the computation, resulting in a bias on the caliper measurement. Ultrasonic caliper data can be recomputed with an updated ultrasonic velocity after acquisition.

There are a number of factors that may result in the ultrasonic measurement giving an invalid reading. These factors include

- insufficient acoustic impedance contrast between mud and formation

- excessive acoustic impedance contrast between the mud and mudcake

- excessive reflections from cuttings suspended in the mud flow

- signal attenuation due to gas in the mud

- signal attenuation due to excessive mud weight

- signal attenuation due to excessive path length (excessive hole diameter)

- loss of reflected signal due to scattering from a rugose borehole wall.

While these effects cannot be corrected, the measurement specifications provide an indication of the range of the measurement as a function of mud weight and type.

4.13.3. Azimuthal ultrasonic caliper applications

The primary applications of the ultrasonic caliper measurement are the evaluation of hole shape for borehole stability determination, hole size for environmental correction of other measurements, and hole volume for calculation of mud and cement volumes.

Other applications include detection of free gas in the mud and evaluation of formation damage. Free gas in the mud changes the impedance contrast between the transducer, mud, and formation, resulting in significant changes in the reflection coefficients at each interface. Free gas can be detected by monitoring changes in the relative reflection amplitudes from the transducer-mud and mud-formation interfaces.

Figure 4-99 compares an ultrasonic image (left image track) with PEF and density images (middle and right image tracks) acquired simultaneously by an EcoScope tool. Note that the dark patches on the ultrasonic image correspond to increased hole diameter as measured by the density caliper (thin black line, left track). The borehole damage can be correlated to changes in formation lithology indicated by the changes in color on the PEF and density images.

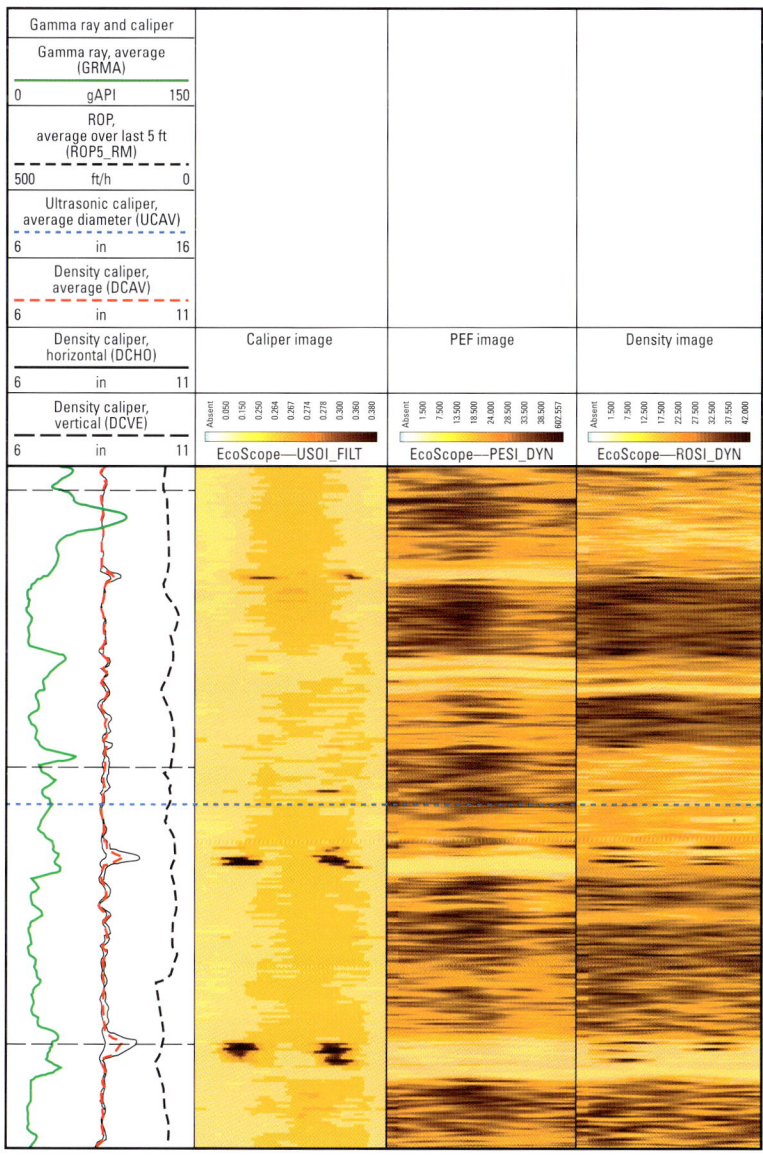

Figure 4-99. Ultrasonic image (left image track), showing dark patches associated with borehole breakout. These correlate with changes in lithology indicated by the PEF image (middle image track) and density image (right image track).

4.13.4. Azimuthal ultrasonic caliper log quality control

The following factors should be considered in quality controlling an ultrasonic caliper log.

- **Ultrasonic velocity through the mud**

  If the user-defined ultrasonic velocity through the mud is too high, the computed hole diameter will be too large and vice versa. The possibility of changes in mud properties over the duration of the acquisition should be taken into account when interpreting ultrasonic measurements. It is common practice for "pills" of mud with properties differing from the original mud to be circulated during drilling.

- **Acoustic impedance**

  The ultrasonic measurement depends on a strong reflection from the borehole wall. If the acoustic impedance of the mud is close to that of the formation, the amplitude of the reflected wave may be too small to be detected. This effect is only likely to occur when heavy mud is used while drilling poorly consolidated sandstone.

- **Mudcake**

  If the difference between the acoustic impedances of the mud and mudcake is large enough to reflect a wave with an amplitude that exceeds the threshold of the ultrasonic sensor, the measurement will yield the ID of the mudcake rather than that of the borehole wall.

- **Gas-cut mud**

  The ultrasonic measurement depends on transmission and reflection of high frequency waves. If there is excessive free gas in the mud, the reflection at the transducer-mud interface will become so strong that insufficient energy will be transmitted through to the mud and onto the mud-formation interface.

- **Mud weight**

  If the mud weight becomes too high, the loss of energy in the mud will attenuate the signal to the point where it can no longer be detected back at the transducer.

- **Excessively large borehole**

  If the distance between the detector and the side of the borehole is too large, insufficient reflected signal will reach the transducer. This signal is closely related to the mud weight. The lighter the mud weight, the larger the borehole that can be measured.

- **Cuttings**

  If there are large quantities of cuttings, the ultrasonic wave will be scattered, and insufficient reflected signal will reach the transducer. Isolated cuttings are generally not a problem because the mud flow quickly removes cuttings from the vicinity of the ultrasonic sensors.

- **Excessive borehole rugosity**

  Excessive borehole rugosity, such as that due to fracturing or breakout, may result in the ultrasonic signal being reflected away at an angle rather than directly back to the transducer. In this case, the signal is lost, and the ultrasonic measurements invalid.

If azimuthal ultrasonic information is to be used, then rotation of the tool over the interval of interest must be verified.

The EcoScope hardware incorporates checks to verify correct performance of the acquisition detectors and electronics. Additional checks to ensure that the measurements fall within the expected range are outlined in the QC logic shown in **Table 4-79**.

Table 4-79. Ultrasonic caliper QC flag logic.

QC_UCAL	Hardware	Measurement
Green	Ultrasonic board status word = 0 (ok) and Ultrasonic processing computes a quality flag for each of the 16 sectors with 0 ≤ red sector flags (absent value) ≤ 2 and 0 ≤ yellow and red sector flags ≤ 4	Ultrasonic caliper[†] ≥ 7.875 in and In water-base mud ultrasonic caliper[†] ≤ 21.25 − 0.75 × mud weight In oil-base mud ultrasonic caliper[†] ≤ 20.25 − 0.75 × mud weight and CRPM ≥ 20 and ROP ≤ 450 ft/h
Yellow	Ultrasonic status word (DUSW) set for • CAN overflow or • Sensor1 or Sensor2 error or • High-voltage error or • US1 or US2 detection ratio error or 2 < Number of red sector flags ≤ 4 or Number of yellow and red sector flags > 4	Ultrasonic caliper[†] < 7.875 in or In water based mud ultrasonic caliper[†] > 21.25 − 0.75 × mud weight In oil-base mud ultrasonic caliper[†] > 20.25 − 0.75 × mud weight and CRPM < 20 and ROP > 450 ft/h
Red	Ultrasonic status word (DUSW) set for • SRAM error or • Board in error or Configuration parameters not received or Number of red sector flags > 4	
White		Ultrasonic caliper measurement is absent value or No new ultrasonic caliper tick for more than 2 ft

[†] Average caliper in recorded mode, available caliper among vertical, horizontal, and average caliper in real time.

4.13.5. Azimuthal ultrasonic caliper channels and parameters

4.13.5.1. Real time

Table 4-80. Azimuthal ultrasonic caliper real-time parameters.

Parameter Name	Description	Unit
DTMUD_DH	Slowness of the ultrasonic signal through mud (real time)	us/ft

Table 4-81. Azimuthal ultrasonic caliper real-time transmitted channels

Transmitted Channel	Description
Azimuthal Average	
UCAV_DH_ECO_RT	Ultrasonic caliper, average diameter, real time, computed downhole
DTMUD_DH_ECO_RT	Slowness of the ultrasonic signal through mud, real time, computed downhole
QC_UCAL_DH_ECO_RT	Ultrasonic caliper quality indicator, real time, computed downhole
Horizontal and Vertical	
UCHO_DH_ECO_RT	Ultrasonic caliper, horizontal diameter, real time, computed downhole
UCVE_DH_ECO_RT	Ultrasonic caliper, vertical diameter, real time, computed downhole
DMHO_DH_ECO_RT	Tool delta movement, horizontal, real time, computed downhole
DMVE_DH_ECO_RT	Tool delta movement, vertical, real time, computed downhole

4.13.5.2. Recorded mode

Table 4-82. Azimuthal ultrasonic caliper recorded-mode parameters.

Parameter Name	Description	Unit
CHI_RM	Caliper high limit from bit size (recorded mode)	in
CLO_RM	Caliper low limit from bit size (recorded mode)	in
SSIZ	Stabilizer size	in
BS_RM	Bit size	in
DTMUD	Slowness of the ultrasonic signal through mud (recorded mode)	us/ft

Table 4-83. Azimuthal ultrasonic caliper recorded-mode channels.

Output Channel	Description
Azimuthal Average	
UCAV	Ultrasonic caliper average diameter
QC_UCAL	Ultrasonic caliper quality indicator
Horizontal and Vertical	
UCHO	Ultrasonic caliper, horizontal diameter
UCVE	Ultrasonic caliper, vertical diameter
Sectors	
UCAL1	Ultrasonic caliper, axis one diameter
UCAL2	Ultrasonic caliper, axis two diameter
UCAL3	Ultrasonic caliper, axis three diameter
UCAL4	Ultrasonic caliper, axis four diameter
UCAL5	Ultrasonic caliper, axis five diameter
UCAL6	Ultrasonic caliper, axis six diameter
UCAL7	Ultrasonic caliper, axis seven diameter
UCAL8	Ultrasonic caliper, axis eight diameter
USOI	Ultrasonic standoff image, 0.1 ft, 16-sector
USOI_FILT	Ultrasonic standoff image, filtered, 16-sector

4.13.6. Azimuthal ultrasonic caliper measurement specifications

Table 4-84. Azimuthal ultrasonic caliper specifications.

Item	Value
Accuracy	0.2 in
Standoff range	
Mud weight = 10 lbm/galUS, water-base mud	3 in
Mud weight = 10 lbm/galUS, oil-base mud	2.5 in
Mud weight = 14 lbm/galUS, water-base mud	1.5 in
Mud weight = 14 lbm/galUS, oil-base mud	1 in
Axial resolution at 90 ft/h	1.2 in

4.14. Continuous near-bit inclination

A single-axis accelerometer, located near the base of the EcoScope tool, provides an inclination measurement as close to the bit as possible in the LWD tools.

4.14.1. Continuous near-bit inclination theory of measurement

Inclination is the angle between a vertical line and the path of the wellbore at that point (**Figure 4-100**). The vertical reference is the earth's gravitational field, as it always points to the center of the earth. An inclination of 0° indicates a vertical well. An inclination of 90° indicates a horizontal well.

Figure 4-100. Inclination, the angle between a vertical line and the path of the wellbore at that point.

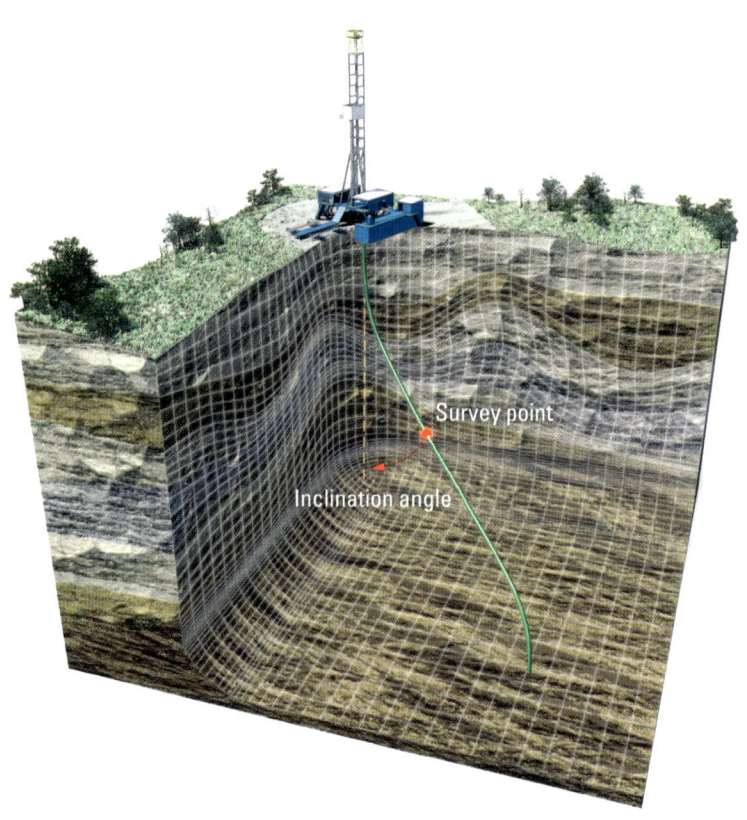

Inclination can be measured using a single accelerometer parallel to the axis of the tool. This accelerometer measures the projection of gravity acceleration along the tool axis (G_x shown in **Figure 4-101**).

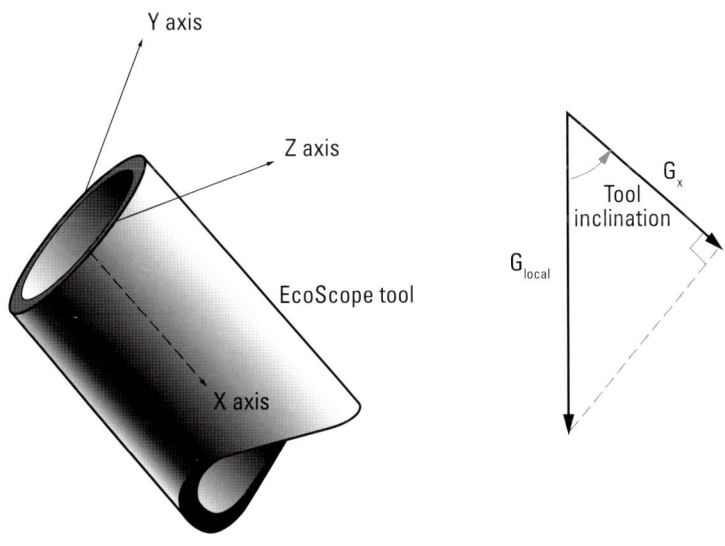

Figure 4-101. Inclination, measured using an accelerometer aligned to the tool axis.

A nominal value for local gravitational field strength, G_{local}, (approximately 32.17 ft/s^2) is programmed into the tool during initialization so that the inclination angle can be derived.

Inclination = $\cos^{-1}(G_x/G_{local})$ (Equation 4-40).

4.14.2. Continuous near-bit inclination environmental corrections

A sophisticated calibration procedure is performed to account for any misalignment between the accelerometer and the axis of the tool. Temperature corrections are performed with the use of measurements made by a temperature sensor built into the accelerometer package.

4.14.3. Continuous near-bit inclination applications

The primary application of the near-bit inclination is in early determination of the drilling tendency of the BHA. The inclination and azimuth of the bit is important information for drillers to be able to steer the well to the desired target. If information about the inclination and azimuth of the well is only available once the MWD survey system has reached a given depth, assumptions must be made about the drilling behavior of the BHA ahead of the MWD tool. Having an inclination measurement closer to the bit at the base of the LWD system provides earlier indication of the behavior of the BHA, facilitating better directional control of the well.

4.14.4. Continuous near-bit inclination log quality control

The continuous near-bit inclination measurement is obtained from a single-axis accelerometer and therefore does not have the QC of a triaxial accelerometer set to verify the magnitude of the axial accelerometer response. The static survey inclination (which is measured with triaxial accelerometers) should be taken as the true inclination of the borehole.

4.14.5. Continuous near-bit inclination channels and parameters

4.14.5.1. Real time

Table 4-85. Continuous near-bit inclination real-time parameters.

Parameter Name	Description	Unit
ToolG	Earth gravity	ft/s^2

Table 4-86. Continuous near-bit inclination real-time transmitted channels.

Transmitted Channel	Description
CNGX	Continuous projection of gravity acceleration along the tool axis from accelerometer near bit

Table 4-87. Continuous near-bit inclination real-time computed channels.

Computed Channel	Description
CNINC	Continuous inclination near bit

4.14.5.2. Recorded mode

Table 4-88. Continuous near-bit inclination recorded-mode parameters.

Parameter Name	Description	Unit
LOCG	Local earth gravitational field strength	ft/s^2

Table 4-89. Continuous near-bit inclination recorded-mode channels.

Output Channel	Description
GX_CAL	X-axis accelerometer, calibrated
INCL_CONT	Continuous inclination

4.14.6. Continuous near-bit inclination measurement specifications

Table 4-90. Near-bit inclination sensor specification.

Item	Value
Range	0 to 180°
Accuracy	0.1° for inclination > 45°
Resolution	0.1°

4.15. Annular pressure

A pressure transducer in the tool measures the pressure of the fluid in the borehole. This measurement allows effective mud weights to be monitored for borehole stability applications.

4.15.1. Annular pressure theory of measurement

The EcoScope tool measures the pressure of the fluid in the annulus between the collar and borehole wall. The pressure measurement is taken below the stabilizer using a strain gauge inside the tool connected to the fluid in the annulus through a port in the collar wall.

The application of pressure deforms an element in the sensor. The value of a resistor deposited on this sensor changes according to the pressure-induced deformation and temperature. The resistance is measured with a Wheatstone bridge configuration, which somewhat corrects for the effect of temperature. However, it is necessary to measure temperature precisely to fully correct the pressure measurement. The temperature measurement is performed with a platinum resistor located on the pressure-sensing element. After residual temperature correction, the output voltage from the Wheatstone bridge is converted to pressure by applying the calibration gain and offset. The pressure measurement is recorded every 2 seconds.

4.15.2. Annular pressure environmental corrections

As outlined above, the pressure gauge has a built-in temperature sensor to ensure that the required temperature correction is applied.

4.15.3. Annular pressure-while-circulating applications

Pressure drives the mud flow from the mud pumps, through the drillpipe and BHA, out through the bit, and back up the annulus between the drillpipe and borehole wall to the surface (**Figure 4-102**). The mud pressure is measured in the standpipe that carries the mud up the side of the rig derrick from where it enters the drillstring. The standpipe pressure (SPP) can be thought of as the output pressure from the mud pumps.

Figure 4-102. Annular and internal mud pressure measurements.

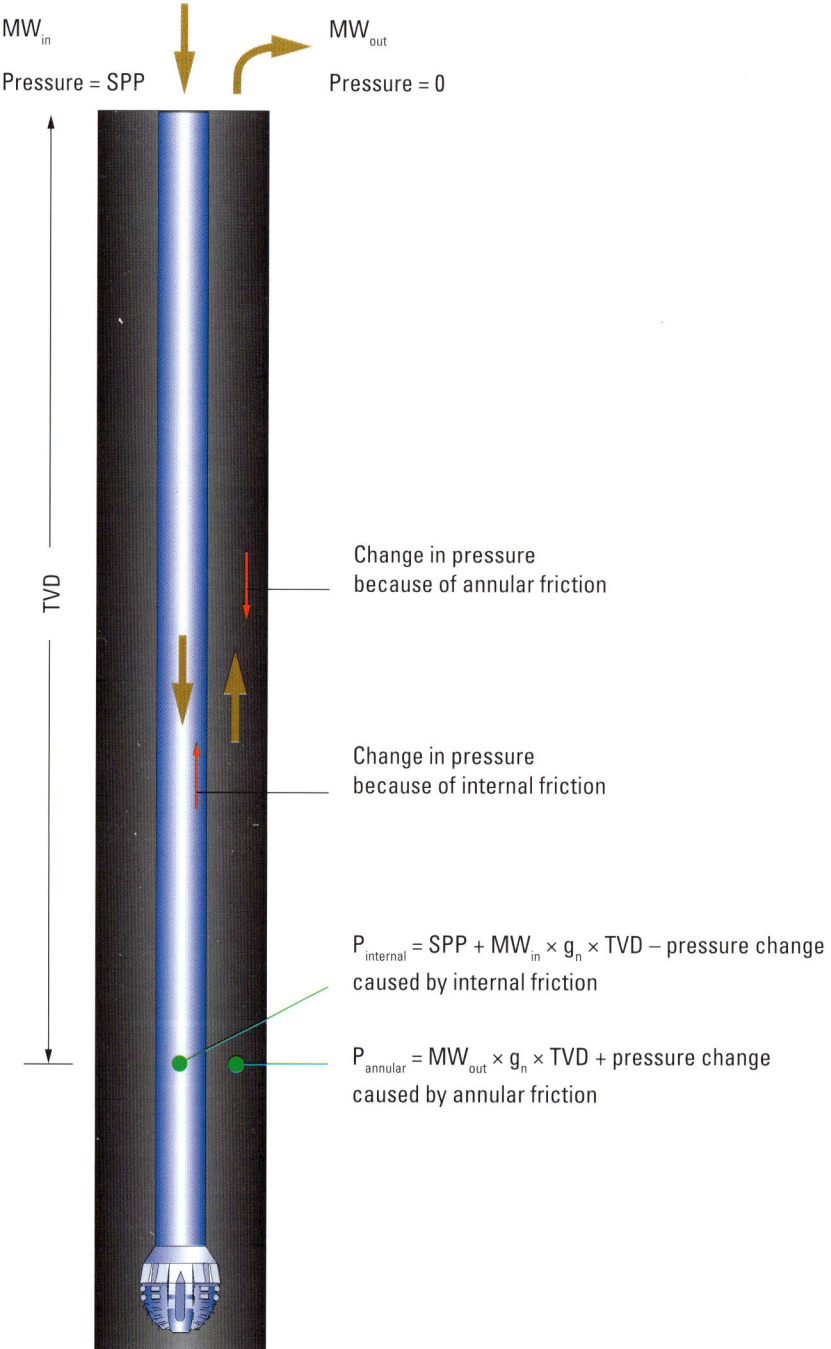

MW$_{in}$

Pressure = SPP

MW$_{out}$

Pressure = 0

TVD

Change in pressure because of annular friction

Change in pressure because of internal friction

P$_{internal}$ = SPP + MW$_{in}$ × g$_n$ × TVD − pressure change caused by internal friction

P$_{annular}$ = MW$_{out}$ × g$_n$ × TVD + pressure change caused by annular friction

As mud travels down the inside of the drillpipe the mud pressure increases according to the equation

Internal hydrostatic pressure = $MW_{in} \times g_n \times TVD$ (Equation 4-41),

where

MW_{in} = mud weight (density) as the mud is pumped into the hole

g_n = acceleration due to gravity.

When the mud is flowing, the pressure is increased by the standpipe pressure applied at the surface and decreased by frictional pressure losses along the inside of the drillstring.

Internal circulating pressure = SPP + internal hydrostatic pressure − internal frictional losses. (Equation 4-42).

Additional pressure losses occur across the BHA, where components such as the MWD mud turbine and the drilling motor extract power from the mud flow. There is also a significant pressure loss across the nozzles in the bit, which accelerate the mud flow to clean the bit and agitate the rock cuttings into the mud flow for transport back to the surface.

As the mud flows back up the annulus between the drillstring and the borehole wall, a strain gauge in the EcoScope tool measures the annular mud pressure.

It is generally easiest to think of the annular pressure in terms of the factors that affect the flow on its return up the annulus to surface.

Mud returning to surface is generally at atmospheric, or zero gauge, pressure as it flows into the mud tanks. The annular hydrostatic pressure at a point in the well is given by

annular hydrostatic pressure = $MW_{out} \times g_n \times TVD$ (Equation 4-43),

where

MW_{out} = mud weight (density) of the mud in the annulus, which is greater than MW_{in} as a result of the rock cuttings that are suspended in the mud returning to the surface.

To overcome the annular frictional forces acting against the flow, the pressure at a given depth must be increased so that the mud arrives at the surface with zero gauge pressure.

Annular circulating pressure = annular hydrostatic pressure + annular frictional losses (Equation 4-44).

One special case where the mud does not return to surface is where mud is circulated out to the sea floor during some offshore operations. The pressure of the mud at the sea floor is equal to the hydrostatic pressure of the seawater at the sea floor, which is given by

seabed hydrostatic pressure = seawater density $\times g_n \times$ seabed TVD (Equation 4-45).

In this special case, the annular circulating pressure is given by

annular circulating pressure = seabed hydrostatic pressure + annular hydrostatic pressure (Equation 4-46).
+ annular frictional losses

The equivalent circulating density of the mud in the annulus includes the effect of cuttings in circulation and frictional effects (Figure 4-101).

ECD = annular circulating pressure $/(g_n \times$ TVD) (Equation 4-47),

where
ECD = equivalent circulating density.

Drillers watch the ECD to ensure that the hole is being cleaned effectively and that cuttings are not building up around the BHA downhole, a situation that could lead to stuck pipe. Monitoring the ECD also allows timely adjustments to the mud density to maintain pressure control of the borehole.

Wellbore pressure control is critical for safe and smooth drilling operations. If the mud pressure is too high, the formation may fracture, resulting in the loss of borehole fluid and subsequent drilling problems. If the mud pressure is too low, then the fluid in the formation may flow into the well and begin migrating to surface, resulting in a kick, or loss of pressure control. Depending on the mechanical properties of the rock, too high or too low of a mud weight may result in a variety of borehole failure modes, most of which are detrimental to efficient drilling.

Well pressure control design involves determining the upper and lower pressure limits within which the well can be drilled safely. These limits are usually expressed as fluid densities. The upper limit is called the fracture gradient and defines the fluid density above which the formation fractures, resulting in mud loss. The lower gradient is usually the pore pressure gradient, below which formation fluid flows into the borehole. During drilling, the ECD must be kept within the safe drilling window.

Figure 4-103 shows an example mud-weight window. The overburden pressure (purple curve) defines the fracture gradient, which is the upper limit of the pressure window. The predrill seismic estimate of the pore pressure (black curve) defines the pressure window's lower limit. The closeness of the two curves indicates that there is little margin for error in mud weight. The pore pressure estimated from real-time resistivity (red curve) generally agrees with the predrill seismic estimate. The annular pressure–derived ECD (blue curve) falls below the pore pressure at two points. Kicks of formation fluid were taken at these points. In the lower part of the well, the ECD briefly exceeds the fracture gradient. Accelerated mud losses were experienced at this point, and the mud weight lowered slightly to remain within the safe mud-weight window. Overall, real-time monitoring of the ECD permitted adjustment of the mud weight to stay within the narrow mud-weight window, allowing the well to be drilled to total depth.

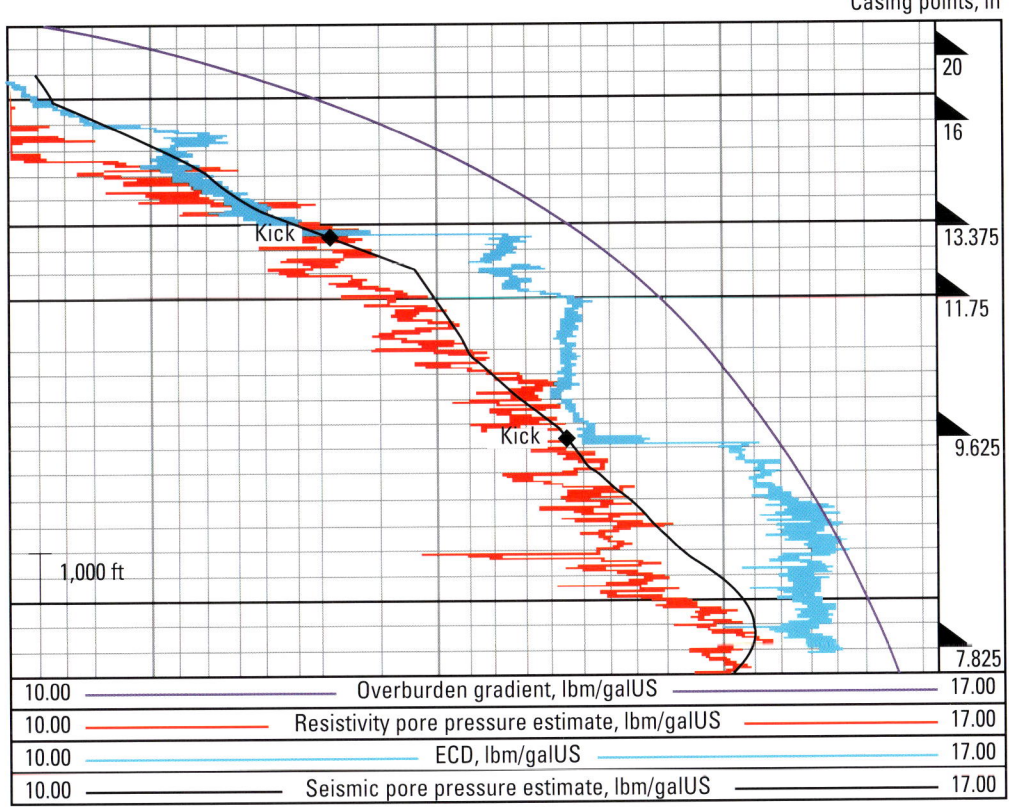

10.00	— Overburden gradient, lbm/galUS —	17.00
10.00	— Resistivity pore pressure estimate, lbm/galUS —	17.00
10.00	— ECD, lbm/galUS —	17.00
10.00	— Seismic pore pressure estimate, lbm/galUS —	17.00

Figure 4-103. Real-time monitoring of ECD, which allows adjustment of the mud weight to remain in the mud-weight window where the formation remains stable. In this example, where the mud weight drops below the lower limit of the mud-weight window, the deepwater well took kicks. Where the mud weight is too high the formation fractured, resulting in accelerated mud losses to the formation.

4.15.4. Annular pressure-during-connection applications

When the mud pumps are off, and there is no mud flow, the frictional forces drop to 0, and the heavier formation cuttings fall out of suspension, while some of the smaller cuttings remain in suspension. The lack of frictional pressure against the flow and the decrease in the cuttings load in the mud both result in a decrease in the annular mud pressure.

A pressure-during-connection test is performed with the EcoScope tool downhole and the pressure sensor powered by the clock battery in the tool. The rest of the EcoScope system is shut down as there is no circulation to provide power from the MWD turbine. The annular hydrostatic pressure is monitored as a function of time after the mud pumps are turned off, as shown in **Figure 4-104**.

Figure 4-104. Pressure profile after the pumps are turned off. The pressure generally decreases and settles to the hydrostatic value. A pressure increase over normal circulating pressure is generally observed when mud circulation restarts.

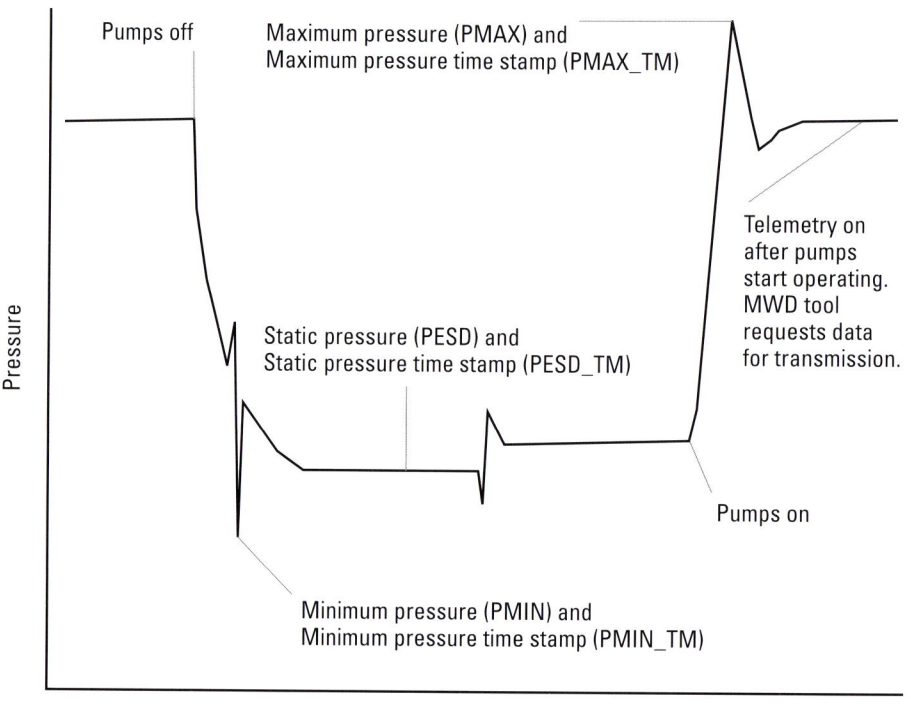

Six measurements are used to quantify the pressure response from the time the pumps are turned off until they are turned on again and the MWD telemetry is ready to transmit data. This period includes the pressure response to both decreasing and increasing mud flow.

The minimum and maximum static pressures (PMIN and PMAX, respectively, in Figure 4-104) and the times at which they occur indicate the behavior of the borehole and formation to rapid changes in mud flow rate and pressure. For example, slow pressure stabilization may be because of fractures closing as the mud flow rate, and therefore the borehole pressure, decreases, or opening as flow rate and pressure increase.

A significant delay between the mud pumps being turned on and the maximum pressure may be an indication of deformation of the borehole called "ballooning" that occurs as the pressure increases as the mud begins to circulate.

The stabilized static pressure and the time at which it occurs (PESD and PESD_TM, respectively, in Figure 4-104) give information about the formation pressure. PESD_TM is the elapsed time from when the stabilized static pressure is measured until the moment when the MWD tool first requests the real-time parameters.

The measured static pressure is converted to the equivalent static density (ESD) of the mud in the annulus by the equation

$$ESD = \text{annular hydrostatic pressure (PESD)} / g_n \times TVD \qquad \text{(Equation 4-48)}.$$

As with the ECD, the ESD must remain above the pore pressure to avoid an influx of formation fluid into the borehole.

A leak-off test (LOT) is an operation performed to test the fracture limit of a formation. The downhole EcoScope pressure sensor is powered by the internal clock battery. The hole is pressure sealed at surface using the blowout preventer (BOP). The mud pressure is slowly increased at surface until the pressure suddenly stops increasing or decreases (**Figure 4-105**). This change in pressure gradient indicates the formation is starting to fracture. The pressure measurements are taken downhole and retrieved with the static pressure data, giving valuable information on the maximum pressure that can be applied to a well before fracturing occurs at the weakest point in the openhole interval.

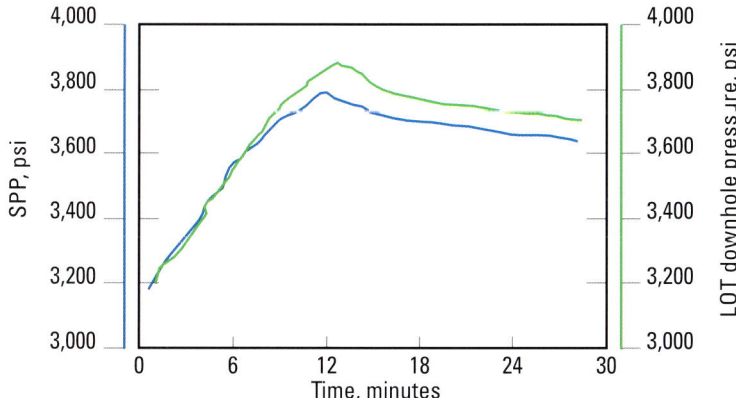

Figure 4-105. A leak-off test. LOT, which determines the maximum pressure that can be applied to an openhole interval before fracturing occurs.

The EcoScope can store up to 3,600 static pressure measurements in memory, normally over a 4-hour period, at a 4-second sampling rate. The duration of acquisition can be adjusted from 2 to 24 hours in steps of 2 hours. The maximum pressure averaged over four samples and time of the maximum is transmitted in real time, as shown in **Table 4-92**.

4.15.5. Annular pressure channels and parameters

4.15.5.1. Real time

Table 4-91. Annular pressure real-time parameters.

Parameter Name	Description	Unit
EHED	Elevation hydraulic head	ft
SEABDEPTH	Water depth	ft
RHO_SEAWATER	Density of seawater	lbm/galUS
SF_FLAG	Return to sea floor	Yes/no (default is no)
FRACP	Fracture pressure gradient	ppg
PPRES	Pore pressure gradient	ppg

Table 4-92. Annular pressure real-time transmitted channels.

Transmitted Channel	Description
DHAP_DH_ECO_RT	Downhole annulus pressure, real time, computed downhole
ESD	
PESD	Hydrostatic pressure (ESD)
PESD_TM	Time between when PESD measured and transmission of the data point
PMAX	Maximum static pressure, real time
PMAX_TM	Maximum static pressure, time delay
PMIN	Minimum static pressure, real time
PMIN_TM	Minimum static pressure, time delay
LOT	
LOTMAXP_DH_ECO_RT	LOT, maximum pressure of four cycles, real time, computed downhole
LOTTMAP_DH_ECO_RT	LOT, time of maximum pressure of four cycles, real time, computed downhole

Table 4-93. Annular pressure real-time computed channels.

Computed Channel	Description
ECD_ECO_RT	Equivalent circulating density, real time
PFPG_ECO_RT	Fracture pressure gradient, real time
PPMW_ECO_RT	Computed pore pressure expressed in mud-weight density, real time
ESD	
PMIN_RT	Minimum static pressure, real time
PMAX_RT	Maximum static pressure, real time
PESD_RT	Static pressure, real time
ESD_MIN_RT	Minimum static density, real time
ESD_MAX_RT	Maximum static density, real time
ESD_RT	Static density, real time
PMIN	Minimum static pressure
PMAX	Maximum static pressure
PESD	Static pressure
ESD_MIN	Minimum static density
ESD_MAX	Maximum static density
ESD	Static density
PRS_P1	Density tadpole graphic
PRS_P2	Density tadpole tail azimuth
PRS_P3	Density tadpole tail azimuth
BB	ESD balloon alarm
K+	ESD kick increasing alarm
K–	ESD kick decreasing alarm
ECD_MW_IN	Equivalent circulating density, mud weight in, 2 Hz

4.15.5.2. Recorded mode

Table 4-94. Annular pressure recorded-mode parameters.

Parameter Name	Description	Unit
EHED	Elevation hydraulic head	ft
SEABDEPTH	Water depth	ft
RHO_SEAWATER	Density of seawater	lbm/galUS
SF_FLAG	Return to sea floor	Yes/no (default is no)
FRACP	Fracture pressure	lbm/galUS
PPRES	Pore pressure	lbm/galUS

Table 4-95. Annular pressure recorded-mode channels.

Output Channel	Description
DHAP	Downhole annulus pressure
TAB_DHAP	Annular pressure time after bit
TICK_DHAP	Annular pressure samples
ECD	
ECD	Equivalent circulating density
PFPG	Fracture pressure gradient
PPMW	Computed pore pressure expressed in mud-weight density

4.15.6. Annular pressure measurement specifications

Table 4-96. Annular pressure sensor specifications.

Item	Value
Range	1 to 25,000 psi
Accuracy	25 psi
Resolution	1 psi

4.16. Annular temperature

The borehole temperature is measured at the same point as the borehole pressure. This measurement provides insight into mud condition, the temperature experienced by the LWD tool, and the operating temperature required for environmental correction of some of the formation measurements.

4.16.1. Annular temperature theory of measurement

The annular temperature is measured using a platinum resistor located on the pressure-sensing element. The temperature data is sampled every 2 seconds.

4.16.2. Annular temperature applications

The annular temperature is used for environmental correction on several measurements including pressure, neutron response, and resistivity. Downhole mud temperature can also be used to diagnose fluid influxes to the borehole. Liquid influxes are likely to increase the wellbore temperature while gas influx is likely to cause cooling as the gas expands as it enters the wellbore.

4.16.3. Annular temperature channels and parameters

4.16.3.1. Real time

Table 4-97. Annular temperature real-time transmitted channels.

Transmitted Channel	Description
DHAT_DH_ECO_RT	Downhole annulus temperature, real time, computed downhole

4.16.3.2. Recorded mode

Table 4-98. Annular temperature recorded-mode channels.

Output Channel	Description
DHAT	Downhole annulus temperature

4.16.4. Annular temperature measurement specifications

Table 4-99. Annular temperature sensor specifications.

Specifications	Value
Range	−71.42 to 312.5 degF
Accuracy	0.89 degF
Resolution	0.17 degF

4.17. Three-axis shock and vibration

Axial, lateral and torsional shock amplitude, shock frequency, and vibrational energy to which the collar is subjected are measured. These measurements allow recognition of unfavorable drilling states and diagnosis of the cause so that remedial action to minimize the consequences of the shock and vibration can be taken.

4.17.1. Shock-and-vibration theory of measurement

Shock-and-vibration data is acquired using three orthogonally mounted accelerometers to measure acceleration along each of the three axes of the tool.

Two magnetometers mounted perpendicular to the axis of the tool measure the frequency of the sinusoids generated as the tool rotates in the earth's magnetic field (**Figure 4-106**). The frequency of the sinusoid is used to determine the rotational speed of the tool.

Figure 4-106. Tool rotation measured by magnetometers mounted perpendicular to the axis of rotation of the tool. They show a sinusoidal response as the tool rotates in the earth's magnetic field, allowing the rpm of the collar to be determined.

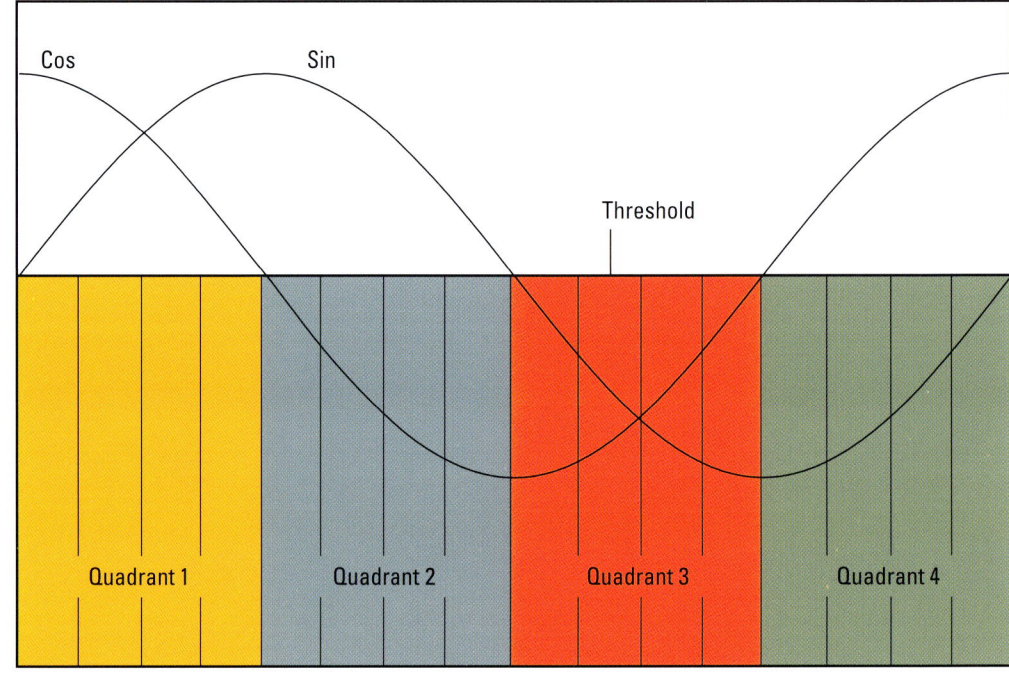

4.17.2. Shock-and-vibration applications

The primary application of shock-and-vibration information is in recognizing the presence and diagnosing the cause of the various mechanical resonance modes in the BHA that result in energy being diverted from cutting rock into damaging the BHA and reducing drilling efficiency.

Shock in a drilling environment is the sudden input of energy when the BHA, bit, or drillstring impacts the borehole. Vibrations are the response of the BHA/bit/drillstring to the shock. Shock and vibration can cause damage to the BHA (collars, stabilizers, connections, downhole tools, or drillbit) or failure of these components. The potential cost when components in the BHA are affected because of shock and vibration can be significant. These costs may include

- extra rig time to trip a failed BHA out of hole and run a replacement

- twisting off connections, leading to a fishing operation or lost-in-hole charge for the BHA

- overgauge hole leading to increased mud and cement volumes

- inability to evaluate the reservoir due to severely degraded formation evaluation sensors and borehole environment.

Monitoring shock-and-vibration measurements, diagnosing the causes of unwanted BHA behavior, and subsequent reduction of shock and vibration delivers improved drilling ROP by ensuring that energy is transmitted smoothly to the cutting surfaces of the bit (**Figure 4-107**) rather than being wasted in damaging the BHA (**Figure 4-108**). This smooth transmission reduces the total time and cost to drill a well.

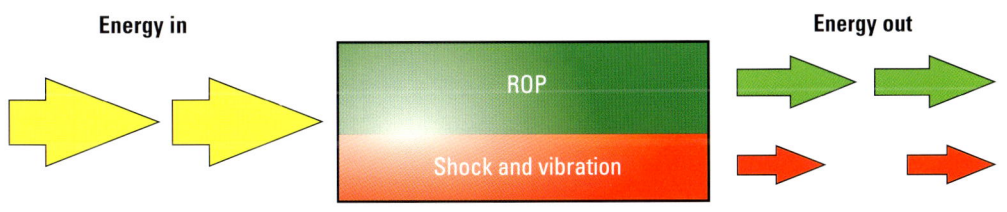

Figure 4-107. Drilling process with normal shock and vibration where energy is mainly used in drilling rock, which maximizes ROP.

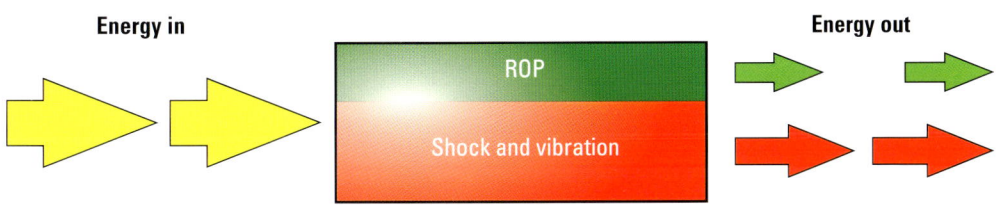

Figure 4-108. Shock and vibration, reducing ROP by diverting energy away from the drilling process and into destructive BHA behavior.

Figure 4-109 displays the following four principal mechanical resonances of a BHA which may result in significant shock and vibration.

- Bit bounce—axial oscillations resulting in repetitive impact on the bit.

- Stick/slip—repetitive rotational acceleration and deceleration of the BHA due to friction against the borehole wall and/or the interaction of the bit with the formation.

- Lateral shocks—sideways bending of the BHA, resulting in impact with the borehole wall.

- Whirl—circular motion of the center of the BHA around the borehole. Forward whirl is the clockwise motion of the center of the BHA around the borehole. As normal drilling rotation is also clockwise, this motion often results in one side of the BHA rubbing against the borehole wall, as indicated by the red dot in the upper panel of **Figure 4-110**. This motion can cause uneven wear of the collar and bit called a "flat spot." Backward whirl is the anticlockwise rotation of the center of the BHA around the borehole, as shown in the lower panel of Figure 4-110.

Figure 4-109. Four main mechanical BHA resonances that can result in significant shock and vibration.

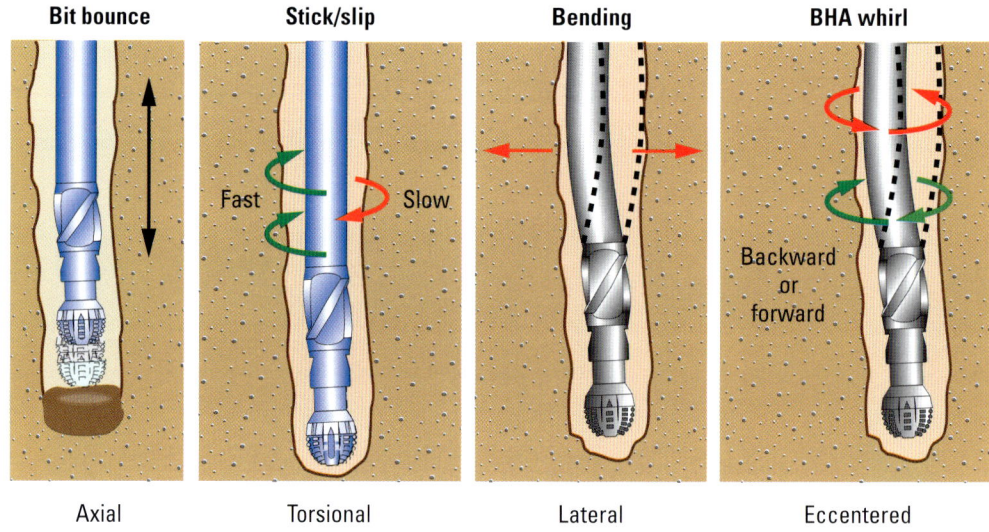

Bit bounce	Stick/slip	Bending	BHA whirl
Axial	Torsional	Lateral	Eccentered

Figure 4-110. Whirling motion of the BHA, which may be clockwise as viewed when looking toward the bit (forward whirl) or anticlockwise (backward whirl).

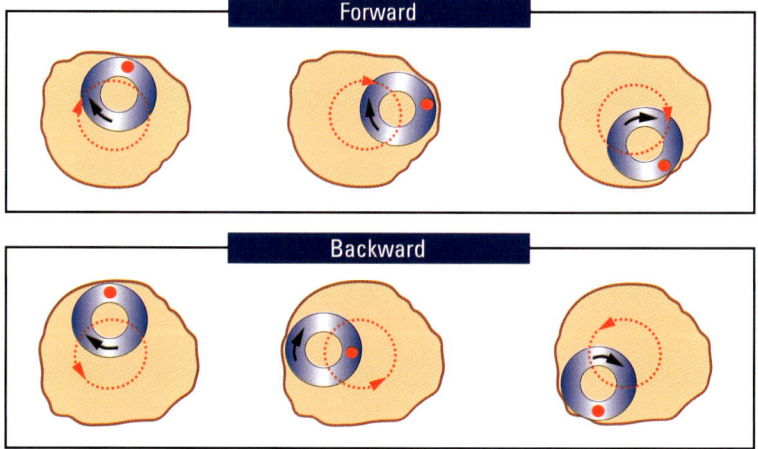

Once drilling, the two main parameters that can be changed at surface to mitigate shocks are the rpm and the weight on bit (WOB). For most BHAs, there is a preferred operating range of WOB and rpm where shocks are minimized (**Figure 4-111**). Outside this stable drilling range, the interaction of the BHA with the borehole environment results in mechanical resonances that divert energy into destructive shocks and vibrations. By monitoring the shocks and vibrations that the BHA experiences as the rpm and WOB are varied, the stable drilling zone can be determined, ROP can be optimized, and shocks can be minimized.

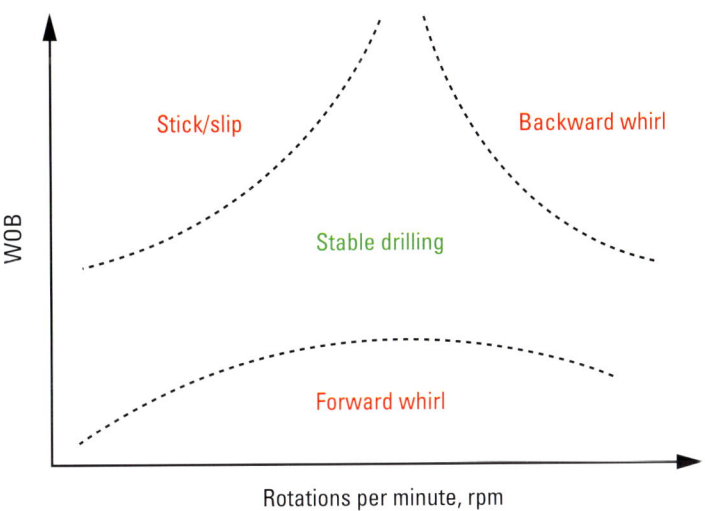

Figure 4-111. WOB and collar rpm, adjusted to maintain the drilling in a stable range where energy is not wasted on shock and vibration.

The EcoScope tool is equipped with triaxial shock sensors that measure the number of shocks per second greater than 50 g_n in each of the three axes. Based on the shock counts/s, a risk factor is determined, as shown in **Table 4-100**.

Table 4-100. EcoScope shock risk.

Shock Risk	Specifications	Value
0	No risk	counts/s < 1
1	Medium risk	1 < counts/s < 5
2	High risk	5 < counts/s < 10
3	Risk of tool failure	counts/s > 10

For example, in **Figure 4-112**, six shocks above the 50-g_n threshold are counted in less than 1 second. This count is converted to a shock risk value that is transmitted to surface in real time, allowing the engineer to determine the kind of drilling condition the tool is undergoing. In this example, the shock risk is high, requiring measures to be taken to reduce the shocks.

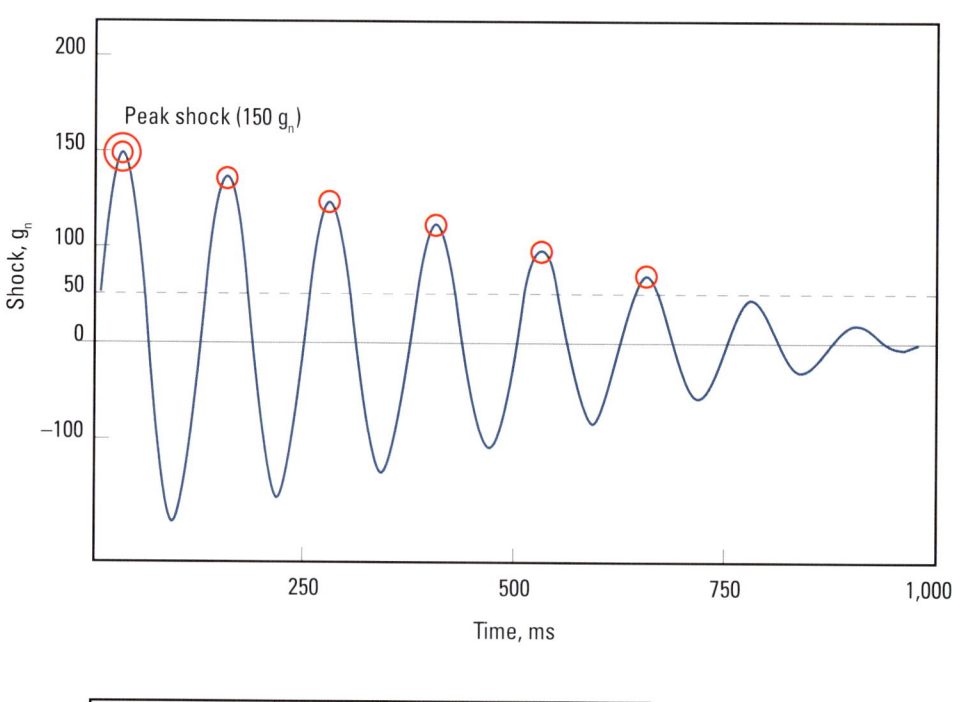

Figure 4-112. Determination of shock risk levels.

Shock Risk	Risk Level	Shock ≥ 50 g_n
2	**High risk**	6

In addition to the shock risk, the amplitude of the largest shock since the previous update was transmitted is measured. This peak shock data gives the magnitude of the shock problem while the shock risk indicates the frequency at which the shocks are greater than 50 g_n.

Prior to each job the tool is subjected to Shock Risk Level 3 as part of the qualification of the EcoScope tool for wellsite operations. This means that the tool is subjected to greater than 10 shocks per second that are over 50 g_n before the tool is qualified for drilling.

4.17.2.1. Vibrations

Vibrations are the "ringing" of the BHA after an impact. The extent of the vibration is indicated by calculating the area under the acceleration curve plotted versus time, as indicated by the green shading in the left panel of **Figure 4-113.** The root mean square (RMS) of the tool acceleration oscillations gives an effective continuous acceleration of the BHA, which is an indication of the vibrational energy in the system. The RMS value is measured along the tool (axial), sideways from the tool (lateral), and around the tool (torsional).

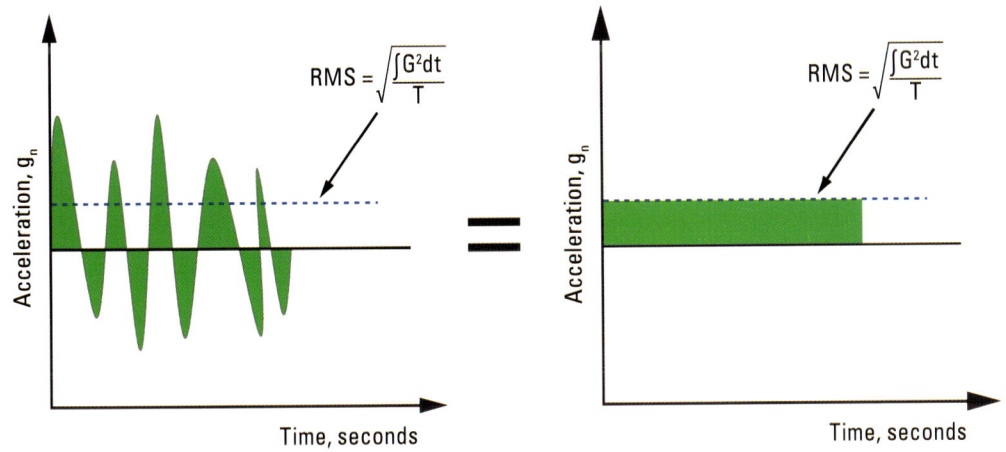

$$RMS = \sqrt{\frac{\int G^2 dt}{T}}$$

$$RMS = \sqrt{\frac{\int G^2 dt}{T}}$$

G = amplitude of the vibrational acceleration expressed in g_n.
dt = the mathematical expression for the integration function of time.
T = the mathematical expression for the total period of time over which the integration is performed.

Figure 4-113. RMS value of the vibrational acceleration, which gives the effective continuous acceleration experienced by the BHA.

4.17.2.2. Stick/slip

Stick/slip describes the situation where a BHA, rather than rotating smoothly at a given rpm, rotationally accelerates, and decelerates because of interactions between the BHA and borehole. This irregular rotation of the BHA decreases drilling efficiency and trajectory control. It increases the stress on the downhole hardware and can lead to severe shocks, both of which increase the probability of downhole equipment failure. Stick/slip can be observed as torsional vibrations or by monitoring the rotation speed of the collar. If the minimum and maximum rpm values separate, it is indicates that stick/slip is occurring.

4.17.2.3. Internal tool temperature

The internal tool temperature is measured on the electronics chassis inside the collar. This is the operating temperature of the tool electronics. These are the most temperature sensitive subsystems so exceeding their rated temperature may result in tool failure.

4.17.3. Shock-and-vibration channels and parameters

4.17.3.1. Real time

Table 4-101. Shock-and-vibration real-time transmitted channels.

Transmitted Channels	Description
SHKL_DH_ECO_RT	Shock level, real time, computed downhole
VIB_LAT_DH_ECO_RT	Transverse (lateral) RMS vibration, real time, computed downhole
VIB_TOR_DH_ECO_RT	Torsional (rotational) RMS vibration, real time, computed downhole
VIB_X_DH_ECO_RT	Axial (X axis) RMS vibration, real time, computed downhole
QC_DRIL_DH_ECO_RT	Drilling measurements quality indicator, real time, computed downhole

4.17.3.2. Recorded mode

Table 4-102. Shock-and-vibration recorded-mode channels.

Output Channel	Description
SHKPK_LAT	Lateral shocks peak amplitude
SHKPK_TOR	Torsional (rotational) shocks peak amplitude
SHKPK_X	Axial shocks peak amplitude
VIB_LAT	Transverse (lateral) RMS vibration
VIB_TOR	Torsional (rotational) RMS vibration
VIB_X	Axial (X axis) RMS vibration
CRPM	Collar rotational speed
SRPM	rpm standard deviation
IRPM	Instantaneous collar rotational speed
IRPM_MAX	Instantaneous collar rotational speed, minimum
IRPM_MIN	Instantaneous collar rotational speed, maximum
TTEM	Internal tool temperature
SHKC_50G	Equivalent integrated 50-g_n shock counter
SHKR_100G	100-g_n torsional (rotational) shock rate
SHKR_200G	200-g_n torsional (rotational) shock rate
SHKR_25G	25-g_n torsional (rotational) shock rate
SHKR_50G	50-g_n torsional (rotational) shock rate
SHKT_50G	Shocks above 50 g_n count rate
SHKX_100G	100-g_n axial shock rate
SHKX_200G	200-g_n axial shock rate
SHKX_25G	25-g_n axial shock rate
SHKX_50G	50-g_n axial shock rate
SHKZ_100G	100-g_n lateral shock rate
SHKZ_200G	200-g_n lateral shock rate
SHKZ_25G	25-g_n lateral shock rate
SHKZ_50G	50-g_n lateral shock rate
SHOCKS_ACC	Total shocks accumulated over tool lifetime
EDT	Lifetime equivalent drilling time

4.17.4. Shock-and-vibration measurement specifications

Table 4-103. Shock sensor specifications.

Item	Value
Range (2-kHz bandwidth)	0 to 500 g_n
Count thresholds (transverse, axial, and rotational shocks)	25, 50, 100, 200 g_n
Resolution (2-kHz bandwidth)	1 g_n

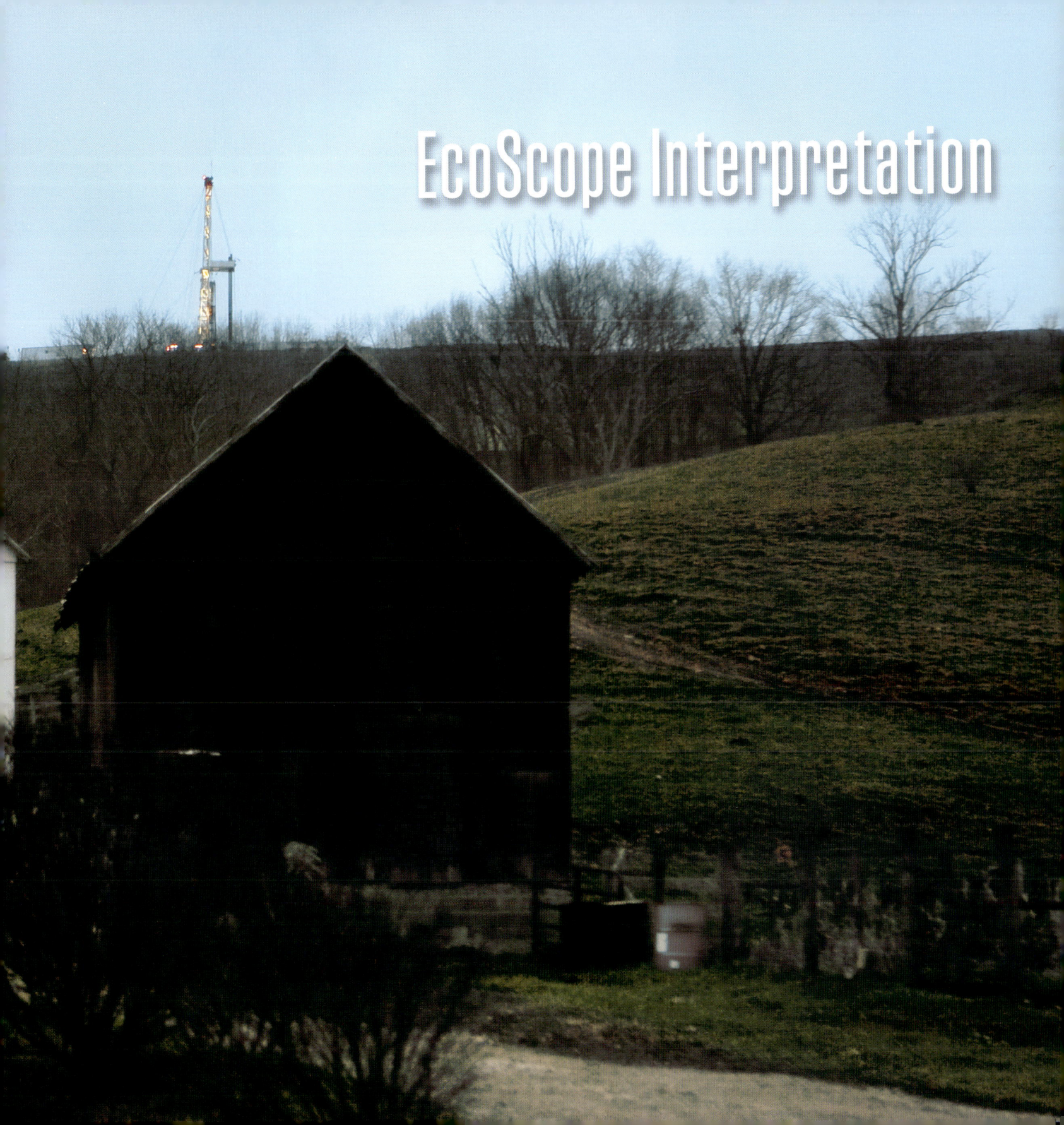

EcoScope Interpretation

EcoScope Interpretation is facilitated by complementary, collocated measurements permitting sophisticated multi-measurement analysis, delivering improved information value and operational efficiency to a wide range of users.

5.1. Introduction

The EcoScope multifunction logging-while-drilling (LWD) service was designed with the objective of increasing operational efficiency and information value for all users, including drillers, acquisition engineers, geologists, petrophysicists, reservoir engineers, and well placement specialists.

The EcoScope system, as with all LWD tools, has the advantage of measuring formation properties before drilling fluids invade deeply (**Figure 5-1**). In addition, the measurements are acquired during the drilling process, so they require minimal additional rig time. The availability of LWD formation information also facilitates well placement in the reservoir based on real-time geological information rather than the assumed geometrical location of the reservoir, which is often inaccurate, particularly in the more complex reservoirs being targeted today.

Figure 5-1. Mud filtrate invasion, creating an altered zone near the wellbore for which the measurements must be corrected. LWD measurements generally do not require as much correction as wireline measurements because the invasion is not as deep soon after the hole has been drilled.

r_i = radius of invasion
R_m = mud resistivity
R_{mc} = mudcake resistivity
R_{xo} = invasion zone resistivity
S_{xo} = invasion zone saturation
R_t = uninvaded formation resistivity
R_w = in situ water resistivity
S_w = formation water saturation
d_i = diameter of invasion
d_j = diameter of transition zone
R_S = shoulder or adjacent bed resistivity
R_{mf} = mud filtrate resistivity
h_{mc} = mudcake thickness

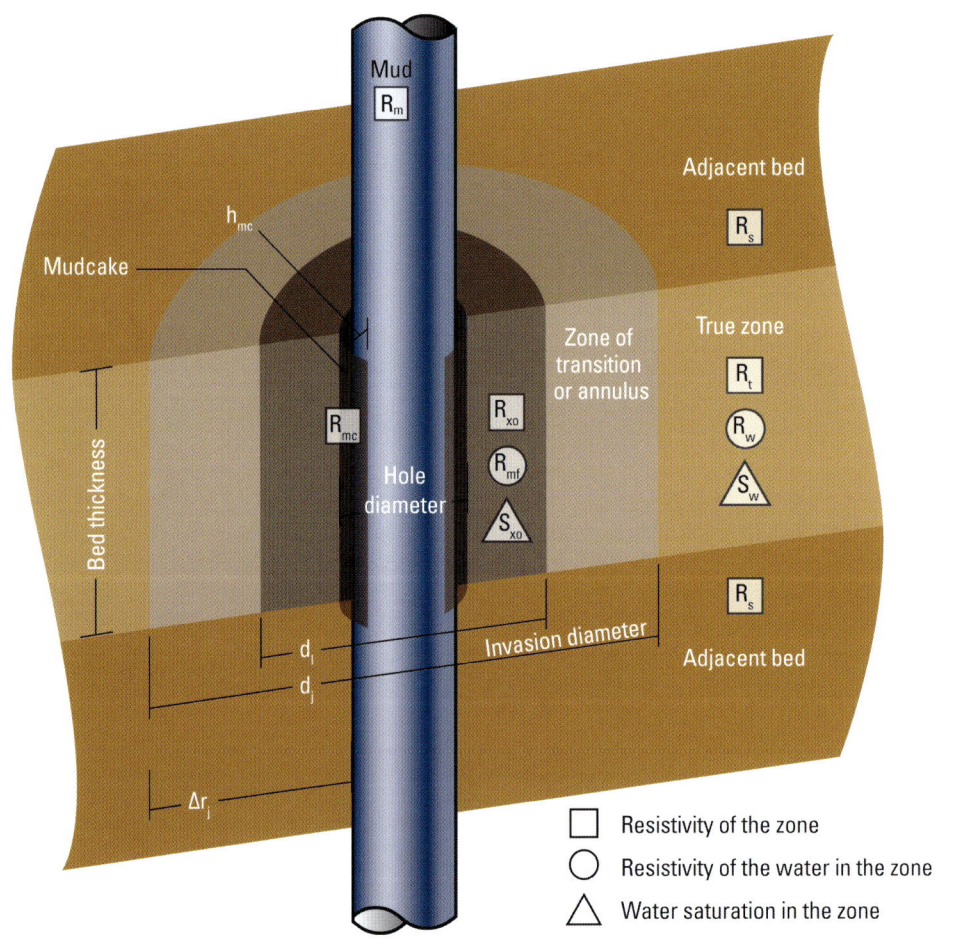

5.2. Resistivity-derived saturation

5.2.1. Archie's equation

One of the main objectives of logging is to evaluate the volume, properties, and producibility of any hydrocarbons in the formation. Since it was first published in 1942, the Archie equation and its variants have been the standard method for evaluating formation water saturation and hence, hydrocarbon saturation. Based on empirical correlation to experimental data, Gus Archie proposed

$$S_w = \sqrt[n]{\frac{a}{\phi^m} \times \frac{R_w}{R_t}}$$ (Equation 5-1),

where

n = saturation exponent

a = empirically derived constant

ϕ = formation porosity

m = cementation exponent.

The classic triple-combo of resistivity, density, and neutron measurements are available from the EcoScope system. They constitute the minimum set of inputs to solve for S_w using Archie's equation.

Density is generally used to derive porosity by

$$\rho_{bulk} = \phi \times \rho_{fluid} + (1 - \phi) \times \rho_{grain}$$ (Equation 5-2a),

where

ρ_{bulk} = bulk density

ϕ = formation porosity

ρ_{fluid} = fluid density

ρ_{grain} = grain density.

Solving for the density porosity, ϕ_d,

$$\phi_d = \frac{\rho_{grain} - \rho_{bulk}}{\rho_{grain} - \rho_{fluid}}$$ (Equation 5-2b),

The neutron and density measurements are often presented on a lithology-compatible scale (e.g., limestone-compatible scale) such that in a freshwater-filled clean formation of the selected lithology, the neutron and density measurements overlay. Any separation between the two curves is an indication that either the fluid in the pore space is not freshwater or the matrix is different from the selected overlay scale. These separations help identify when the ρ_{fluid} or ρ_{grain} terms in the above ρ-ϕ equation should be reviewed to obtain the correct porosity.

The photoelectric factor (PEF) and its volumetric equivalent, U, add information to help evaluate the formation mineralogy so that the correct grain density can be determined. The grain density is critical in the transformation from density to porosity. The EcoScope spectroscopy measurement significantly improves formation mineralogy determination by delivering formation elemental composition information. This can be directly transformed into formation grain response parameters such as grain density, thereby facilitating accurate porosity determination.

The resistivity measurements are used to evaluate the true formation resistivity. Because of possible invasion of mud filtrate into the formation during drilling, modern resistivity tools measure the formation resistivity at multiple depths of investigation (DOI) to characterize and correct for the near-wellbore invasion. It has been found that for LWD measurements close to the bit, the invasion is often minimal, allowing direct formation measurements.

In addition to porosity, determined using a combination of the density, neutron, and photoelectric measurements and a formation resistivity representative of true formation resistivity, the Archie parameters a, m, and n need to be known (e.g., from core analysis) or assumed (the default values are 1, 2, and 2, respectively). The formation water resistivity at downhole conditions also needs to be known. R_w is generally calculated based on downhole temperature, pressure and water salinity, which has been determined from water samples.

Archie's equation can then be solved to find the proportion of the pore space filled with water, otherwise known as the water saturation, S_w. The remaining pore space is assumed to be filled with hydrocarbon. Hence, hydrocarbon saturation is found by using

$$S_{hc} = 1 - S_w \qquad \text{(Equation 5-3)},$$

where

S_{hc} = hydrocarbon saturation.

The volume of hydrocarbon per unit volume of formation, V_{hc}, is given by

$$V_{hc} = \phi \times S_{hc} \qquad \text{(Equation 5-4)}.$$

5.2.2. Other resistivity-based saturation equations

Archie's equation does not account for the excess conductivity associated with clay minerals in the formation. Many modifications of the Archie equation have been proposed, including the Simandoux, Indonesia, Nigeria, dual water, Waxman-Smits, and Sen-Goode-Sibbit. All of these techniques require accurate evaluation of the volume of clay in the formation and the conductivity of the clay material. The spectroscopy measurement made by the EcoScope service yields quantitative clay volumes needed for accurate evaluation of hydrocarbon volumes using these equations.

5.3. Sigma-derived saturation

As outlined in Section 4.11, LWD sigma (Σ) measurements facilitate resistivity-independent saturation evaluation in saline formation waters by delivering the sigma measurement before mud filtrate invasion sweeps the hydrocarbons away from the near wellbore investigated by the sigma measurement.

Sigma is a volumetric measurement related to the chlorine (Cl) content of the formation. Because chlorine generally occurs dissolved in the formation water, sigma can be used to derive resistivity-independent water saturation as long as the formation water is sufficiently saline to produce a usable sigma contrast between the water and hydrocarbons.

Typical sigma values for common formation components are shown in **Figure 5-2**.

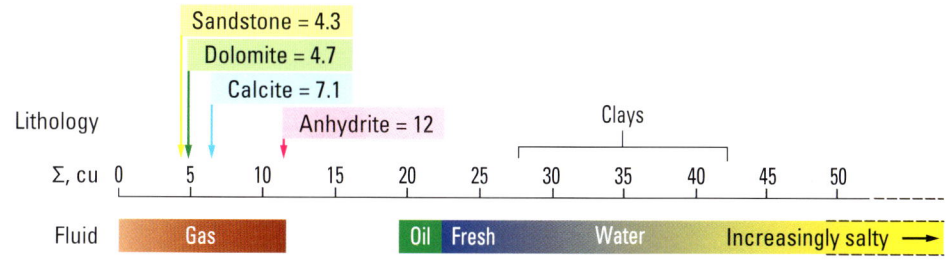

Figure 5-2. Typical sigma values for common formation components. There is a large difference between values for hydrocarbons and those for saline water.

There is a large difference between the sigma values for hydrocarbons and salty water, allowing S_w to be derived according to

$$S_w = \frac{(\Sigma_{bulk} - \Sigma_{grain}) + \phi \times (\Sigma_{grain} - \Sigma_{hc})}{\phi \times (\Sigma_{water} - \Sigma_{hc})}$$

(Equation 5-5),

where

Σ_{bulk} = bulk sigma
Σ_{grain} = grain sigma

Σ_{hc} = hydrocarbon sigma
Σ_{water} = water sigma.

Σ_{bulk} is measured by the EcoScope tool. Σ_{grain} can be determined if the formation mineralogy is known or calculated directly from the elemental composition information available from the EcoScope spectroscopy measurement. Σ_{water} can be measured directly on water samples or calculated from the water salinity. Σ_{hc} can be calculated from the hydrocarbon properties, temperature, and pressure.

Derivation of water saturation from sigma is a particularly useful technique in formations where the traditional resistivity-based methods of estimating water saturation fail to provide reliable results, such as in some low-resistivity pay (LRP) zones.

The combination of resistivity-independent water saturation with a resistivity method allows an additional unknown in the system of equations to be determined. For example, if one of the Archie parameters is not well known, then the in situ value can be derived from the combination of a sigma saturation and resistivity measurement.

Alternatively, if there is significant uncertainty in the formation water salinity, the in situ salinity can be determined by solving for the water salinity at which sigma and the resistivity-derived saturations agree. In this way, a simultaneous solution for both the water salinity and saturation is obtained.

5.4. EcoView interpretation system

5.4.1. Introduction

The EcoView integrated petrophysical interpretation system is an integrated data visualization tool and answer product designed to extract value from the EcoScope suite of measurements and to deliver robust, automated petrophysical interpretation in a wide variety of geological environments.

The EcoView system presents a selection of EcoScope measurements relevant to drillers, such as annular pressure and temperature, shock and vibration, and borehole caliper. These measurements reduce the risks associated with drilling and assist in real-time optimization of the drilling process.

The three-dimensional (3D) visualization component of the EcoView product is designed to give an overview of the borehole trajectory and its position in 3D space. In addition, because it is closely linked to the interpretation component, it gives the user the ability to gain a better understanding of the relationship between the petrophysical (log) data and the borehole geometry, which in turn aids in the interpretation of logging data and the position of the well in the geological sequence.

The petrophysical interpretation process is automated to an unprecedented level through the use of complementary advanced logging measurements to provide results that could previously only have been delivered by an experienced log analyst.

These three components represent the drilling, well placement, and formation evaluation functionalities the of EcoScope service.

The EcoView system can be used at the wellsite or in an office environment. It is a personal computer (PC) (Windows® operating system)-based application that runs as a stand-alone unit in either real-time or postacquisition mode to facilitate timely decision making (**Figure 5-3**).

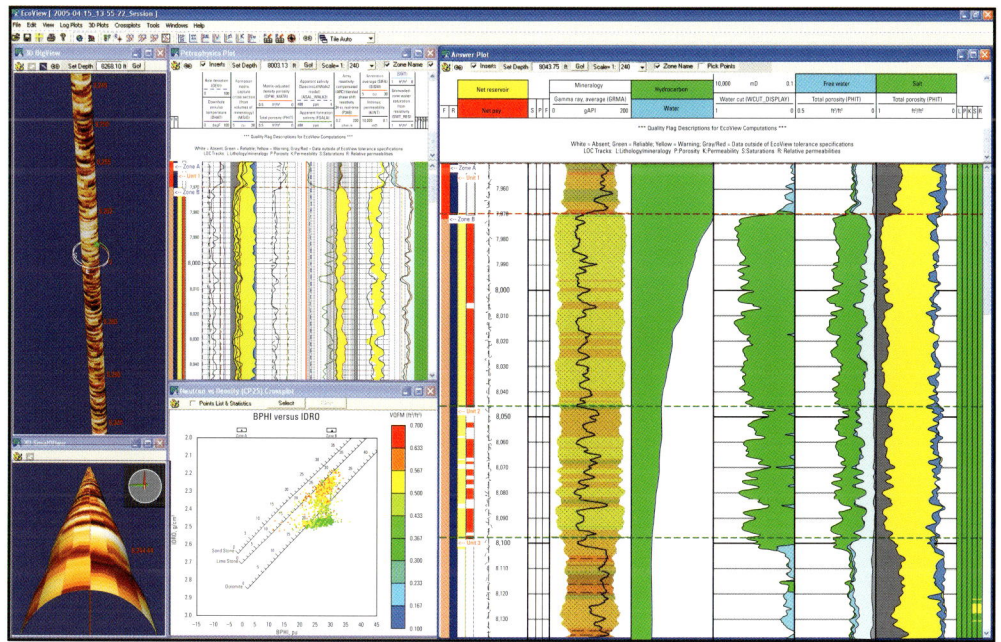

Figure 5-3. EcoView interpretation system, which allows the user to visualize the interpretation of all the measurements in a single workspace that combines two-dimensional (2D) and 3D information.

5.4.2. Three-dimensional visualization and log display

EcoView 3D visualization is presented in two windows. The EcoView big view (**Figure 5-4**) provides 3D visualization of the entire borehole trajectory with the option to render any of the EcoScope images wrapped around the borehole, along with planes representing the dip and azimuth of features picked from the data.

Figure 5-4. EcoView big view, providing for 3D visualization of EcoScope image data wrapped around the borehole trajectory. Dips can be picked on the image data to be displayed as planes (green surface) so that the intersection between the formation feature and the borehole can be visualized in 3D.

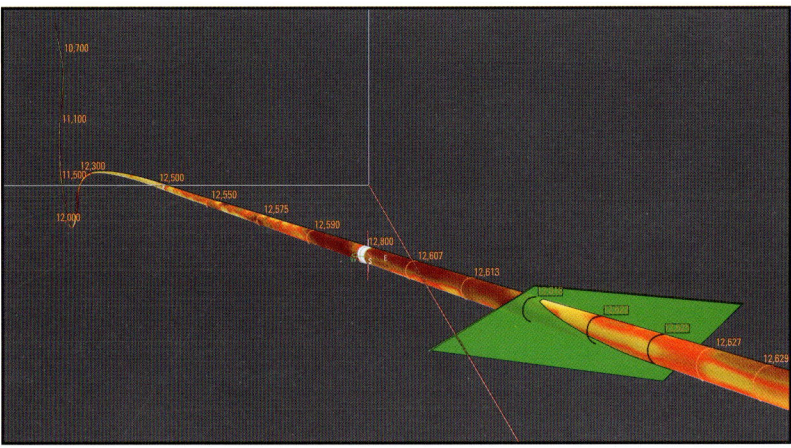

The EcoView small view (**Figure 5-5**) displays a short interval (50 ft) of the borehole, including the borehole shape measured by the EcoScope calipers.

Figure 5-5. The EcoView small view, displaying borehole shape and image over a 50-ft interval.

The EcoView 3D navigation technique (shown in **Figure 5-6)** is centered on the point of interest (POI). The POI remains in the center of the window at all times, avoiding the problem of becoming lost in the 3D environment. The POI navigation technique facilitates rapid identification and inspection of both large features at the scale of the entire well and smaller features at the scale of the borehole diameter. Once identified, synchronization between the 2D and 3D views permits rapid identification of the corresponding log responses. In synchronized mode, the center of the 2D plot is kept at the same measured depth (MD) as the POI in the 3D views. A change in the POI in any of the windows will be reflected in all the windows. The ability to review data interactively in synchronized 2D and 3D environments helps the user better understand the impact of wellbore and formation geometry on log responses.

Navigation in the 3D environment is achieved by using three mouse buttons: left allows rotation around the POI, middle moves the POI along the trajectory, and right zooms in and out. The POI is represented by two perpendicular circles. The red and pink circle defines the vertical and north-south (red-pink) orientation. The white and green circle defines the horizontal plane with the compass quadrants marked.

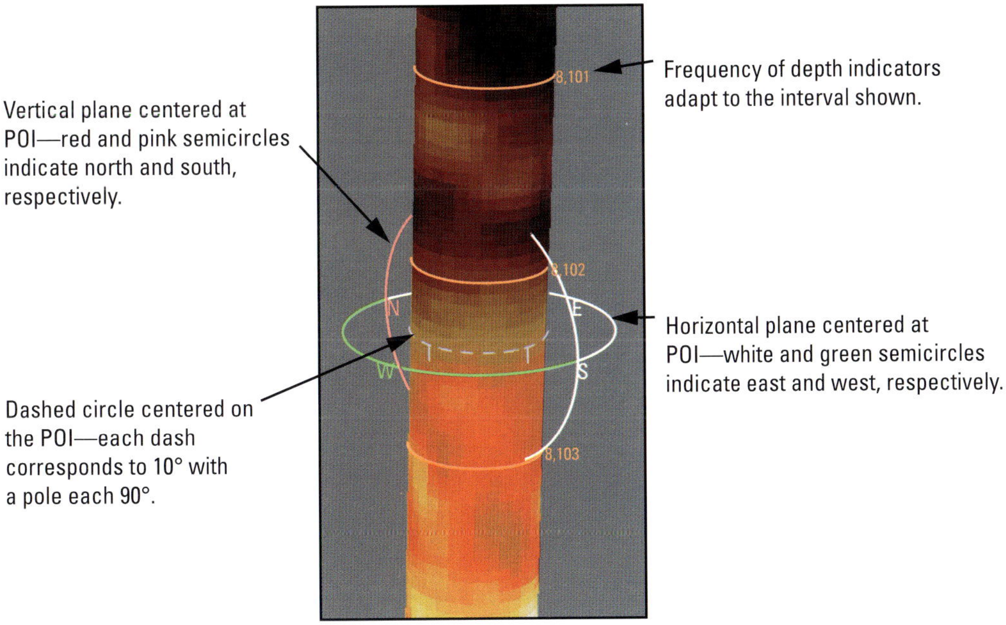

Vertical plane centered at POI—red and pink semicircles indicate north and south, respectively.

Frequency of depth indicators adapt to the interval shown.

Horizontal plane centered at POI—white and green semicircles indicate east and west, respectively.

Dashed circle centered on the POI—each dash corresponds to 10° with a pole each 90°.

Figure 5-6. Ecoview POI, which permits simple navigation around the Ecoview 3D environment because it is always centered on the borehole and remains in the middle of the viewable area.

The EcoScope tool was designed to address the needs of multiple user groups, including drillers, geologists, petrophysicists, reservoir engineers, and field acquisition personnel. Interactive 2D log displays and crossplots are presented with default formats to guide the various groups of users to the data they require. The formats can be modified or user-defined, and presentations can be created as required.

5.4.3. Drilling data display

Several drilling-related measurements are made by the EcoScope system. This information, along with formation-related measurements relevant to drilling operations, is visualized on the Driller's Display window (**Figure 5-7**).

Figure 5-7. EcoView Driller's Display, showing EcoScope data relevant to drilling operations.

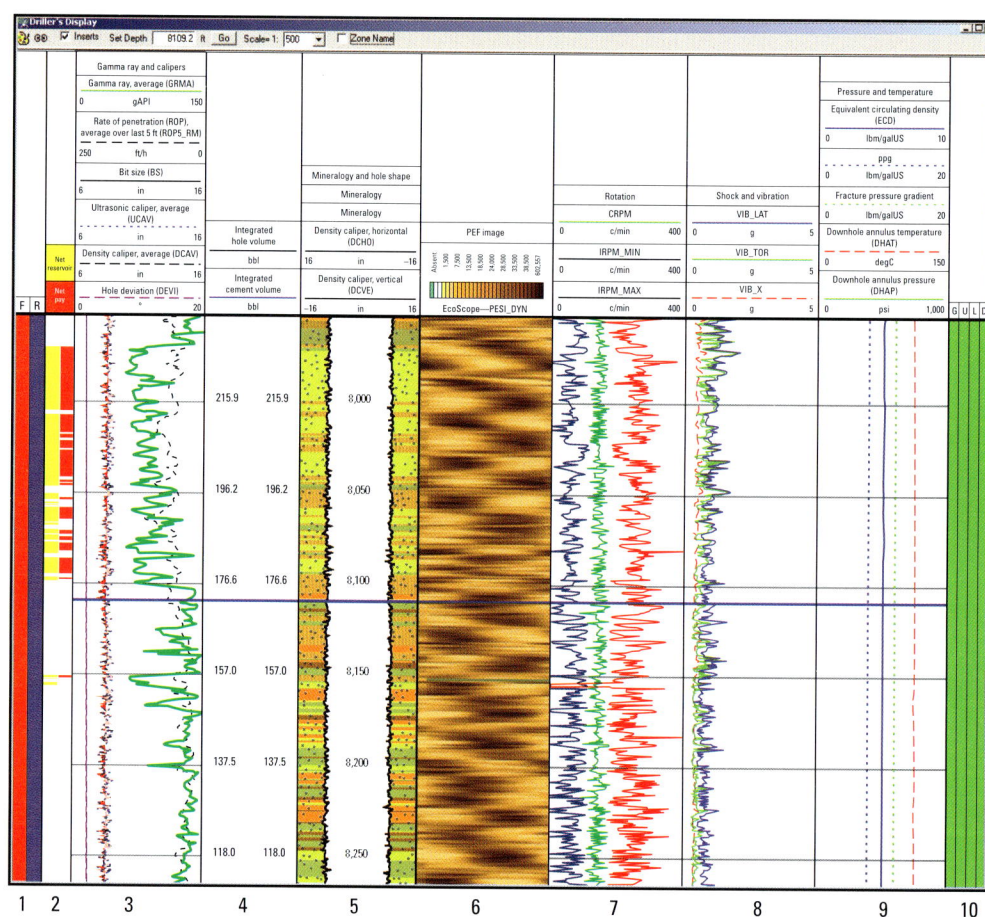

Figure 5-7 shows the following tracks.

- Track 1—zones where formation (F) and reservoir (R) interpretation parameters have been changed.

- Track 2—net reservoir (yellow) and net pay (red) intervals.

- Track 3—hole deviation, caliper, gamma ray, and ROP data.

- Track 4—caliper data converted to integrated hole volume (left) and integrated annular cement volume (right).

- Track 5—depth numbers plotted between the horizontal and vertical hole diameters, plotted to create a schematic of the borehole drilling through the lithology (as indicated by the colors and shading). This allows quick visual correlation among hole condition, bottomhole assembly (BHA) behavior, and formation lithology.

- Track 6—dynamically normalized PEF image, with the shallow nature of the measurement providing an image that is sensitive to borehole rugosity. The alternating light and dark patches suggest some gouging or spiraling on the inner surface of the borehole).

- Track 7—average, minimum, and maximum rpm of the EcoScope collar, which should overlay in smooth drilling conditions. The separation indicates significant stick/slip.

- Track 8—vibration experienced by the tool. The stick/slip motion noted from the rpm behavior is shown by the torsional vibration response (green line). The high lateral vibration response (blue line) suggests that the BHA is also impacting the side of the borehole as part of the BHA motion in the borehole.

- Track 9—mud temperature, pressure, and equivalent mud-circulating density, along with the allowable mud-weight window, as defined by the pore pressure gradient (blue dotted line) and fracture pressure gradient (green dotted line).

- Track 10—log quality control (LQC) indicators for the gamma ray, ultrasonic caliper, lithology, and density measurements, respectively.

5.4.4. Image data display

The EcoScope system acquires four images simultaneously. These are presented in the EcoView image analysis window (**Figure 5-8**).

Figure 5-8. EcoView image analysis window, presenting the four images acquired by the EcoScope tool.

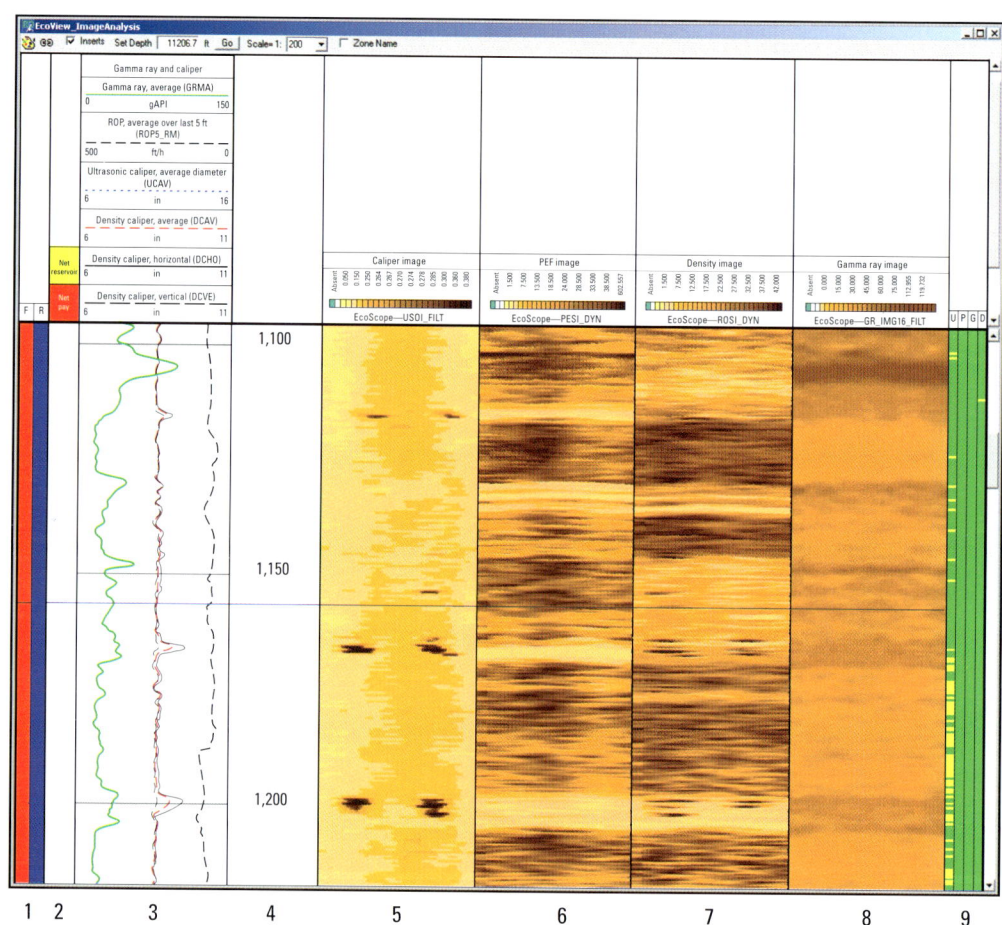

The following tracks are shown in Figure 5-8.

- Tracks 1 and 2—zones of different interpretation parameters, net reservoir (yellow), and net pay (red) intervals.

- Track 3—gamma ray caliper and ROP information.

- Track 4—MD.

- Track 5—image derived from the ultrasonic measurement. This indicates the shape of the borehole and the tool position in the borehole. Darker colors suggest greater standoff between the tool and borehole wall. In this example, the pairs of very dark patches indicate drilling-induced breakout of the borehole wall. Note that these correlate with the lighter colored layers seen on the photoelectric and density images and increased borehole caliper measurements.

- Track 6—PEF image, which responds primarily to formation lithology.

- Track 7—density image, which responds to the combination of porosity, fluid density, and lithology (in the form of grain density).

- Track 8—gamma ray image. Note the correlation between the gamma ray image and the average gamma ray shown in Track 3.

- Track 9—quality control (QC) flags for the ultrasonic, photoelectric, density, and gamma ray images, respectively.

Each of these images responds to different formation properties, so comparison of their responses enhances the interpretation of each. Any of this image information can be visualized in 3D in big and small view windows. The horizontal blue line indicates the 3D POI synchronization depth.

5.4.5. Petrophysical interpretation

The EcoView system follows the interpretation sequence outlined in **Figure 5-9**.

Figure 5-9. Petrophysical interpretation sequence that the EcoView system follows.

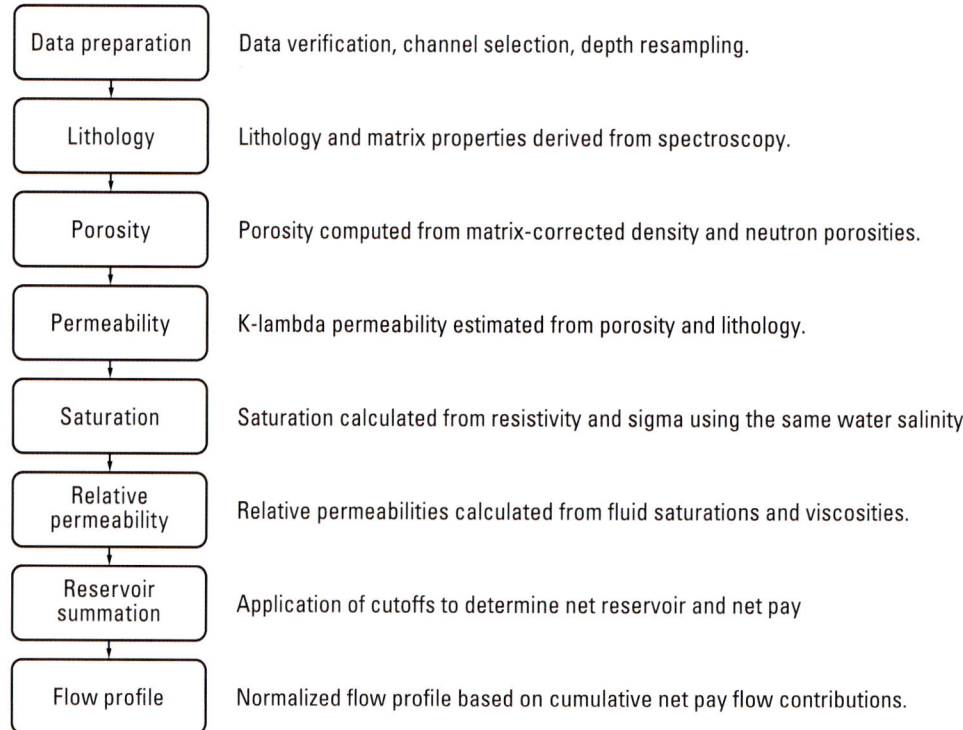

Data preparation	Data verification, channel selection, depth resampling.
Lithology	Lithology and matrix properties derived from spectroscopy.
Porosity	Porosity computed from matrix-corrected density and neutron porosities.
Permeability	K-lambda permeability estimated from porosity and lithology.
Saturation	Saturation calculated from resistivity and sigma using the same water salinity
Relative permeability	Relative permeabilities calculated from fluid saturations and viscosities.
Reservoir summation	Application of cutoffs to determine net reservoir and net pay
Flow profile	Normalized flow profile based on cumulative net pay flow contributions.

The four user inputs required are

- formation water salinity—used to calculate water resistivity and sigma

- hydrocarbon capture cross section—used for sigma saturation calculation

- hydrocarbon viscosity—used for flow profile calculations

- clay cation exchange capacity (CEC)—used for resistivity-based saturation calculation.

If the spectroscopy-derived dry weight elemental concentrations are transformed to clay volume before calculating the grain sigma, rather than calculating grain sigma directly from the elements (the default procedure), then the user is also required to input

- clay capture cross section—used in the conversion of mineral volumes to grain sigma for use in the sigma saturation calculation.

Several calculators and crossplots, examples of which are shown in **Figure 5-10**, help identify appropriate user-defined parameters.

All other information required for the interpretation is available from the suite of EcoScope measurements.

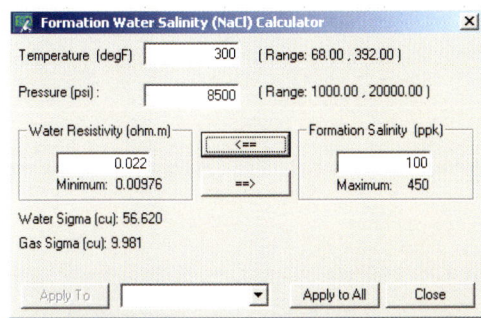

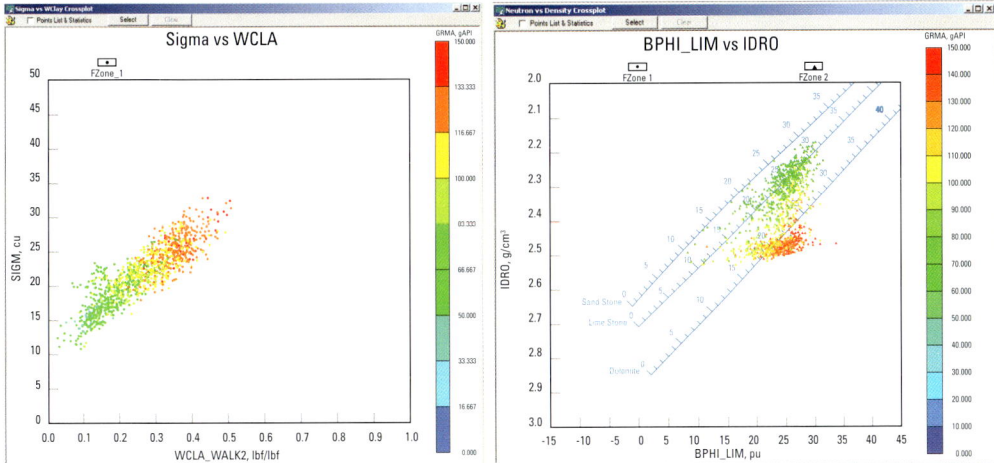

Figure 5-10. Various calculators (upper) and crossplots (lower) available to assist in parameter selection.

The EcoView petrophysics plot (**Figure 5-11**) displays the data in approximately the same sequence as the processing logic.

Figure 5-11. EcoView petrophysics plot.

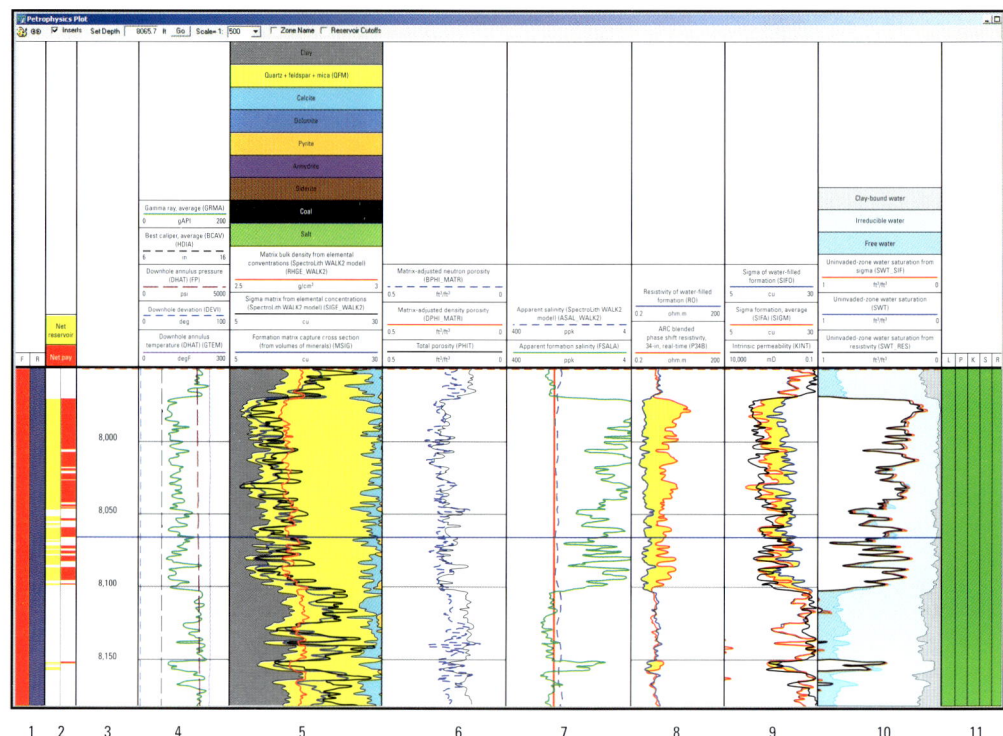

The following tracks are represented in Figure 5-11.

- Track 1—zones where formation (F) and reservoir (R) interpretation parameters have been changed, indicated by shading.

- Track 2—net reservoir (yellow) and net pay (red) intervals.

- Track 3—depth.

- Track 4—gamma ray, caliper, borehole temperature, pressure, and deviation data.

- Track 5—lithology and grain properties derived from spectroscopy.

- Track 6—porosity corrected for the lithology effect based on the spectroscopy results.

- Track 7—salinity track with interactive formation water salinity selection (red line). Two independent salinity indicators are provided: (1) the apparent water salinity (green curve) needed to match either the measured true resistivity or sigma and (2) the salinity computed from the chlorine/hydrogen (H) ratio derived from the spectroscopy measurements (blue dashed curve).

- Track 8—comparison of the measured formation resistivity (red curve) with the resistivity value expected if the formation were filled with water (blue curve) of the salinity selected in Track 7. Yellow shading indicates zones where the measured resistivity reads higher than the water-filled formation resistivity, suggesting the presence of hydrocarbons.

- Track 9—measured sigma (red curve) compared with the sigma value expected if the formation were filled with water (blue curve) of the salinity selected in Track 7. Yellow shading indicates zones where measured sigma reads lower than the water-filled formation sigma, suggesting the presence of hydrocarbons. The intrinsic permeability computed from porosity and mineralogy is also presented.

- Track 10—comparison of water saturation derived from the resistivity and sigma techniques.

- Track 11—five LQC flags that follow the same processing sequence of lithology (L), porosity (P), permeability (K), saturation (S), and relative permeability (R).

The output from the petrophysical analysis is a crossverified saturation from the resistivity and sigma techniques, along with QC flags to assist in validation of the interpretation.

EcoView processing goes beyond evaluating the static hydrocarbons in place to estimate the dynamic flow potential of the formation, which is then displayed in the answer plot (**Figure 5-12**).

Figure 5-12. EcoView answer plot.

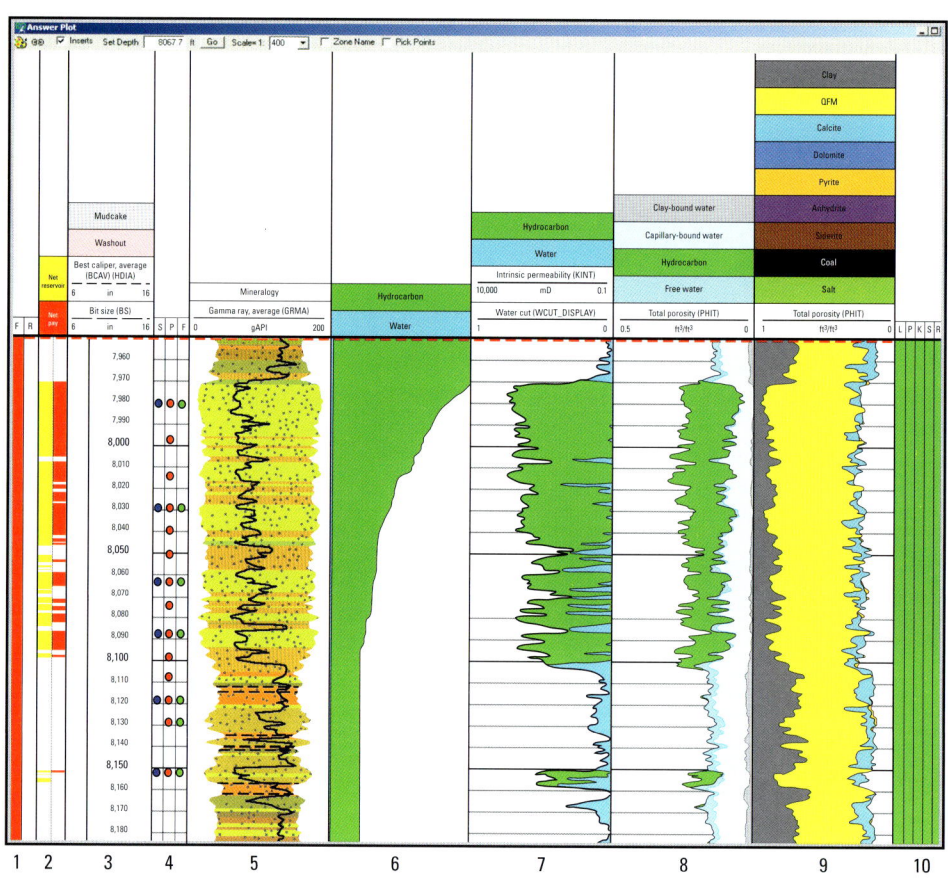

Figure 5-12 shows the following tracks.

- Tracks 1, 2, and 3—identical to those on the petrophysics plot.
- Track 4—interactive selection of depths for subsequent acquisition of sidewall cores (S), pressure points (P), and fluid samples (F). Tables listing the selected depths (**Figure 5-13**) are automatically populated with relevant formation properties to aid in equipment preparation and interpretation of the subsequently acquired data.

- Track 5—qualitative lithology column that is computed from a ternary diagram of quartz-feldspar-mica, calcite, and dolomite, modified based on the fraction of clay present in the formation, as shown in **Figure 5-14**.

- Track 6—normalized cumulative flow profile predicted if the net pay intervals (red shading in Track 2) are perforated.

- Track 7—intrinsic permeability (thick black line) subdivided into the relative permeability to hydrocarbon and water. Using the hydrocarbon viscosity (user-input) and water viscosity (computed from salinity), the relative mobilities of hydrocarbon and water are computed, and hence, the total flow rate is partitioned into hydrocarbon and water flow rates.

- Track 8—fluid volumes

- Track 9—porosity-corrected lithology volumes for comparison with the relative permeability and flow profiles.

- Track 10—five LQC flags, identical to those in the petrophysics plot.

Pressure/Fluids	Sidewall Core		Insert		Delete		☑ Pick Points		
No.	Depth, ft	S	VCLA (ft³/ft³)	PHIT (ft³/ft³)	KINT (mD)	SWT (ft³/ft³)	SWI (ft³/ft³)	RHOB (g/cm³)	Comments
1	7976.0		0.055	0.260	510.124	0.181	0.181	2.670	
2	7997.0		0.112	0.213	69.372	0.410	0.353	2.682	
3	8020.5		0.158	0.246	52.185	0.537	0.455	2.698	

Pressure/Fluids	Sidewall Core			Insert		Delete		☑ Pick Points	
No.	Depth, ft	P	F	PHIT (ft³/ft³)	KINT (mD)	MT (mD/cP)	SWT (ft³/ft³)	Hole Size (in)	Comments
1	7976.0			0.260	510.124	41.784	0.181	8.487	
2	7988.0			0.231	114.770	8.004	0.303	8.510	
3	7997.0			0.213	69.372	4.141	0.410	8.438	
4	8009.5			0.275	207.553	13.783	0.336	8.439	
5	8020.5			0.246	52.185	2.792	0.537	8.453	

Figure 5-13. Selected sidewall core (upper) and formation fluid pressure and sampling depths (lower), with relevant formation properties listed automatically.

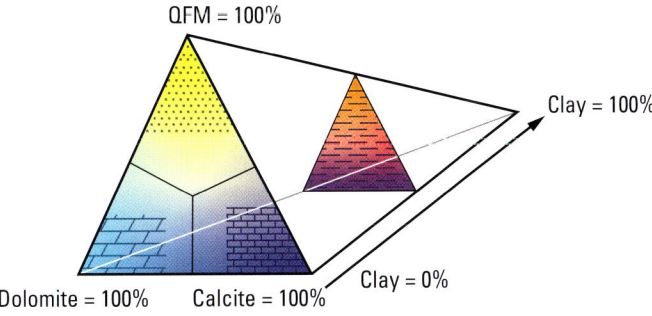

QFM = 100%

Clay = 100%

Dolomite = 100%   Calcite = 100%   Clay = 0%

Figure 5-14. A lithology ternary diagram, modified by the clay content to generate the lithology column shown in Figure 5-12.

5.4.5.1. Data preparation

Prior to launching, the EcoView system ensures that the required measurement types are available to perform the processing. The minimum dataset must include spectroscopy, density, neutron, and resistivity measurements. Data preparation follows the sequence shown in **Figure 5-15**. The EcoScope tool can provide multiple data channels for several of the measurements based on the processing that has been applied to the raw data to compute the desired formation properties. Where multiple channels exist for a measurement type, a selection hierarchy is applied, as displayed in the upper panel of **Figure 5-16**. For example, the density measurement used in the EcoView computations could be from image-derived density (IDD) processing, the bottom quadrant density, the average gamma-gamma density, the upper quadrant density, or the neutron-gamma density (NGD). The EcoView tool scans the input Digital Log Information Standard (DLIS) or real-time file to determine which channels are available and automatically selects the channel highest in the selection hierarchy. Figure 5-16 shows all the density channels in green, indicating that they are all available in the input file. In this case, the IDD channel, IDRO, will be used. The user can override the automatic selection if desired.

Figure 5-15. Sequence of data preparation operations.

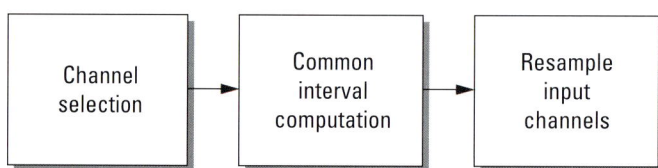

Note that the EcoView software does not perform any environmental corrections on the data. The input channels are used directly in the EcoView computations, so input data must be appropriately corrected prior to processing in the EcoView system.

Figure 5-16. EcoView data channel validation and selection window.

Data Validation

Algorithm Options:

Resistivity Data Input source	ARCWizard	Blended	34 inch 2 MHz	28 inch 2 MHz	40 inch 2 MHz	GVR Average	GVR Bottom
Uninvaded Zone	RT_ARC	P34B	P34H	P28H	P40H	RES_BD	RES_BD_DN
Density Data Input Source	Imaged Derived	Bottom Quadrant	Average	Upper Quadrant	Neutron Gamma		
Bulk Density	IDRO	ROBB	RHOB	ROBU	RHON		
Photoelectric Effect Factor	IDPE	PEB	PEF	PEU			
Neutron Data Input Source	Average	Bottom Quadrant	Upper Quadrant				
Neutron Porosity	BPHI	BPHB	BPHU				
Capture Cross Section	SIFA	SIFB	SIFU				
Hole Size Option	Density Caliper	Ultrasonic Caliper	Resistivity Caliper				
Hole Size	DCAV	UCAV	CALE_ARC				

☑ Detail

Channel Name	Start Depth	Stop Depth	Channel Name	Start Depth	Stop Depth	Channel Name	Start Depth	Stop Depth
IDRO	9000.0 ft	9078.5 ft	IDPE	9000.0 ft	9078.5 ft	ROBB	9000.0 ft	9078.5 ft
PEB	9000.0 ft	9078.5 ft	RHOB	9000.0 ft	9078.5 ft	PEF	9000.0 ft	9078.5 ft
ROBU	9000.0 ft	9078.5 ft	PEU	9000.0 ft	9078.5 ft	RHON	9000.0 ft	9078.5 ft
PEFX			BPHI	9000.0 ft	9078.5 ft	SIFA	9000.0 ft	9078.5 ft
BPHB	9000.0 ft	9078.5 ft	SIFB	9000.0 ft	9078.5 ft	BPHU	9000.0 ft	9078.5 ft
SIFU	9000.0 ft	9078.5 ft	RT_ARC			P34B	9000.0 ft	9078.5 ft
P34H	9000.0 ft	9078.5 ft	P28H	9000.0 ft	9078.5 ft	P40H	9000.0 ft	9078.5 ft
RES_BD			RES_BD_DN			BCAV	9000.0 ft	9078.5 ft
DCAV	9000.0 ft	9078.5 ft	UCAV	9000.0 ft	9078.5 ft	CALE_ARC		
BS			ASAL_WALK2			DHAT	9000.0 ft	9078.5 ft
ATMP			TTEM			TEMP		
DHAP	9000.0 ft	9078.5 ft	APRS			WCLA_WALK2	9000.0 ft	9078.5 ft
WQFM_WALK2	9000.0 ft	9078.5 ft	WCAR_WALK2	9000.0 ft	9078.5 ft	WANH_WALK2	9000.0 ft	9078.5 ft
WPYR_WALK2	9000.0 ft	9078.5 ft	WSID_WALK2	9000.0 ft	9078.5 ft	WCOA_WALK2		
WEVA_WALK2			DWSI_WALK2	9000.0 ft	9078.5 ft	DWCA_WALK2	9000.0 ft	9078.5 ft
DWFE_WALK2	9000.0 ft	9078.5 ft	DWSU_WALK2	9000.0 ft	9078.5 ft	DWTI_WALK2	9000.0 ft	9078.5 ft
DWGD_WALK2	9000.0 ft	9078.5 ft	DWAL_WALK2	9000.0 ft	9078.5 ft	DWK_WALK2		
PPG	9000.0 ft	9078.5 ft	TVDE			TIME		
QC_GR			QC_UCAL			QC_SPEC		
QC_RHOB	9000.0 ft	9078.5 ft	QC_PEF			QC_SIGM	9000.0 ft	9078.5 ft
QC_RES			QC_BPHI					

All data needed for EcoView has been found.   [OK]   [Ignore]   [Cancel]

In the lower panel, the available channels are highlighted in green, and the available data interval is indicated. Where the data is streaming into the EcoView system in real time, the interval is updated as new data arrives. In this example, the EcoView processing has sufficient information to be able to run, despite the channels highlighted in red not being available.

The common top and bottom depths of all input channels are then determined, defining the interval over which the subsequent algorithms will work.

All input channels are then resampled at a 6-in sampling rate.

5.4.5.2. Lithology computation

The formation lithology and grain properties

- grain density

- grain thermal neutron response

- grain epithermal neutron response

- grain capture cross section

are derived from the spectroscopy measurement and are the foundation of the EcoView processing. **Figure 5-17** shows the sequence of lithology computation operations.

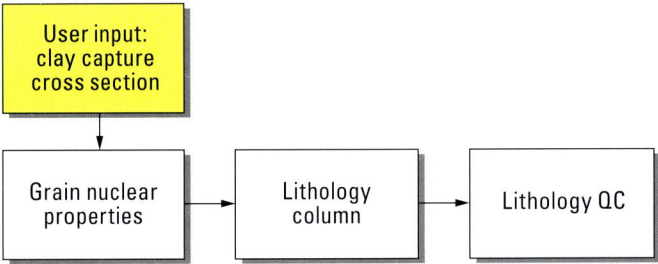

Figure 5-17. Sequence of lithology computation operations.

The EcoView system computes the grain properties directly from the elemental yields. In the case of grain sigma, a second method is available. Using the lithology and porosity information (available after the porosity computation is performed in the next step), the volumetric contribution of each of the minerals is summed to give the grain sigma from volumes. In this case, the sigma value of clay is required (user-input). The user can select either grain sigma from elements or from mineral volumes as the input to subsequent computations.

The lithology information is used directly in the EcoView processing. Where magnesium (Mg) has been detected in the spectral data and included in the lithology model, the volume of dolomite is driven by the spectroscopy results. Where magnesium has not been used in the spectral model, PEF is used to split the spectroscopy-derived carbonate volume into calcite and dolomite.

As shown in Figure 5-14, for presentation purposes, a qualitative lithology column is computed from a ternary diagram of QFM, calcite, and dolomite, modified based on the fraction of clay present in the formation. This is generally displayed with the width varied as a function of clay volume such that cleaner intervals are wider, as seen in Track 5 of Figure 5-12.

5.4.5.3. Porosity computation

Nuclear properties—density, hydrogen index (HI), and sigma—of the formation water are computed as a function of salinity, temperature, and pressure. **Figure 5-18** shows the sequence of porosity computations.

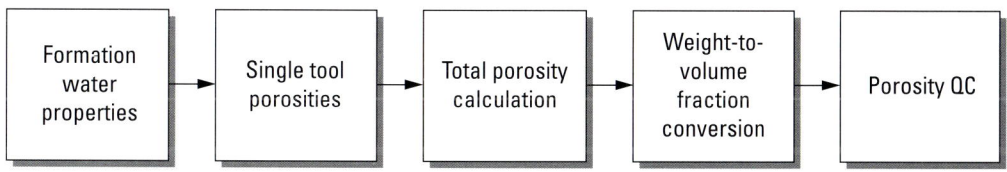

Figure 5-18. Sequence of porosity computation operations.

For each of the measurements of bulk density, thermal neutron porosity, epithermal neutron porosity, and sigma, it is possible to compute a grain-corrected single tool porosity estimate based on the assumption that the formation is water filled. These computations all take the general form of

$$\phi_z = \frac{Z_g - Z_b}{Z_g - Z_f} \qquad \text{(Equation 5-6),}$$

where

ϕ_z = grain corrected single tool porosity from measurement Z (v/v)
Z_g = grain response for measurement Z
Z_b = bulk value of Z measured by the tool
Z_f = fluid response for measurement Z.

Based on the assumption that the fluid in the pores is formation water at the user-specified salinity, Z_f is computed based on the water salinity, temperature, and pressure. In each case, Z_g is derived from spectroscopy.

The density porosity is the single most reliable porosity estimate available, except when light hydrocarbons are present. Therefore, the density porosity is taken as the total porosity except where the neutron porosity reads lower than the density porosity, suggesting the presence of light hydrocarbons. In this case, the average of the matrix-corrected density and neutron porosity is used.

With the availability of porosity and matrix density information, the weight fraction mineralogy derived from the spectroscopy is converted into the more traditional volume fraction format using

$$V_i = \frac{W_i}{\rho_i} \times \rho_g \times (1 - \phi_t)$$

(Equation 5-7),

where

V_i = volume fraction of mineral i (v/v)

W_i = weight fraction of mineral i (w/w)

ρ_i = density of mineral i (g/cm³)

ρ_g = formation grain density from spectroscopy (g/cm³)

ϕ_t = total porosity (v/v).

5.4.5.4. Permeability estimation

Absolute permeability is computed from the available mineralogy and porosity data using the mineral form of the k-Lambda algorithm. This algorithm modifies a porosity-permeability transform with the permeability effect of the surface area of each of the minerals in the formation. This surface area is computed by summing the weight proportions of the minerals, as derived from the spectroscopy multiplied by the specific surface area per gram of the minerals. **Figure 5-19** shows the sequence of permeability estimation operations.

Figure 5-19. Sequence of permeability estimation operations.

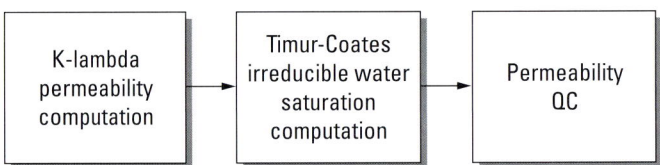

Once permeability has been computed, it is possible to compute irreducible water saturation and volume of irreducible water using an inverse form of the Timur-Coates algorithm. Traditionally, the Timur-Coates algorithm is used to transform porosity and bound- and free-fluid volumes derived from magnetic resonance measurements into permeability. In this inverse application, the absolute permeability derived from the k-Lambda algorithm is used to solve for the bound- and free-fluid volumes. This irreducible bound water saturation is used with the total water saturation and clay-bound water saturation to partition the water volume into bound- and free-water fractions.

Clay-bound water associated with the volume of clay derived from the spectroscopy is displayed in gray in Track 10 of Figure 5-11. The light blue shading indicates the remaining irreducible water. The sum of these two equals the total bound fluid derived from the inverse Timur-Coates algorithm. The darker blue shading indicates producible water.

5.4.5.5. Saturation computation

Resistivity-derived total water saturation is computed using the Waxman-Smits equation. The sequence of operations in shown in **Figure 5-20**. This computation requires the input of two user parameters: formation water salinity and CEC. The CEC is used in conjunction with the spectroscopy-derived weight fraction of clay to compute the excess conductivity caused by the presence of the clay. The water salinity is transformed to water resistivity, which is combined with porosity to compute water saturation and water-filled formation resistivity. Water-filled formation resistivity should overlie the measured true resistivity in all water-filled formations, regardless of lithology, and show separation in the presence of hydrocarbons, thereby providing a direct visual indication of hydrocarbon-bearing zones.

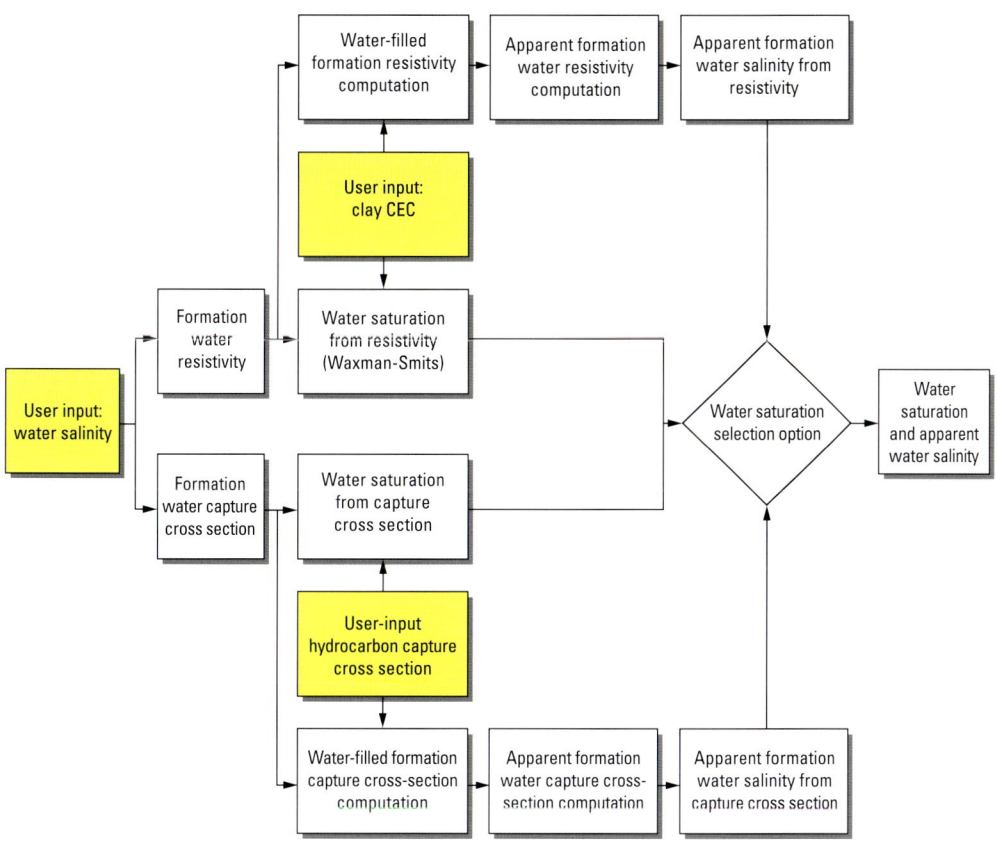

Figure 5-20. Sequence of saturation computation operations.

To aid in water salinity selection, the water resistivity required to match the measured true resistivity is computed, converted to the equivalent water salinity, and displayed in the interactive salinity selection track.

The sigma-derived total water saturation calculation requires two user-defined parameters: (1) formation water salinity and (2) hydrocarbon sigma. The user-defined formation water salinity is transformed to water sigma, which is then used with the hydrocarbon sigma and porosity to compute the total hydrocarbon saturation. Sigma of the formation when completely water-filled is also calculated. It should overlay the measured true formation sigma in all water-filled formations, regardless of lithology, and show separation in the presence of hydrocarbons, providing a direct visual indication of hydrocarbon-bearing zones similar to the resistivity overlay.

Both the resistivity- and sigma-derived water saturations are presented on the petrophysics plot for comparison. A selection toggle allows the user to define which of the saturations is to be used for subsequent processing. This toggle also defines whether the resistivity-derived apparent salinity or sigma-derived apparent salinity is presented on the salinity track of the petrophysics plot.

5.4.5.6. Relative permeability estimation

Total and irreducible water saturation data are used to compute relative permeability to hydrocarbons and water. The calculated relative permeabilities are then combined with the absolute permeability to compute effective permeabilities to the two fluids. The water viscosity is computed based on the water salinity, temperature, and pressure, and the hydrocarbon viscosity is input by the user. The effective permeabilities are divided by the fluid viscosities to obtain the effective mobilities of the hydrocarbon and water. This sequence of operations is shown in **Figure 5-21**.

Figure 5-21. Sequence of operations involving relative permeability estimation, reservoir summation, and flow profile estimation.

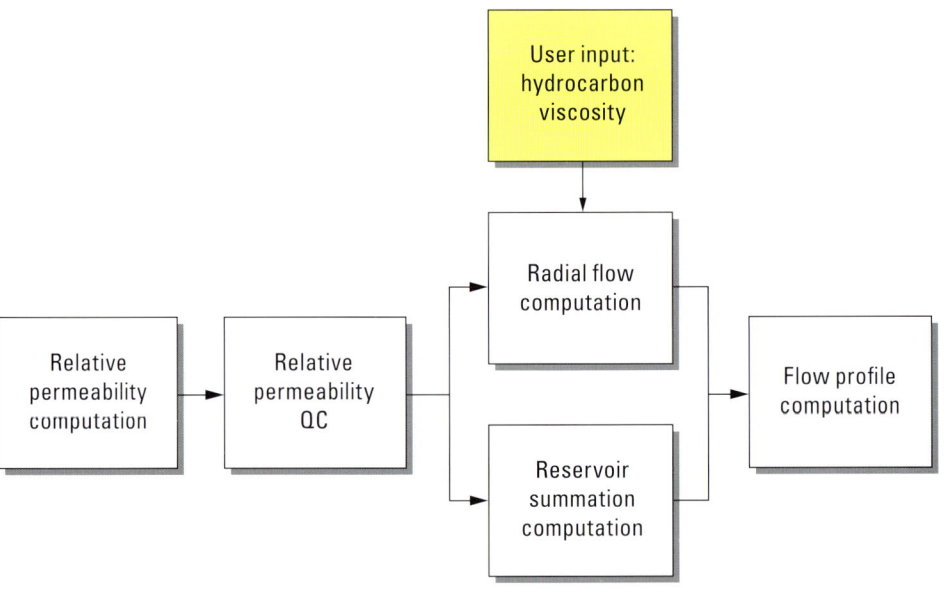

5.4.5.7. Reservoir summation

Reservoir summation is performed based on the final computed values of volume of clay, porosity, permeability, and total water saturation. Reservoir summation is a two-level process.

1. Net reservoir is identified based on the volume of clay being less than, and porosity and permeability values exceeding, user-input cutoffs.

2. Net pay is identified based on those net reservoir intervals where water saturation is less than a user-input cutoff.

The cumulative thickness percentage that passes the thresholds applied to the various properties during net reservoir and net pay calculations is displayed on interactive threshold selection crossplots, examples of which are shown in **Figure 5-22**. The orange thresholds can be adjusted interactively on the plot, allowing the impact on the percentage that is accepted to be assessed.

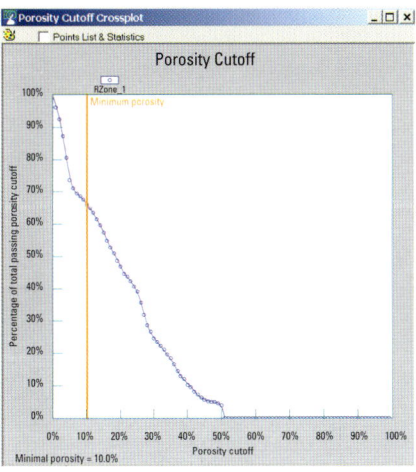
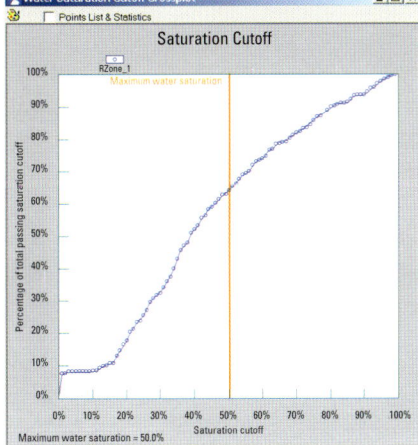

Figure 5-22. Interactive cutoff selection crossplots for porosity and water saturation. Similar crossplots are provided for clay volume and permeability cutoff evaluation.

For net reservoir intervals, the thickness and arithmetic average porosity and permeability are computed. For net pay intervals, the thickness and arithmetic average porosity, permeability, and water saturation are computed, with the arithmetic average water saturation being a porosity-weighted average (in fact, bulk volume water is averaged and then converted back to an average water saturation using the average porosity).

Reservoir summation is not performed over intervals where the interpretation results are of questionable quality, as defined by the relative permeability computation QC indicators.

As shown in **Figure 5-23**, a summary report is generated containing information for each continuous interval of net reservoir. This information includes

- total net reservoir thickness
- average porosity in net reservoir
- average permeability in net reservoir

- total net pay thickness
- average porosity in net pay
- average permeability in net pay
- average water saturation in net pay.

Figure 5-23. Example reservoir summation report.

Reservoir Summation Report

Well: Example
Date: 09-Apr-2010
Time: 17:33:12

RESERVOIR SUMMATION CUTOFFS

No.	Zone Name	Top, ft	Bottom, ft	Maximum Volume of Clay, ft³/ft³	Minimum Porosity, ft³/ft³	Minimum Permeability, mD	Maximum Water Saturation, ft³/ft³
1	Upper zone	8,328.0	8,789.8	0.250	0.100	5.000	0.500
2	Lower zone	8,789.8	8,945.0	0.250	0.100	1.000	0.300

NET RESERVOIR SUMMARY

No.	Zone Name	Top, ft	Bottom, ft	Net Reservoir Thickness, ft	Average Porosity, ft³/ft³	Porosity Thickness, ft³/ft³	Average Permeability, mD	Permeability Thickness, mD.ft
1	Upper zone	8,328.0	8,789.8	149.5	0.268	40.087	129.420	19,348.219
2	Lower zone	8,789.8	8,945.0	58.0	0.213	12.364	19.700	1,142.609
	Total	--	--	207.5	0.253	52.451	98.751	20,490.828

NET PAY SUMMARY

No.	Zone Name	Top, ft	Bottom, ft	Net Reservoir Thickness, ft	Average Porosity, ft³/ft³	Porosity Thickness, ft³/ft³	Average Permeability, mD	Permeability Thickness, mD.ft	Average Water Saturation, ft³/ft³	HCPV Thickness, ft³/ft³.ft
1	Upper zone	8,328.0	8,789.8	125.0	0.265	33.125	116.244	14,530.506	0.238	25.229
2	Lower zone	8,789.8	8,945.0	48.0	0.221	10.628	22.886	1,098.507	0.190	8.604
	Total			173.0	0.253	43.753	90.341	15,629.013	0.225	33.833

5.4.5.8. Flow profile estimation

Absolute permeability is used in a radial flow variation of Darcy's law to compute the downhole flow rate at each depth level. In addition, using the relative permeability data at each depth and knowledge of the formation water viscosity (computed from salinity) and hydrocarbon viscosity (user-input), the relative mobilities of hydrocarbon and water are computed, and hence, the total flow rate is partitioned into hydrocarbon and water flow rates.

Once the downhole phase flow rates have been computed at each depth, an indicative flow profile is computed. This would normally be done over intervals open to the wellbore. As a simplistic solution, zones identified as net pay from the reservoir summation computation are used as open intervals in this process.

Then, over these intervals, the downhole phase flow rates are cumulated to produce cumulative flow profiles. The computed flow profile is then normalized to the cumulative total flow to produce an indicative flow profile on a zero-to-one scale. The relative permeability and normalized flow profile information is presented on the answer plot, an example of which is shown in Figure 5-12.

More details on the EcoView computations can be obtained from the help file embedded in the software.

5.5. Formation evaluation challenges

5.5.1. Porosity in complex lithologies

No single measurement directly delivers true formation porosity. Formation density is often used to derive porosity based on

$$\phi_d = \frac{\rho_{grain} - \rho_{bulk}}{\rho_{grain} - \rho_{fluid}}$$

(Equation 5-8).

Porosity can be determined provided ρ_{fluid} and ρ_{grain} are known; however, they are often unknown, so additional measurements sensitive to the fluid (neutron and magnetic resonance) and lithology (neutron response, PEF, and spectroscopy) are required. Knowledge of the formation lithology allows ρ_{grain} to be determined, which is a critical input to obtain accurate porosity.

For example, for a measured formation ρ_{bulk} of 2.6 g/cm^3, the computed ϕ_d varies from 3 to 13.5 pu, depending on the assumed formation lithology (**Figure 5-24**). Heavy minerals add further complexity to this evaluation.

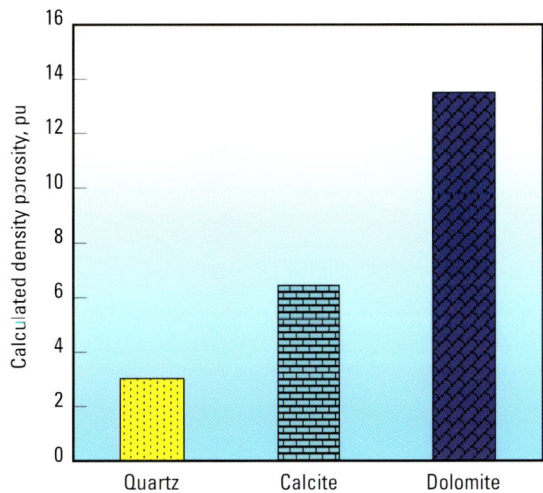

Figure 5-24. Density porosity calculated for various lithologies, indicating the importance of accurate lithology determination.

Spectroscopy delivers the elemental composition of the formation, permitting the volumetric proportions of the various minerals in the formation to be identified. Knowing the volumetric proportion of calcite cement in sandstone, for example, allows more accurate and higher porosity determination. Similarly, the presence of heavy minerals such as pyrite or siderite will result in underestimation of the porosity if their presence is not identified and quantified. Partial dolomitization of calcite and distributed anhydrite in carbonates are other common examples where porosity is often underestimated.

5.5.2. Carbonate lithology identification in heavy mud

The photoelectric measurement delivers critical lithology information, particularly in carbonates where it is used to distinguish calcite from dolomite. However, because of its shallow DOI, the photoelectric measurement often suffers from borehole effects. This can render the measurement unusable in rugose boreholes and heavy mud systems. Where PEF measurement degradation is caused by poor contact between the tool and formation, image-derived density processing (Section 4.4.2.2) may be able to improve data quality. However, if image data is not available because the tool was not rotating across the interval of interest, or if high-PEF mud solids like barite have invaded the formation, it may not be possible to improve the PEF measurement. Distinguishing calcite from dolomite may become difficult under these circumstances.

Figure 5-25 (Track 2) shows an example of how, in favorable borehole and mud conditions, the PEF measurement can be used to distinguish calcite from dolomite. This requires light mud and a smooth borehole wall to ensure good contact between the tool and formation.

The spectroscopy measurement is significantly less sensitive to the borehole than the PEF measurement. Spectroscopy does not require contact between the tool and formation, and spectroscopy processing explicitly excludes elements commonly found in borehole fluids during the conversion from the detected elemental composition to formation mineralogy, as outlined in Section 4.10.

Under favorable conditions such as the slow logging typical of LWD operations, EcoScope spectroscopy is able to quantify the proportion of magnesium in the formation, allowing the proportion of dolomite to be distinguished from calcite using spectroscopy alone. Figure 5-25 (Track 3) shows how EcoScope spectroscopy has identified the dolomite intervals. This is in agreement with the result derived from the PEF seen in Track 2.

5.5.3. Formation evaluation without chemical nuclear sources

The continuing drive to minimize the environmental impact of oilfield operations favors a transition to formation evaluation without the use of chemical nuclear sources, but density and neutron measurements are key inputs to established formation evaluation workflows. The pulsed neutron generator (PNG) in the EcoScope system permits both neutron and density measurements to be acquired without chemical nuclear sources in the hole, significantly reducing risk to both personnel and the environment without compromising formation evaluation.

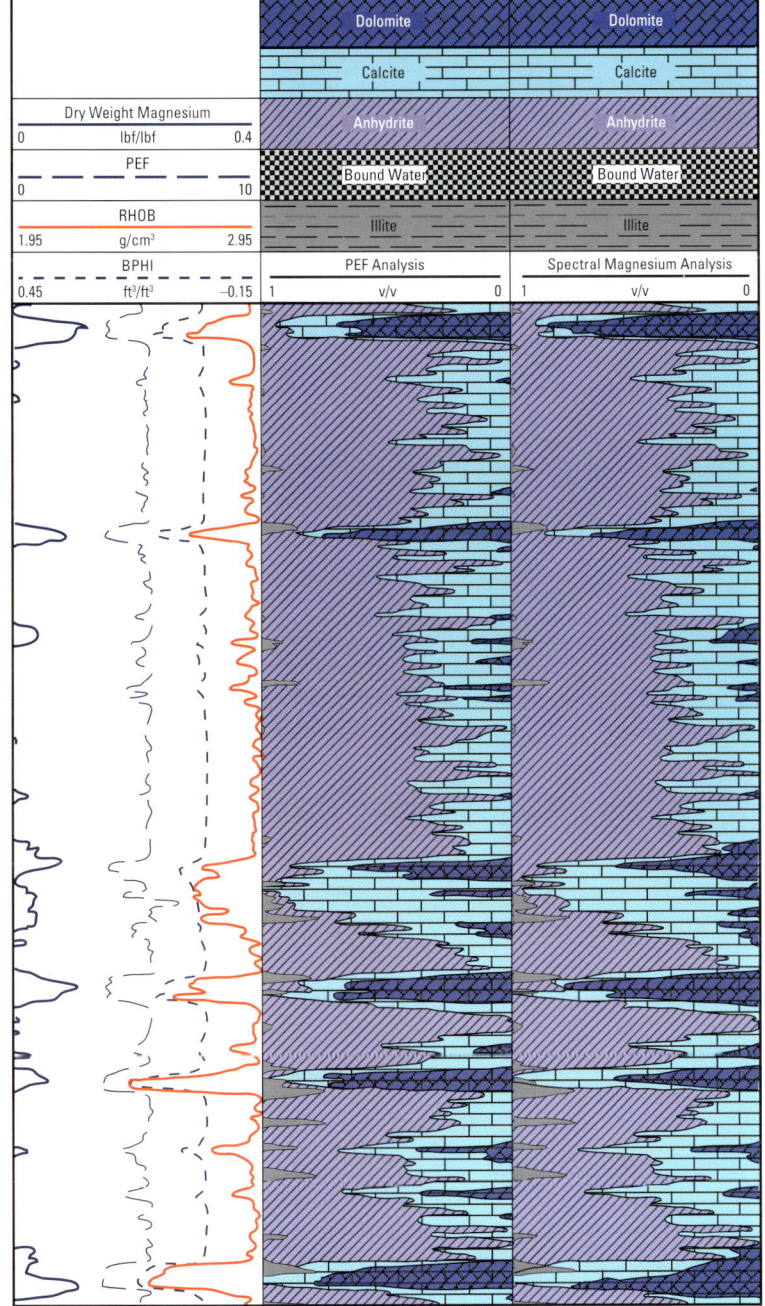

Figure 5-25. Comparison of formation lithology, derived using the PEF (left) and spectroscopy (right). This example is from the Middle East.

Figure 5-26 demonstrates the feasibility of formation evaluation without chemical nuclear sources. The PNG has already replaced the chemical neutron source for the thermal neutron porosity (TNPH) and HI-based best thermal neutron porosity (BPHI) measurements. The agreement between the conventional gamma-gamma density (RHOB, red curve, Track 5) and the NGD (RHON, black curve, Track 5) derived from the PNG and associated detectors indicates that the remaining chemical source for the conventional density can be removed and formation evaluation performed without the need for chemical nuclear sources.

Figure 5-26. Agreement between the NGD and the conventional gamma-gamma density measurement, demonstrating the feasibility of formation evaluation without chemical nuclear sources. This example is from deepwater Nigeria.

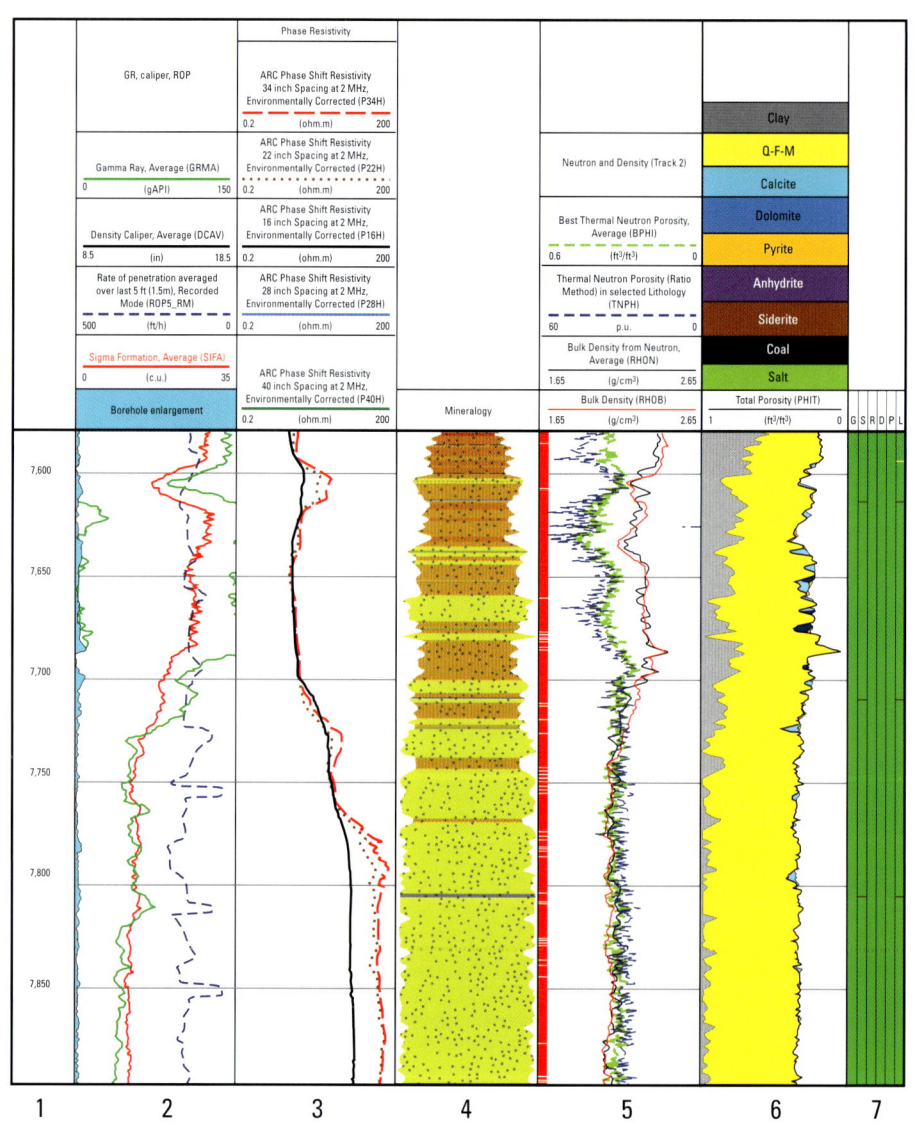

5.5.4. Formation density evaluation while sliding or in poor hole condition

The conventional gamma-gamma density measurement requires good contact between the formation and the tool. If the standoff between the detector windows and the formation becomes greater than can be compensated by the spine-and-ribs technique, the conventional density may become unusable. Such a situation may occur in enlarged or rugose boreholes or while sliding when drilling with a motor if the detectors are not correctly oriented to face the borehole wall.

The sourceless NGD is almost insensitive to borehole contact because the high-energy neutrons and gamma rays involved in the measurement readily penetrate the borehole and collar wall. In contrast, the low-energy gamma rays in the conventional gamma-gamma density measurement require special low-density windows in the collar that must be facing the formation to allow the measurement gamma rays from the formation into the detectors.

Figure 5-27 compares the response of the NGD (green curve, right track) with the gamma-gamma density (red curve) in both gauge hole (0960 to 1,110 ft) and overgauge hole (approximately 0950 ft).

While showing agreement when the hole is in gauge, the conventional density measurement is severely degraded by poor detector window contact with the borehole wall caused by the hole enlargement at approximately 0950 ft. The NGD is significantly less affected because (1) it does not require windows or orientation toward the formation and (2) it is a much deeper measurement than the conventional gamma-gamma density.

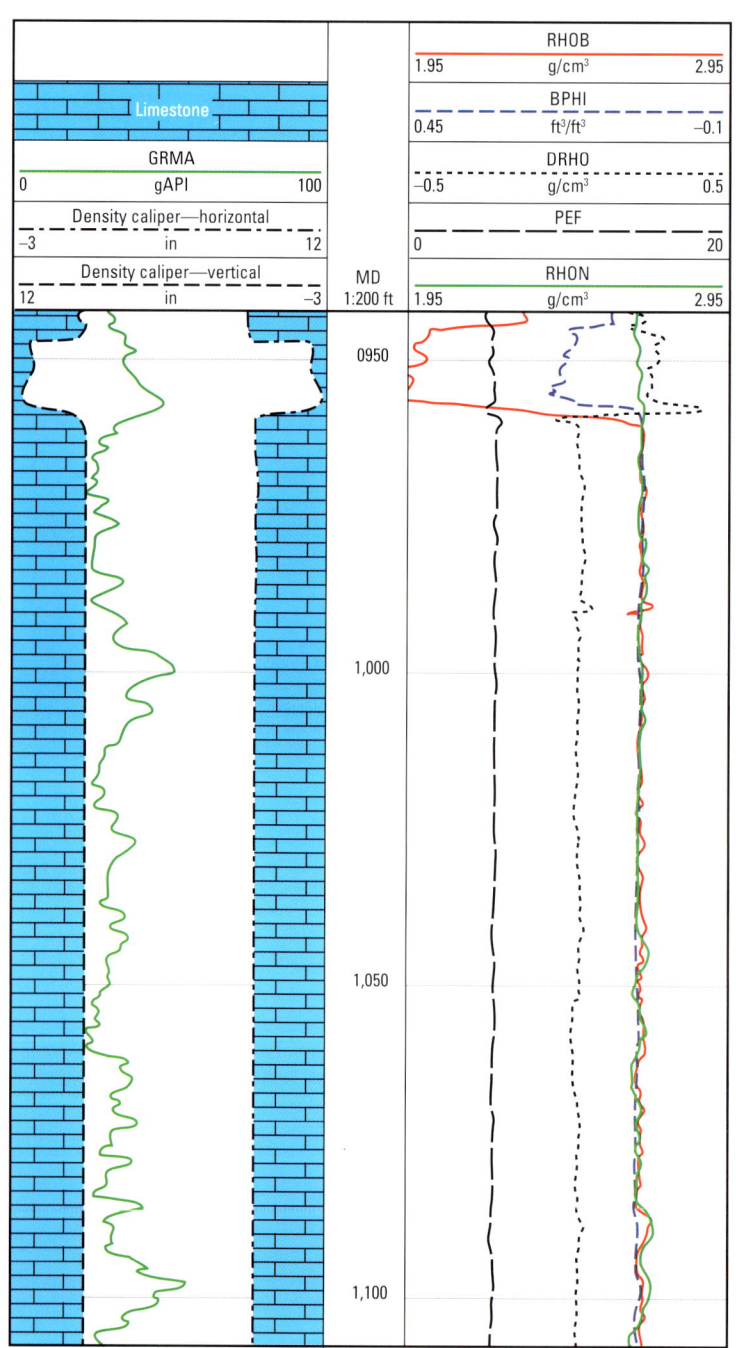

Figure 5-27. Overgauge hole at approximately 0950 ft, demonstrating the relative insensitivity of the NGD to hole condition and tool orientation. This example is from the Middle East.

5.5.5. Clay volume evaluation in radioactive formations

The term shale refers to fine-grained, fissile, detrital sedimentary rock formed by consolidation of clay- and silt-sized particles. Clay minerals (as opposed to clay-sized particles) are generally alumino-silcates. Shale may contain clay minerals and clay-sized particles that may not be clay minerals. Because potassium (K) and thorium (Th), two of the three naturally occurring radioactive elements, are commonly associated with silicate minerals including feldspars, micas and clays, the gamma ray measurement has traditionally been used to approximate the the proportion of shale in a formation. A high gamma ray reading was assumed to indicate high shale content while a low gamma ray reading indicated a clean formation. The "clean" and "shale" thresholds were generally open to significant interpretational uncertainty.

The third naturally occurring radioactive element, uranium (U), commonly exists as water-soluble salts that can be dissolved and transported by water moving through the formation. When these uranium salts deposit in otherwise clean formations, the associated high gamma ray reading is easily misinterpreted as indicating the presence of shale in the formation. Where this occurs, the assumed link between high gamma ray response and shale content is invalid, resulting in inaccurate shale volume estimations if the gamma ray approximation is applied.

Spectroscopy measurements permit evaluation of the formation clay content through elemental composition analysis, resulting in a more direct and accurate assessment of the proportion of alumino-silcates (the clay minerals). The associated silt-sized material is often composed of silicate minerals, which appear on the spectroscopy analysis under the QFM group.

Figure 5-28 shows an example where the gamma ray correlates with the spectroscopy-derived clay content in the interval below 000 ft, but suggests the presence of more clay than determined by the spectroscopy in the upper interval. The density-neutron crossover supports the spectroscopy-derived clay volume, showing a clear shale separation in the lower interval, but overlying in the upper interval, suggesting that the formation is relatively clean despite the high gamma ray reading.

Note that spectroscopy delivers clay volume, not shale volume. The use of spectroscopy avoids overestimation of the clay volume and underestimation of the porosity and hydrocarbon potential of the reservoir, as would be the case if the gamma ray technique were used.

Figure 5-28. An example of a high gamma ray reading (upper section) which does not correlate with the formation clay content. Below 000 ft the gamma ray correlates with the clay content. Spectroscopy provides more reliable clay content evaluation than the traditional GR correlation.

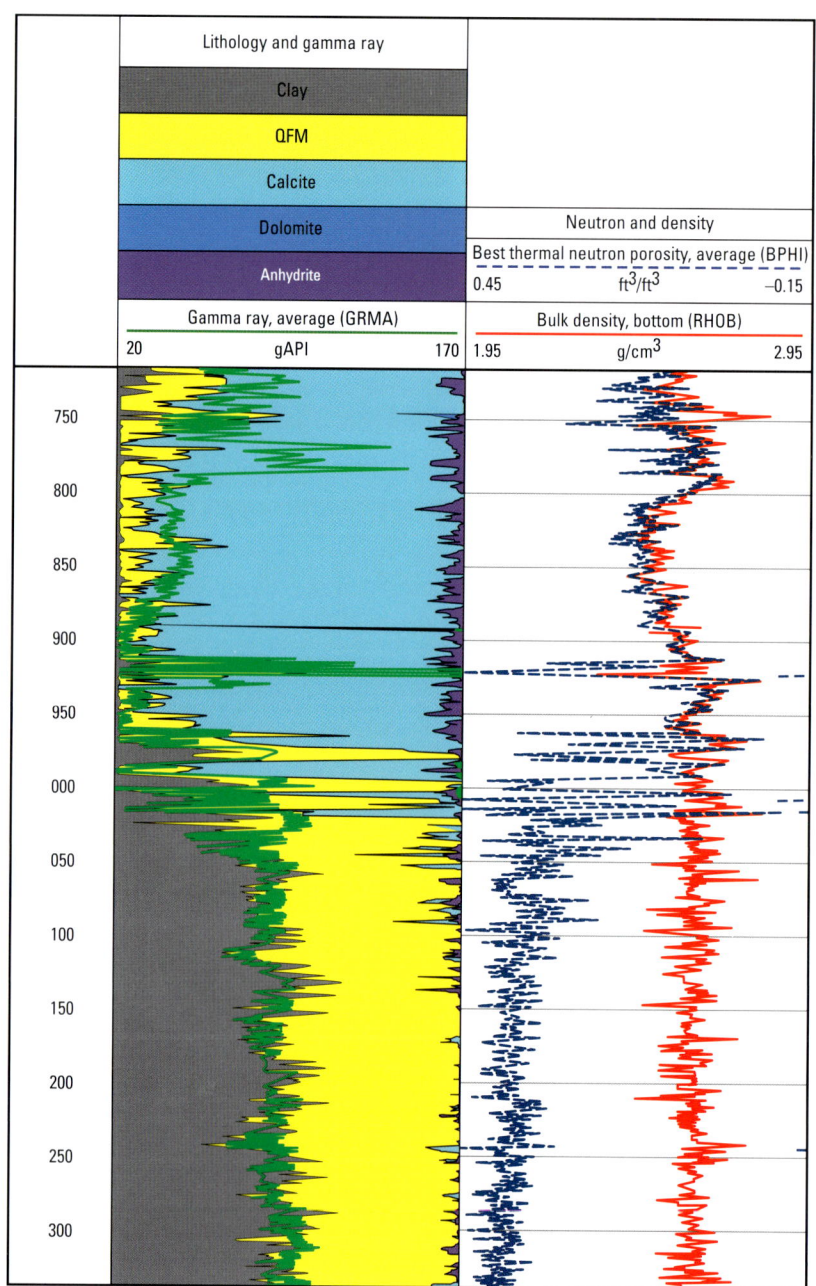

5.5.6. Water saturation evaluation in high-angle wells

As wells logged with LWD tools are typically either at high angles or are horizontal, one of the most common complications encountered in LWD formation evaluation is shoulder bed effect on the propagation resistivity measurements. In a well drilled perpendicular to the formation layering, as shown in **Figure 5-29**, the measure current loops traverse both layers in parallel so that the resistivity measured is the parallel sum of the layer resistivities.

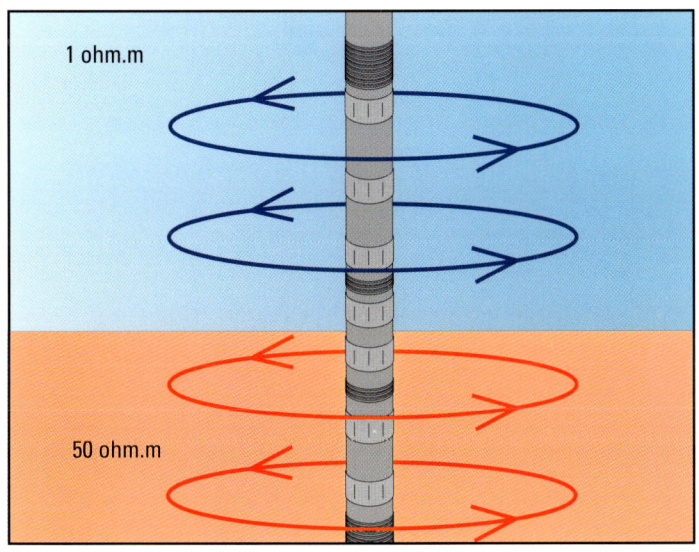

Figure 5-29. Propagation resistivity tool, perpendicular to layering. The resistivity response is the parallel sum of the layer resistivities. In this example, the measured resistivity would be 1.96 ohm.m.

In wells drilled closer to parallel to the formation layering, the measure currents may encounter more than a single layer in their DOI (**Figure 5-30a**). This leads to electromagnetic averaging of the layer resistivities and, if the difference in layer resistivities is sufficient, charge buildup at the interface between the layers, resulting in polarization horns on the resistivity responses (**Figure 5-30b**). When the elevated resistivities due to polarization are applied to Archie's equation, they result in misleadingly high hydrocarbon saturation. The sigma measurement responds volumetrically and does not suffer from polarization effects, resulting in more accurate water saturation evaluation (**Figure 5-30c**).

Figure 5-30a. LWD propagation measure currents, creating charge buildup at the interface between layers of different resistivity in high-angle and horizontal wells.

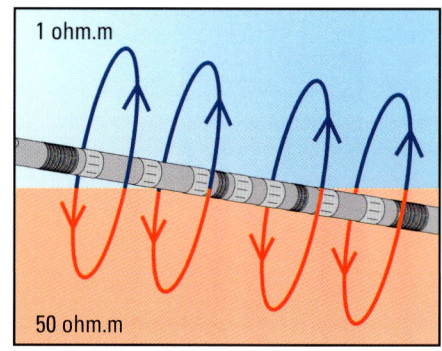

Figure 5-30b. A propagation resistivity polarization horn, caused by oscillating charges interfering with the tool transmitter signal as the propagation resistivity tool passes through the interface.

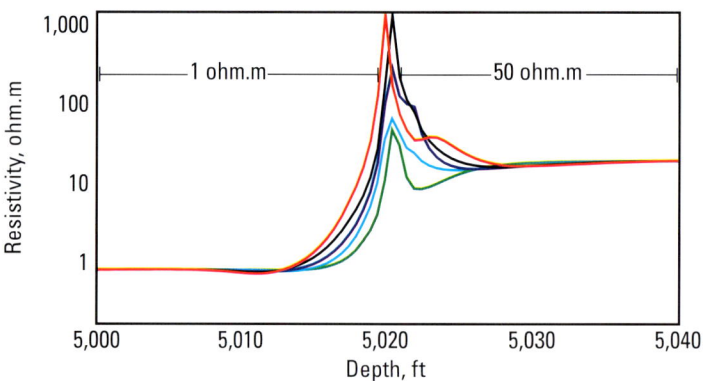

Figure 5-30c. EcoScope sigma measurement, showing no polarization effect. Because the sigma measurement is volumetric, it provides accurate saturation evaluation, even in laminated intervals.

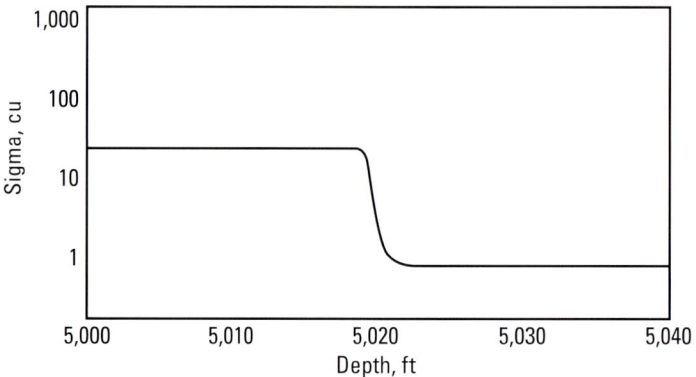

Figure 5-31 shows a well that deviates from vertical to horizontal. In the well's upper, near-vertical interval, the saturations (Track 2) derived from resistivity (green line) and from sigma (black line) agree well because there is no polarization effect on the resistivity. In the well's lower, high-angle section, polarization effects cause high resistivity readings that mimic the presence of hydrocarbons. The saturation derived from the EcoScope sigma response does not suffer from these effects and accurately indicates that minimal hydrocarbon is present. Dark, curved sections on the density image (Track 3) show the location of the layers that are causing the resistivity polarization.

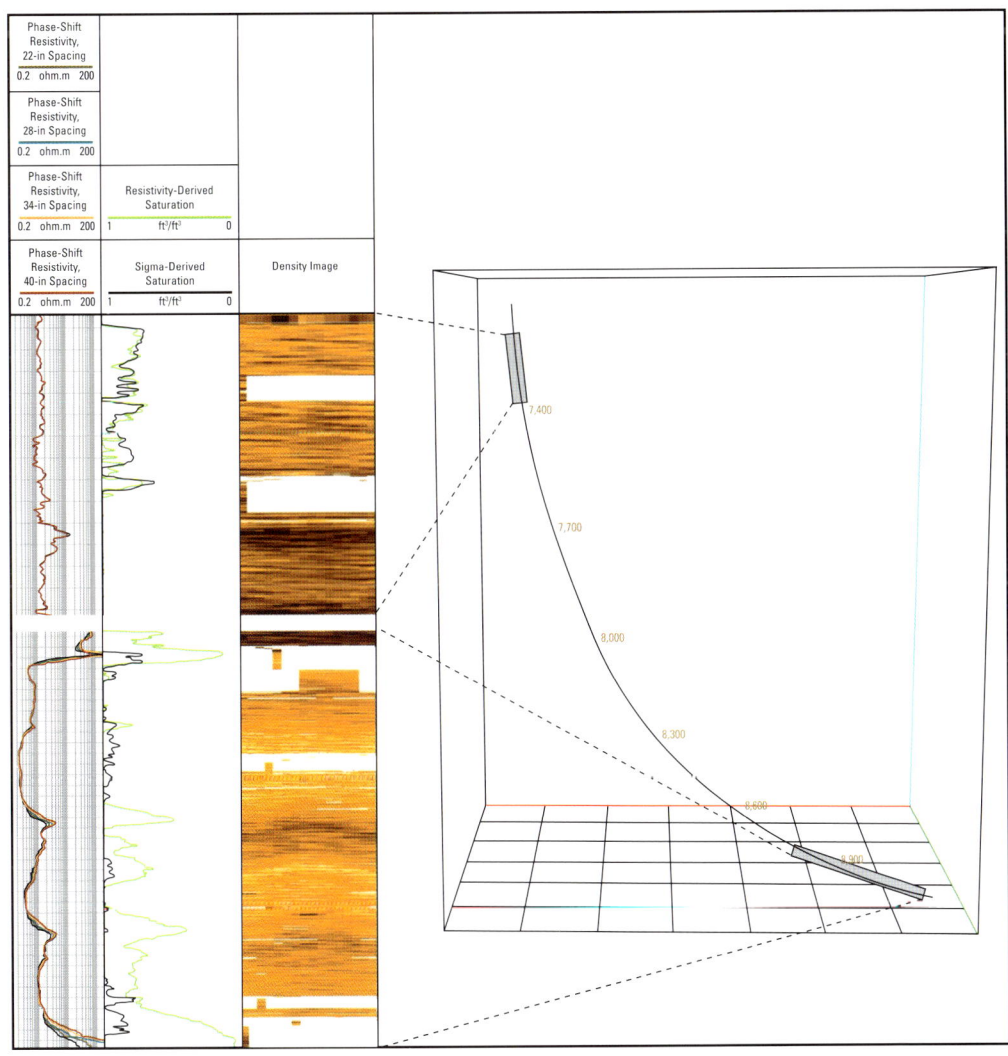

Figure 5-31. Polarization effects, becoming more pronounced as the well deviates closer to parallel with the formation layering.

There are numerous examples of intervals that have been interpreted as hydrocarbon-bearing based on the resistivity response that subsequently produce only water when put on production. In many cases, this is because of shoulder effects on the resistivity measurements that have not been accounted for during the interpretation.

For the sigma measurement to yield valid saturations, the acquisition must be made in a formation with sufficiently saline water and sufficient porosity and before significant mud filtrate invasion of the formation. Details of the water salinity and porosity required can be found in Section 4.11.3.

5.5.7. Fluid saturation in anisotropic formations

5.5.7.1. Resistivity solution in a bimodal formation

Formation resistivity anisotropy is often caused by thin layers of differing resistivities that result in different measured resistivities depending on the direction that the measure currents traverse the layering (**Figure 5-32**).

Figure 5-32. Propagation resistivity array of an EcoScope tool oriented perpendicular to formation layering, measuring volume-weighted parallel resistivity (R_h, left), and a tool parallel with the layering, measuring a combination of volume-weighted parallel resistivity and volume-weighted series resistivity (R_h and R_v, right).

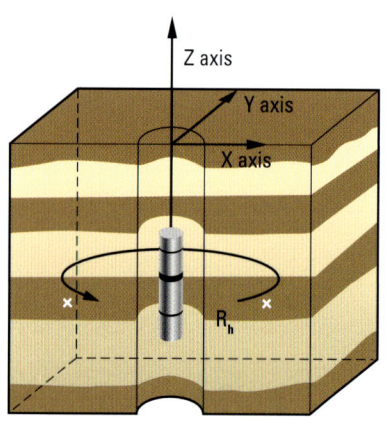
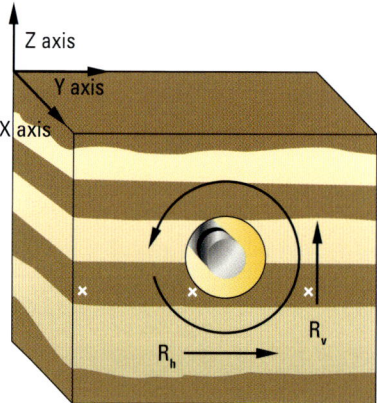

Perpendicular to layers; propagation tools measure R_h

Parallel to layers; propagation tools measure a combination of R_h and R_v

For well deviations parallel to the formation layering within approximately 40°, the EcoScope propagation resistivity measurements can be used to determine the volume-weighted parallel resistivity and volume-weighted series resistivity of the formation. Details of the propagation resistivity response to an anisotropic formation can be found in Section 4.3.2.5.

To transform the anisotropic resistivities into water saturation, a model of the formation layering must be applied. The simplest model assumes a bimodal sequence of isotropic layers (Figure 5-33 left). In this simplest case, there are four unknowns, including

* isotropic resistivity and volumetric proportion of the lower-resistivity layer

* isotropic resistivity and volumetric proportion of the higher-resistivity layer.

The volumetric proportions of the two layers must sum 1 (**Figure 5-33**, right).

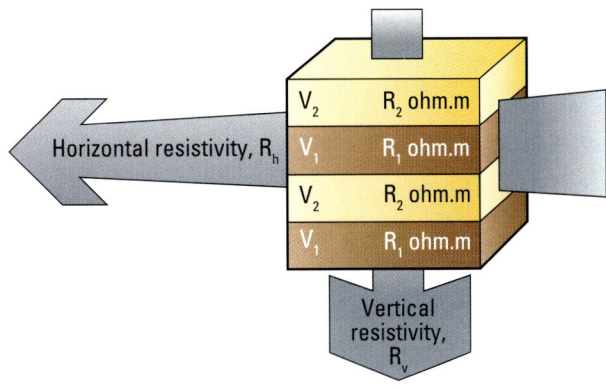

Figure 5-33. A bimodal model consisting of a repeating sequence of two isotropic layers, generally assumed in transforming the measured R_h and R_v into the individual layer resistivities, R_1 and R_2, using Equations 5-9a, b, and c.

The isotropic resitivities of the lower and higher resistivity layers (R_1 and R_2, respectively) are transformed by

$$\frac{1}{R_h} = \frac{V_1}{R_1} + \frac{V_2}{R_2}$$

(Equation 5-9a),

where

V_1 = volumetric proportion of the lower-resistivity layer in the volume of measurement
V_2 = volumetric proportion of the higher-resistivity layer in the volume of measurement

$$R_v = V_1 \times R_1 + V_2 \times R_2$$

(Equation 5-9b).

$$V_1 + V_2 = 1$$

(Equation 5-9c).

Equations 5-9a, b, and c describe the three known parameters, but the bimodal model has four unknowns. Consequently, information about one of the layer proportions or resistivities must be obtained from another source. The EcoScope system provides three possible sources of additional information to solve this problem.

If the layers have differing lithologies—alternating clean sand and clay, for example—then spectroscopy can be used to derive the volumetric proportion of each in the formation. The corresponding isotropic layer resistivities can then be calculated.

If the layers are sufficiently thick to be seen on one of the EcoScope images in a high-angle well, then the volumetric proportions can be calculated from the image (**Figure 5-34**). Because layers are detected over several feet of MD when encountered by a high-angle well, layers too thin to be detected in a vertical well may be visible in a high-angle well. For example, a horizontal, 1-in-thick layer is detected over almost 6 ft of MD in a well inclined at 89°.

Figure 5-34. Layers too thin to be identified in a vertical well. These may be visible in a high-angle well because they are encountered by the wellbore over several feet of MD.

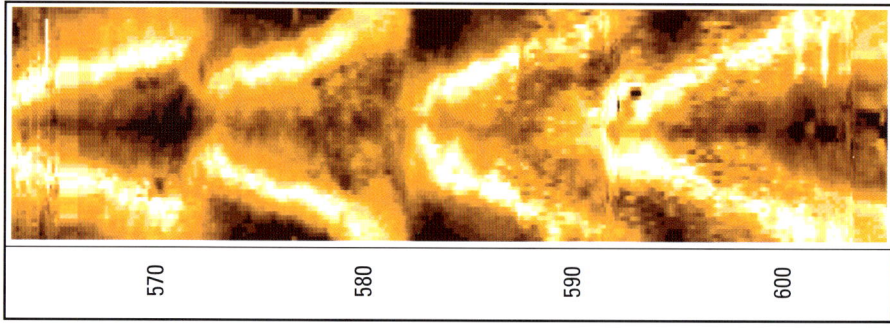

If the resistivity of one of the layers is known, such as when the shale in a sand-shale sequence has a consistent resistivity that can be identified in a thicker shale layer, then the system of equations can be solved for the resistivity of the second layer and the volumetric proportions of each using

$$R_2 = R_h \frac{R_v - R_1}{R_h - R_1}$$ (Equation 5-10a)

$$V_2 = \frac{R_v - R_1}{R_2 - R_1}$$ (Equation 5-10b)

$$V_1 = 1 - V_2$$ (Equation 5-10c).

After deriving the isotropic resistivities of the two layers and their volumetric proportions, S_w is calculated for each using Archie's equation or similar. This requires that the layer porosity be known.

If the separate layer densities can be identified, individual layer density porosities can be calculated.

If the neutron response of one of the layers is known, such as a consistent shale neutron signature in a sand-shale sequence, the neutron response of the other layer can be calculated based on the volumetric neutron response equation

$$\phi_n = V_1 \times \phi_{n1} + V_2 \times \phi_{n2}$$ (Equation 5-11),

where

ϕ_n = measured neutron response

ϕ_{n1} = neutron response of the lower-resistivity layer

ϕ_{n2} = neutron response of the higher-resistivity layer.

Hence,

$$\phi_{n2} = (\phi_n - V_1 \times \phi_{n1}) / V_2$$ (Equation 5-12).

In the absence of any additional information enabling the individual layer porosity to be determined, the common practice is to assume that the two layers have the same porosity.

To calculate total water saturation of the formation, the resistivity-derived saturation for each of the layers is proportioned according to the porosity and volume of each layer:

$$S_{wt} = (S_{w1} \times \phi_1 \times V_1 + S_{w2} \times \phi_2 \times V_2) / (\phi_1 \times V_1 + \phi_2 \times V_2)$$ (Equation 5-13),

where

S_{wt} = formation total water saturation

S_{w1} = water saturation in the lower resistivity layer

ϕ_1 = porosity of the lower-resistivity layer

S_{w2} = water saturation in the higher-resistivity layer

ϕ_2 = porosity of the higher-resistivity layer.

The bimodal formation model solution to water saturation evaluation in resistivity-anisotropic formations is built on the assumptions that

- a bimodal model is a good representation of the formation layering. If the formation has three types of layers, for example, then there are at least six unknowns, requiring that additional information about three of the formation parameters be acquired from elsewhere (a bimodal model applied in a trimodal formation will yield inaccurate results)

- the validity of the volumetric partitioning or the validity that one of the layer resistivities is known and unchanging

- the layers are resistivity-isotropic

- the porosity of the individual layers can be accurately extracted.

When evaluating water saturation in a resistivity-anisotropic formation, care must be taken to validate the interpretive assumptions applied.

5.5.7.2. Volumetric sigma solution

In formations with sufficient porosity and water salinity, the resistivity-independent saturation evaluation available with a preinvasion sigma measurement circumvents many of the complications of resistivity evaluation in anisotropic formations.

Because sigma is volumetric, it responds to water saturation and porosity in each of the layers according to their volumetric proportion, delivering the volumetric water saturation equivalent to that derived from Equation 5-13 without the application of formation layering, resistivity, or porosity assumptions.

5.5.8. Low-resistivity pay evaluation

LRP in sand-shale sequences is often associated with fine layering, which creates an anisotropic resistivity environment. Two possible solutions for formation evaluation in anisotropic formations are outlined in Section 5.5.7.

Resistivity anisotropy is not the only cause of LRP. For example, some carbonates exhibit LRP behavior without any measurable resistivity anisotropy or visible layering. **Figure 5-35** shows a photograph of a carbonate core from a Middle East reservoir that exhibits LRP behavior. A wide range of pore sizes can be seen in close proximity to each other.

Figure 5-35. Carbonate reservoir core section, showing the wide spectrum of pore sizes and the lack of clear depositional structure found in some carbonates.

The pore size spectrum is generally divided into three loosely defined groups: macropores for the largest, mesopores for the intermediate pore sizes, and micropores for the smallest pores, as shown schematically in **Figure 5-36**.

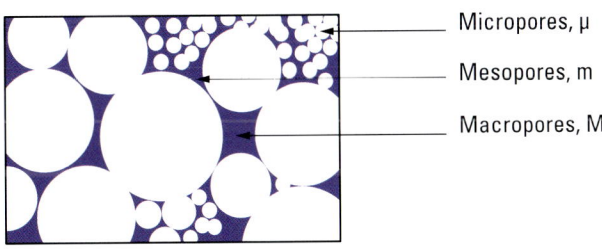

Micropores, μ

Mesopores, m

Macropores, M

Figure 5-36. Schematic of a water-filled (blue) carbonate, showing a range of pore sizes.

Unlike clastic rocks where the microporous intervals (e.g., silt) and macroporous (e.g., sandstone) are often found in distinct layers, carbonates generally do not have such an ordered structure because of the nature of their deposition and subsequent diagenesis. Despite the lack of laminations, path-of-least-resistance effects are still present in carbonates. As shown in **Figure 5-37**, the presence of a continuous path of water-filled micropores in close proximity to the larger hydrocarbon-filled pores creates paths of least resistance, which short the resistivity measure currents. When the resulting low measured resistivity is applied to Archie's equation, the hydrocarbon saturation is underestimated. When put on production, the hydrocarbon in the large pores tends to flow. To explain hydrocarbon production from isotropic low-resistivity zones, as observed in these types of carbonates, requires a nonresistivity evaluation solution providing accurate saturation evaluation.

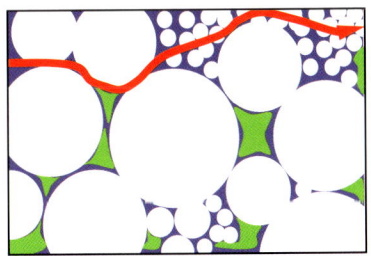

Figure 5-37. The presence of water-filled (blue) micropores, in close proximity with larger pores filled with hydrocarbon (green). The water-filled micropores create paths of least resistance, which short the measure currents (red line) around the hydrocarbon-bearing pores.

In favorable conditions (porous formations with saline water), the sigma measurement acquired prior to significant invasion with the EcoScope tool can be used to perform resistivity-independent saturation evaluation. The volumetric nature of sigma means that it does not suffer from the path-of-least-resistance effects that complicate the resistivity response.

Figure 5-38 compares the resistivity- and sigma-derived saturations in a Middle East reservoir sequence. Several different combinations of n and m (Tracks 1 and 2) are used to compute water saturation from Archie's equation. These are compared to the linear sigma solution for water saturation (black line). Total porosity, along with hydrocarbon volumes from both resistivity and sigma solutions, is shown in Track 3, while lithology and fluid volumes are presented in Tracks 4 and 5.

In the upper panel, the sigma saturation and the resistivity saturations agree for $m = n = 2$. In this gas zone, the low viscosity of the hydrocarbon and the high buoyancy pressure appear to have disrupted the majority of the conductive microporous paths for the measure current, resulting in a formation that obeys the traditional Archie equation.

The middle panel has been segmented to display only undisturbed reservoir zones. The sigma and resistivity saturations agree for $m = n = 2$ at the top of the oil zone because of the high buoyancy pressure injecting oil into the smaller pores, breaking up the conductive paths that would otherwise allow the measure current to bypass the more resistive oil-filled large pores.

Toward the bottom of the middle panel, the discrepancy between the resistivity- and sigma-derived saturations increases. The lower buoyancy pressure on the oil results in higher water saturation. The patches of conductive water-filled smaller pores become electrically continuous, presenting a path of least resistance to the measure currents. Consequently, the measured formation resistivity in the lower part of the reservoir is less than Archie's equation would predict. Sigma delivers a volumetric saturation evaluation unaffected by path-of-least-resistance effects.

The lower panel shows two LRP zones. The wide spread of responses seen in Track 1 demonstrates the considerable sensitivity to m. Track 2 shows a similar sensitivity to n. In both zones, the standard values of m and n do not give a saturation match. The volume of hydrocarbon indicated in Track 3 by sigma is significantly higher than that from a resistivity analysis. When production was tested, both zones flowed high percentages of oil.

Resistivity-independent saturation evaluation using preinvasion sigma data from the EcoScope tool eliminates saturation uncertainty associated with resistivity effects and Archie parameter selection, uncertainties that are particularly significant in LRP intervals.

The following tracks are shown in Figure 5-38.

- Tracks 1 and 2—several different combinations of m and n respectively are used to compute water saturation from Archie's equation. These are compared with the linear sigma solution for water saturation.

- Track 3—total porosity, along with hydrocarbon volumes from both resistivity and sigma solutions.

- Tracks 4 and 5—lithology and fluid volumes.

Despite agreeing toward the top of the reservoir, the resistivity answer (Track 4) was pessimistic, while the sigma-based water saturation (Track 5) more accurately predicted the observed production from the zone.

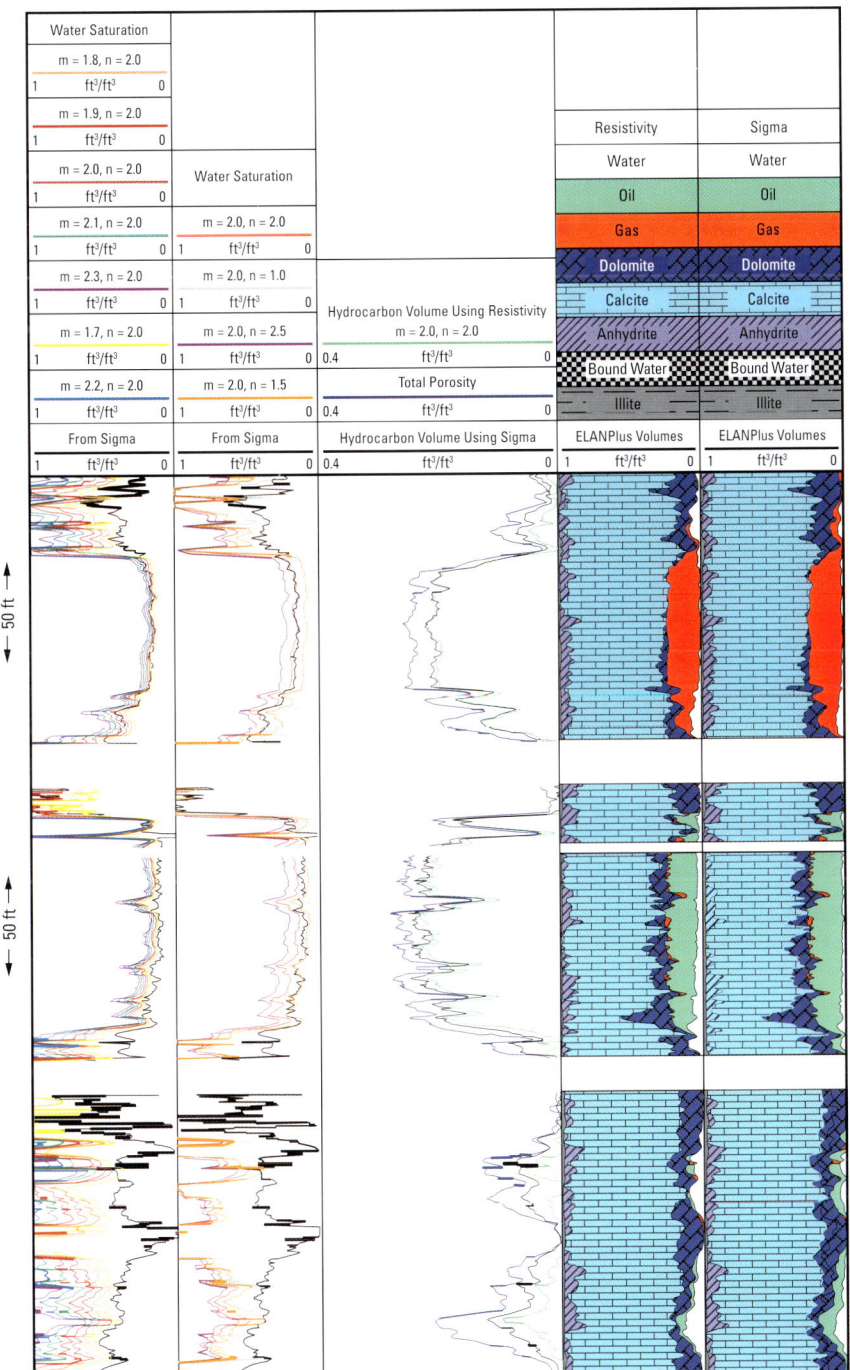

Figure 5-38. Finding productive LRP using sigma.

5.5.9. Solving for Archie's parameters m and n

Porosity and resistivity measurements are generally used in conjunction with a water resistivity to derive a fluid saturation estimate through the Archie equation or equivalent. This requires that the Archie parameters are known. Similarly, porosity and sigma are generally used in conjunction with a water sigma value to derive a fluid saturation estimate. These independent saturation techniques are shown in the left two columns of **Table 5-1**.

Table 5-1. Inputs and outputs from the Archie and sigma equations, with the Archie saturation set equal to the sigma saturation to derive an in situ variable m.

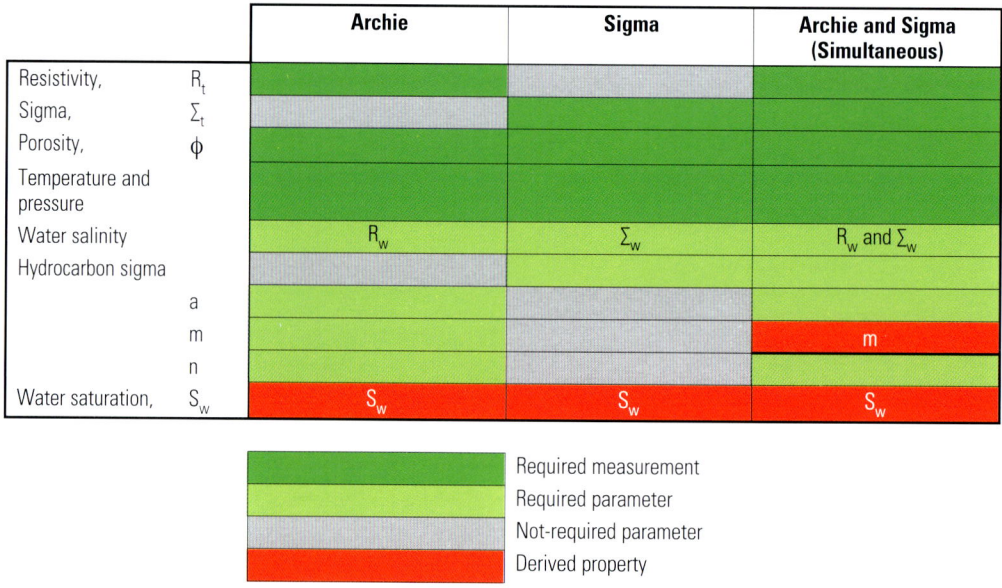

		Archie	Sigma	Archie and Sigma (Simultaneous)
Resistivity,	R_t			
Sigma,	Σ_t			
Porosity,	ϕ			
Temperature and pressure				
Water salinity		R_w	Σ_w	R_w and Σ_w
Hydrocarbon sigma				
	a			
	m			m
	n			
Water saturation,	S_w	S_w	S_w	S_w

Required measurement
Required parameter
Not-required parameter
Derived property

Where there is significant uncertainty in one of the Archie parameters m or n, the resistivity-derived saturation equation can be used to find the m or n value that delivers the same saturation as the sigma evaluation. The right column of Table 5-1 shows the additional information from the sigma saturation being used to solve for variable m. The same methodology could be used to solve for variable n or both by linking the two variables, such as by setting m = n.

Figure 5-39 shows the same intervals as Figure 5-38. The Archie parameters m and n are set equal to each other (i.e., m = n), and the value derived for which the resistivity-derived water saturation equals the sigma-derived water saturation. In the gas zone the m = n value is close to the classic value of 2, indicating that this interval is behaving in the simplest expression of Archie's equation. As explained in the previous section, this is because the low viscosity gas enters the smaller pores and disrupts the paths of least resistance, resulting in tortuous electrical measurement current paths.

At the top of the oil interval the m = n value is close to 2, but it decreases down the oil column as water saturation increases, creating paths of least resistance and lowered measurement current tortuosity. Similarly, low m = n values are observed in the LRP zones.

The information available from preinvasion sigma allows an additional unknown to be derived. In this case, m has been set equal to n. In solving for the single parameter, trends are observed consistent with the fluid and pore interactions known to cause difficulties with conventional saturation estimation in these formations.

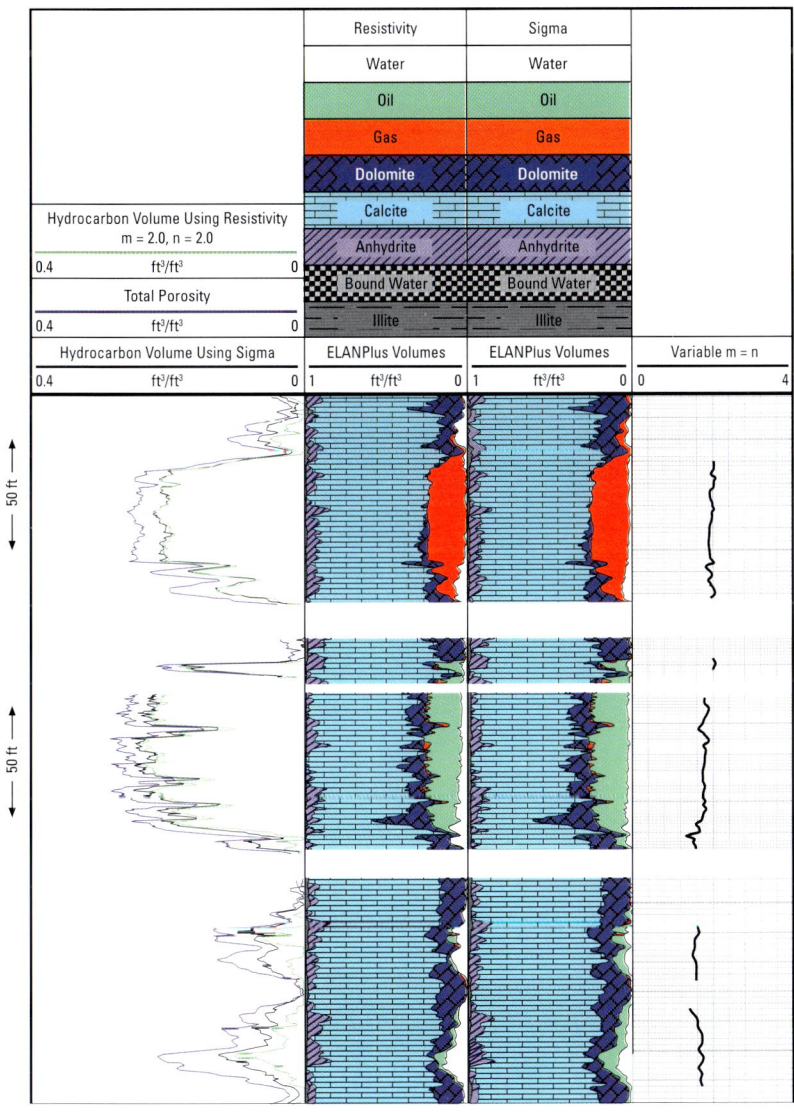

Figure 5-39. Solving for Archie's parameters, using the additional information available from sigma.

5.5.10. Unknown water salinity

With increasing use of water injection for reservoir pressure support and hydrocarbon displacement, the problem of unknown water salinity because of mixing of injection and connate waters is becoming more common.

Where the salinity of the formation water is unknown, neither resistivity nor sigma measurements alone can be used to determine the water saturation. However, a combination of these two measurements allows both water resistivity and saturation to be computed simultaneously.

Both water resistivity and sigma are primarily controlled by the water salinity. Where the water salinity is known, resistivity- and sigma-derived saturations can be calculated independently (**Table 5-2**, two left columns). Where the salinity is unknown, they can be used together to solve for the water salinity in addition to the saturation (Table 5-2, right column). The availability of all the required measurements in the EcoScope tool permits in situ evaluation of water saturation and salinity.

Table 5-2. Inputs and outputs from the Archie and sigma equations, with simultaneous solution of the two equations allowing evaluation of in situ water salinity and saturation.

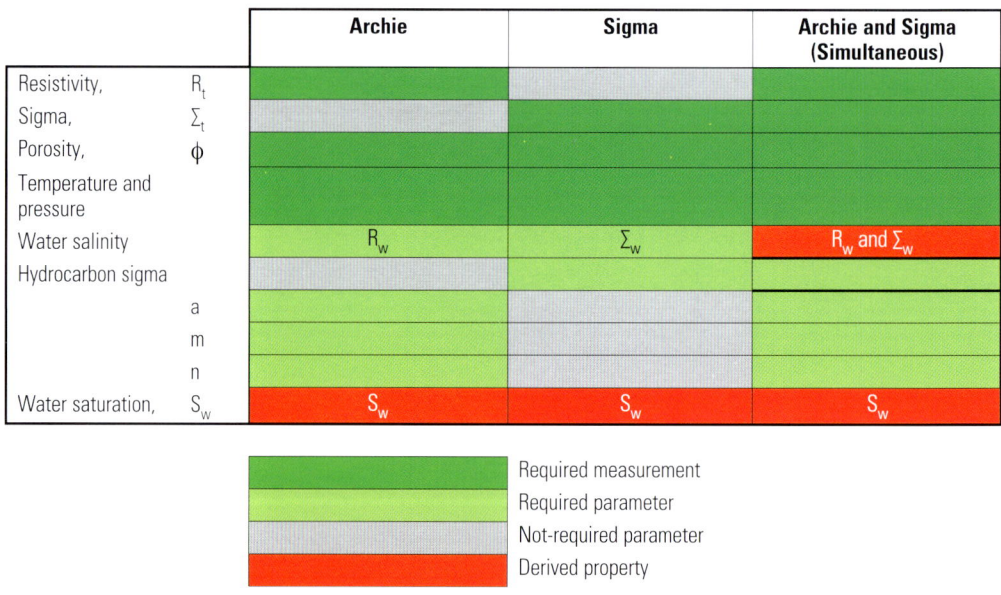

		Archie	Sigma	Archie and Sigma (Simultaneous)
Resistivity,	R_t			
Sigma,	Σ_t			
Porosity,	ϕ			
Temperature and pressure				
Water salinity		R_w	Σ_w	R_w and Σ_w
Hydrocarbon sigma				
	a			
	m			
	n			
Water saturation,	S_w	S_w	S_w	S_w

- Required measurement
- Required parameter
- Not-required parameter
- Derived property

The collocated resistivity and sigma measurements provided by the EcoScope service ensure that the measurements used for these independent saturation evaluations are made at the same depth in the borehole, at the same time, and under the same conditions. Consequently, they can be compared with each other with the confidence that all environmental conditions are identical.

In circumstances in which there is confidence in Archie's parameters a, m, and n, the saturation derived from resistivity and sigma should match if the water salinity (and hence, water resistivity and water sigma) used is correct. If the salinity used is incorrect, the effect on the sigma-derived saturation will be greater than that on the resistivity-derived saturation.

Figure 5-40 shows resistivity- and sigma-derived saturations in a zone under active seawater injection by a nearby well. Both saturations show lower water saturation than expected. The original water salinity was 230 ppk, but with mixing of the seawater, the in situ salinity is expected to be lower.

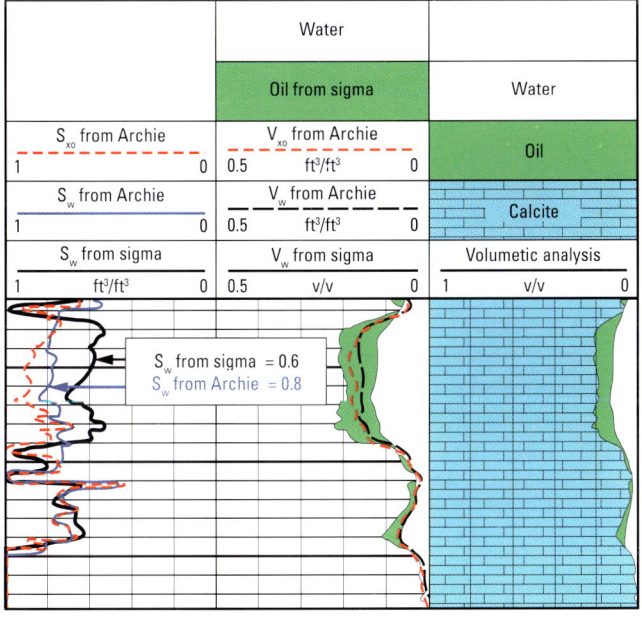

V_{xo} volumetric proportion of water in the invaded zone
V_w volumetric proportion of water in the uninvaded zone

Figure 5-40. Resistivity and sigma-derived saturations, showing unexpectedly low water saturation when assuming that the water salinity remains 230 ppk in a zone under active water injection by a nearby well.

If the in situ water salinity were 230 ppk, the resistivity- and sigma-derived saturations should match. The reduced in situ water salinity due to seawater mixing results in a higher water resistivity and thus, a higher measured formation resistivity, mimicking lower water saturation. Reduced salinity also reduces the water sigma and thus, the measured formation sigma, also mimicking lower water saturation. However, the effect is more pronounced on the sigma response, as shown in **Figure 5-41**.

Figure 5-41. Crossplot of apparent Archie water saturation versus apparent sigma water saturation for a clean 20-pu limestone. The crossplot shows the simultaneous solution of apparent resistivity- and sigma-derived saturations to derive both the in situ water salinity and true water saturation.

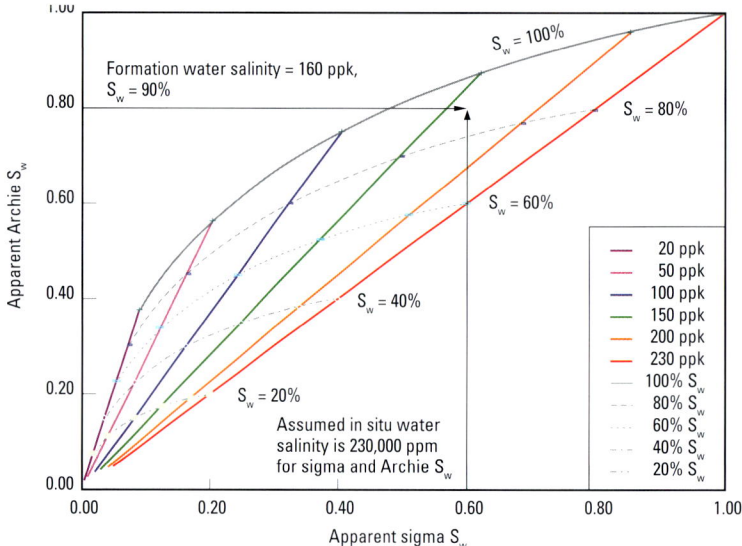

This crossplot, valid for a 20-pu, oil-filled limestone, shows that the resistivity- and sigma-derived saturations will match if the in situ water salinity is 230 ppk (red diagonal line). If the water salinity drops to 200 ppk (orange line), even at 100% water saturation (solid gray curve), the resistivity-derived apparent water saturation will drop to approximately 95% if the original water salinity of 230 ppk is used in the calculation. In this same situation, the sigma-derived apparent saturation will drop to approximately 85% as the impact of a salinity change is greater on the sigma measurement than on the resistivity. The difference in sensitivity to the water salinity allows the two measurements to be used to solve for the in situ water salinity in addition to the true water saturation.

The solution to the differing saturations observed in Figure 5-40 is indicated by the black arrows on Figure 5-41. Crossplotting the resistivity-derived saturation of 80% with the sigma-derived saturation of 60% yields the simultaneous solutions that the in situ water salinity is 190 ppk and the actual water saturation is 90%. This is consistent with 10% residual oil after seawater sweeps the layer. The same technique can be used to determine sweep efficiency by evaluating the residual oil saturation after injection.

The EcoView interpretation platform offers interactive salinity selection, along with automatically updated resistivity- and sigma-derived saturation evaluations. This facilitates salinity selection based on the saturation responses and additional salinity indications. **Figure 5-42** presents an example from the Middle East[vii] in which the water salinity was originally expected to be 250 ppk.

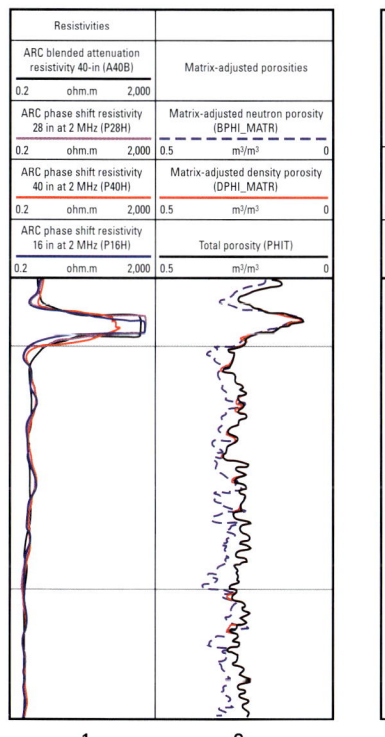
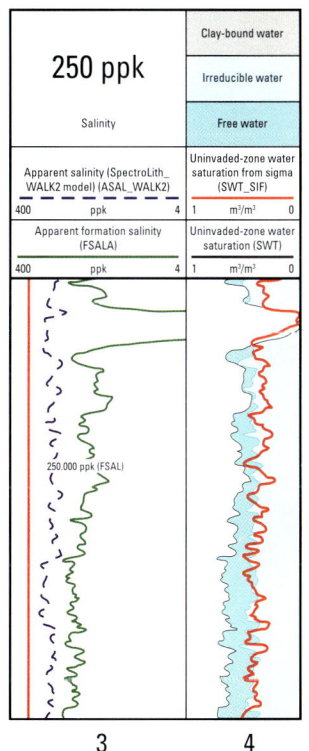
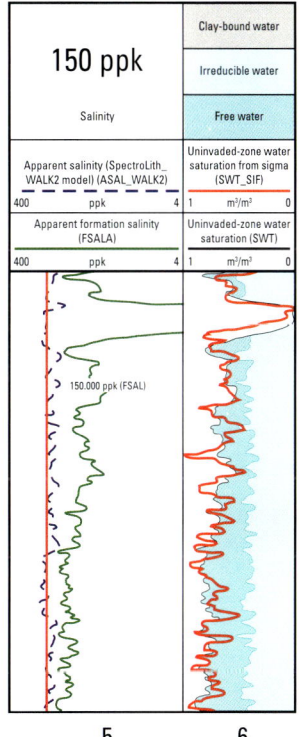

Figure 5-42. Interactive salinity selection in the EcoView interpretation package, permitting salinity indicators to be taken into account in addition to the resistivity- and sigma-derived saturation overlay.

Track 1 of Figure 5-42 displays the overlay of the propagation resistivity measurements as evidence that the sigma measurement has been acquired prior to significant invasion, a prerequisite for its application to uninvaded formation evaluation. Track 2 shows the matrix-adjusted density and neutron porosities. The neutron reading slightly higher than the density porosity is indicative of saline water in the pores of this clean formation.

When the assumed salinity of 250 ppk is applied (Tracks 3 and 4) the resistivity- and sigma-derived saturations do not overlay, nor is the selected salinity in agreement with either of the salinity indicators. Reducing the water salinity to 150 ppk produces a match between the two independently derived saturations (Track 6). A 150-ppk salinity also overlays the water salinity derived from the chlorine/hydrogen ratio derived from the spectroscopy analysis of the elements in the formation (Track 5, blue dashed line). The convergence with the independent salinity indicator and overlay of the two saturations suggest that the in situ water salinity is 150 ppk, not the originally anticipated 250 ppk.

The EcoScope system helps evaluate in situ water salinity through independent saturation measurements with differing salinity sensitivities and through spectroscopic analysis of the relative proportions of chlorine and hydrogen in the formation. All of this information is presented in the EcoView interpretation platform to facilitate consistent and crossvalidated interpretation, resulting in improved saturation evaluation.

5.5.11. Formation evaluation close to the bit

Measurements close to the bit provide critical information for real-time decisions. The entire suite of EcoScope measurements is made in the 16 ft of the tool closest to the bit.

Before drilling a vertical pilot hole, an operator in the Middle East predicted that the vertical distance between the deepest potential pay zone and the water-producing interval would be 26 ft (**Figure 5-43**). The integrated measurements of the EcoScope service were close enough to the bit to supply critical data to the driller and asset team to successfully geostop the well just 2 ft above the water layer. A full evaluation of the complex LRP carbonate reservoir was accomplished within 21 ft of total depth (TD). Currently, only the EcoScope service can make this many measurements so close to the bit.

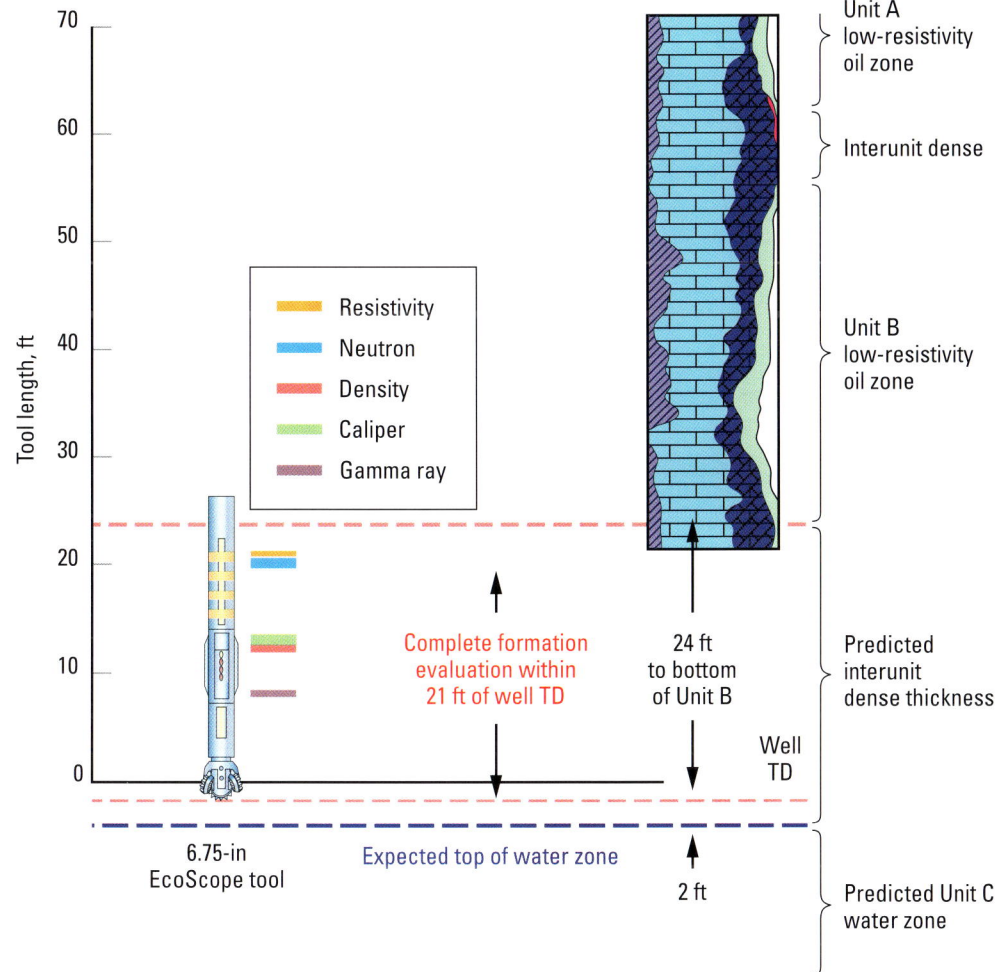

Figure 5-43. The integrated EcoScope measurements, providing complete formation evaluation close to the bit. In this Middle East case, the operator stopped drilling just 2 ft above a water-producing layer and was still able to evaluate the low-resistivity oil zone 21 ft above TD.

Appendix

Nomenclature

$\Delta\rho$	Density correction, 4.4.2.1
Σ	Sigma, 4.6.2.4, 4.7.1, 4.10.3, 4.11.3, 4.11.6, 5.3, 5.5.9, 5.5.10
A	Area through which the measurement current passes, 4.3.1.2
a	Empirically derived constant, 4.11.3, 5.2.1
α	Neutron microscopic cross section (except capture) for elements other than H, 4.7.1
β	Neutron microscopic cross section (except capture) for elements other than H, 4.7.1
β_{adn}	TNPH sensitivity factor, 4.8.1
β_{Eco}	Density sensitivity, 4.8.1
ε	Dielectric constant, 4.3.1, 4.3.2.7, 4.3.3
ε_r	Relative dielectric constant, 4.3.1, 4.3.2.7, 4.3.6.2
F	Nominalization factor that compensates for the fact that Cl, H, and the tool background are eliminated from the analysis and that the yields are relative and divided by their sensitivities, 4.10.1
G_{local}	Local gravitational field strength, 4.14.1
g_n	Acceleration due to gravity, 4.15.3
G_x	Projection of gravity acceleration along the tool axis, 4.14.1, 4.14.5.1
HI_{bulk}	Hydrogen index of the formation, 4.7.1, 4.10.3
HI_{fluid}	Hydrogen index of the fluid, 4.7.1, 4.10.3
K	Permeability, 5.4.5, 5.4.5.4
k_d	Empirically derived density caliper coefficient, 4.12.1
k_u	Empirically derived volumetric PEF caliper coefficient, 4.12.1
L	Characteristic measurement length, 4.3.1.2
m	Cementation exponent, 4.11.3, 5.2.1
MW	Mud weight, 4.12.1
Mw_{in}	Mud weight (density) as the mud is pumped into the hole, 4.15.3
Mw_{out}	Mud weight (density) of the mud in the annulus, 4.15.3
n	Saturation exponent, 4.11.3, 5.2.1
$N(t)$	Count rate at time, t, 4.11.1
$N(x)$	Number of neutrons per second after traveling a distance x, 4.7.1
$N(x)_{density\text{-}corrected}$	Measured neutron count rate corrected for the density effect, 4.7.1
N_0	Number of source neutrons per second, 4.7.1
N_{bulk}	Measured bulk neutron response, 4.10.3
N_{fluid}	Neutron response to the fluid, 4.10.3
N_{grain}	Neutron response to the formation grain, 4.10.3
N_i	Initial count rate , 4.11.1
R	Formation resistivity, 4.3.1, 4.3.1.2, 4.3.2.3, 4.11.3
r	Resistance seen by the tool, 4.3.1.2
r	Radius, 4.3.2.3
R_0	Water-filled formation resistivity, 4.3.2.4
R_1	Isotropic resistivity value of the lower resistivity layer, 4.3.2.4, 4.3.2.5, 5.5.7.1
R_2	Isotropic resistivity value of the higher resistivity layer, 4.3.2.5, 5.5.7.1

r_h	Borehole radius, 4.3.2.3	
R_h	Horizontal resistivity, 4.3.2.5, 4.3.3, 5.5.7.1	
r_i	Radius of invasion, 4.3.2.3, 5.1	
R_{lower}	Resistivity of the lower layer in the volume of measurement, 4.3.2.4	
R_m	Mud resistivity, 4.3.2.1, 5.1	
R_{mc}	Mudcake resistivity, 5.1	
$R_{measurement}$	Measured resistivity, 4.3.2.4	
R_t	True resistivity, 4.3.1.3, 4.3.2.1, 4.3.2.3, 4.3.2.4, 4.3.2.7, 4.3.3, 4.11.3, 5.1, 5.2.1, 5.5.9, 5.5.10	
R_{upper}	Resistivity of the upper layer in the volume of measurement, 4.3.2.4	
R_v	Vertical resistivity, 4.3.2.5, 4.3.3, 5.5.7.1	
R_w	In situ water resistivity, 4.11.3, 5.1	
R_{xo}	Invaded zone resistivity, 4.3.2.3, 4.3.3, 5.1	
σ	Formation conductivity, 4.3.1	
S_{hc}	Hydrocarbon saturation, 5.2.1	
S_i	Sensitivity of element i to the prompt neutron capture measurement, 4.10.1	
Standoff$_{density}$	Density standoff, 4.12.1	
Standoff$_U$	Volumetric photoelectric standoff, 4.12.1	
S_w	Formation water saturation, 4.3.1, 4.10.3, 4.11.3, 5.1, 5.2.1, 5.3, 5.5.7.1, 5.5.9, 5.5.10	
S_{w1}	Water saturation in the higher resistivity layer, 5.5.7.1	
S_{w2}	Water saturation in the lower resistivity layer, 5.5.7.1	
S_{wt}	Total formation water saturation, 5.5.7.1	
S_{xo}	Invaded zone saturation, 5.1, 5.5.10	
τ	Decay constant of the quasiexponential decay of the neutrons or the associated gamma rays in us, 4.6.2	
U	Volumetric photoelectric effect, 1, 4.4.2.2, 4.5.1, 4.5.3, 4.5.4, 4.5.5.2, 4.12.1, 5.2.1	
U_{barite}	Volumetric photoelectric effect of barite	
U_{bulk}	Bulk volumetric photoelectric effect, 4.5.3	
U_{fluid}	Volumetric photoelectric effect of the fluid in the pores, 4.5.3	
U_{matrix}	Volumetric photoelectric effect of the formation matrix	
v	Amplitude, 4.13.1	
V_1	Volumetric proportion of the lower-resistivity layer in the volume of measurement, 4.3.2.5, 5.5.7.1	
V_2	Volumetric proportion of the higher-resistivity layer in the volume of measurement, 4.3.2.5, 5.5.7.1	
V_{hc}	Volume of hydrocarbon per unit volume of formation, 5.2.1	
V_i	Volume fraction of mineral i, 5.4.5.3	
V_{lower}	Volumetric proportion of the lower layer in the volume of measurement, 4.3.2.4	
V_{upper}	Volumetric proportion of the upper layer in the volume of measurement, 4.3.2.4	

W	Standoff weighting factor, 4.12.1	
W_i	Weight fraction of mineral i, 4.10.1, 5.4.5.3	
x	Distance traveled in the material, 4.7.1	
X_i	Oxide association factor used to convert the element to its appropriate order or oxide and related elements, 4.10.1	
Y_i	Relative yield of element i, 4.10.1	
Z_b	Bulk value of Z measured by the tool, 5.4.5.3	
Z_f	Fluid response for measurement Z, 5.4.5.3	
Z_g	Matrix response for measurement Z, 5.4.5.3	
ρ	Material density (atom concentration of other elements), 4.7.1	
ρ_b	Increasing formation density, 4.4.2.1	
ρ_{bulk}	Bulk density of the formation, 4.4.3, 4.5.1, 4.7.3, 4.10.3, 4.12.1, 5.2.1, 5.5.1	
ρ_e	Electron density, 4.5.1, 4.5.3	
ρ_{fluid}	Density of the fluid in the pores, 4.4.3, 4.7.3, 4.10.3, 5.2.1, 5.5.1	
ρ_g	Density of mineral g, 5.4.5.3	
ρ_{grain}	Grain density, 4.4.3, 4.7.3, 4.10.1, 4.10.3, 5.2.1, 5.5.1	
ρ_i	Density of mineral i, 5.4.5.3	
$\rho_{long-spacing}$	Density computed from the long-spacing gamma-gamma density detector, 4.4.2.1, 4.12.1	
$\rho_{short-spacing}$	Density computed from the short-spacing gamma-gamma density detector, 4.4.2.1	
ρ_{matrix}	Density of the formation matrix, 4.4.3	
ρ_{mud}	Mud density, 4.4.2.1, 4.12.1	
Σ_{bulk}	Bulk sigma, 4.10.3, , 4.11.3, 5.3	
Σ_{fluid}	Fluid sigma, 4.11.3	
Σ_{grain}	Grain sigma, 4.10.3, 4.11.3, 5.3	
Σ_{hc}	Hydrocarbon sigma, 4.10.3, 4.11.3, 5.3	
Σ_{water}	Water sigma, 4.10.3, 4.11.3, 5.3	
ϕ	Formation porosity, 4.4.3, 4.5.3, 4.10.3, 4.11.3, 5.2.1	
ϕ_1	Porosity of the lower-resistivity layer, 5.5.7.1	
ϕ_2	Porosity of the higher-resistivity layer, 5.5.7.1	
ϕ_d	Density porosity, 5.2.1, 5.5.1	
ϕ_n	Measured neutron response, 5.5.7.1	
ϕ_{n1}	Neutron response of the lower-resistivity layer, 5.5.7.1	
ϕ_{n2}	Neutron response of the higher-resistivity layer, 5.5.7.1	
ϕ_t	Total porosity, 5.4.5.3	
ϕ_z	Matrix-corrected single tool porosity from measurement Z, 5.4.5.3	

Acronyms

CEC — Cation exchange capacity, 5.4.5, 5.4.5.5

DC — Direct current, 4.3.1.2, 4.3.2.7

DOI — Depth of investigation, 1.1, Fig. 1-6, 2.2, 3.3.1, 4.3.1.2, 4.3.2.3, Fig. 4-23, 4.3.2.6, 4.3.4, 4.4.1, 4.4.2.1, 4.4.2.3, 4.7.1, 4.7.2, 4.8.1, 4.8.2, 4.9.3, 4.11.1, 4.11.3, Fig. 4-93, 4.11.4, 5.2.1, 5.5.2, 5.5.6

DLIS — Digital Log Information Standard, 5.4.5.1

DLS — Dogleg severity, 3.3.8

ECD — Equivalent circulating density, 4.15.3, 4.15.4, 4.15.5.1, 4.15.5.2, 5.4.3

ESD — Equivalent static density, 4.15.4, 4.15.5.1

EVR — Enhanced vertical resolution, 4.4.5.2

FH — Full hole, Table 3-1

FNP — Field neutron plug, 4.6.1

HI — Hydrogen index, 1, 1.1, 2.2, 2.3, 4.6, 4.7, 4.7.1, 4.7.2, 4.7.3, 4.7.4, 4.7.5, 4.7.6, 4.8, 4.8.1, 4.8.2, 4.8.3, 4.9.1, 4.10.3, 4.11.2, 4.11.4, 5.4.5.3

HPHT — High-pressure, high-temperature, 3.2.3

ID — Inner diameter, Table 3-1, 4.7.1, 4.13.4

IDD — Image-derived density, 4.4.2.2, 4.4.5.1, 4.5.2, 4.5.4, 4.5.5.1, 4.5.5.2, 4.9.3, 5.4.5.1, 5.5.2

JNOC — Japan National Oil Corporation, ii

JOGMEC — Japan Oil, Gas and Metals National Corporation, ii

LCM — Lost circulation material, Table 3-2

LOT — Leakoff test, 3.3.9, 4.1.1, 4.15.4, 4.15.5.1

LQC — Log quality control, 1.3, 4.2.4, 4.3.5, 4.4.4, 4.5.4, 4.7.4, 4.8.4, 4.9.4, 4.10.4, 4.11.4, 4.12.3, 4.13.4, 4.14.4, 5.4.3

LRP — Low-resistivity pay, 1.1, 4.11, 4.11.1, 4.11.3, 5.3, 5.5.8, 5.5.9, 5.5.11

LWD — Logging while drilling, 1, 1.1, 1.1.2, 1.3, 2.1, 2.2, 3.1, 3.3.4, 4.1, 4.1.2, 4.2.1, 4.2.4, 4.3.5, 4.4, 4.4.2.1, 4.4.2.3, 4.4.4, 4.5.4, 4.7.1, 4.8.1, 4.11.1, 4.11.2, 4.14, 4.14.3, 4.16, 5.1, 5.2.1, 5.3, 5.5.2, 5.5.6

MBHC — Mixed borehole compensation, 4.3.1.1

MD — Measured depth, 1.1, 4.4.2.3, 5.4.2, 5.4.4, 5.5.7.1

MWD — Measurement while drilling, 1, 3.3.10, 3.3.7, 3.4.2, 4.6.1, 4.14.3, 4.15.3, 4.15.4

NGD — Neutron-gamma density, 1, 2.2, 4.4, 4.6, 4.9, 4.9.1, 4.9.2, 4.9.3, 4.9.4, 4.9.5, 4.9.6, 5.4.5.1

OD — Outside diameter, Table 3-1, 3.4.1, 4.2.5.1, 4.3.6.1

PC — Personal computer, 5.4.1

PEF — Photoelectric factor, 1, 1.1, 1.2.1, 1.2.2, 1.3, 2.2, 2.3, 4.4.2.2, 4.5, 4.5.1, 4.5.2, 4.5.3, 4.5.4, 4.5.5, 4.5.5.1, 4.5.5.2, 4.5.6, 4.9, 4.9.1, 4.10.5.1, 4.10.5.2, 4.10.6, 4.12, 4.12.1, 4.12.3, 4.13.3, 5.2.1, 5.4.3, 5.4.4, 5.4.5.1, 5.4.5.2, 5.5.1, 5.5.2, Fig. 5-27

PEI — Photoelectric image, 1.1

PESD — Pressure used for equivalent static density calculation, 4.15.4, 4.15.5, 6.11.1

PNG — Pulsed neutron generator, 1, 1.1, 2.1, 3.1, 3.2.4, 3.3.3, 4.2.2.1, 4.6.1, 4.6.2, 4.7.1, 4.7.2, 4.8.1, 4.8.2, 4.9, 4.9.1, 4.9.4, 4.10, 4.10.4, 4.11.1, 4.11.2, 4.11.3, 4.11.4, 5.5.3

POI — Point of interest, 5.4.2, 5.4.4

QC — Quality control, 1.2.3, 1.3, 1.3.1, 2.2, 4.3.3, 5.4.4

QFM — Quartz-feldspar-mica, 4.10.1, 4.10.4, 4.10.5.1, 4.10.5.2, 4.10.6, 5.4.5, 5.4.5.2, 5.5.5

RMS — Root mean square, 4.17.2.1, 4.17.3.1, 4.17.3.2

ROP — Rate of penetration, 1.2.1, 2.1, 3.2.4, 3.3.4, 3.3.5, 4.2.4, 4.10.4, 4.17.2, 5.4.3, 5.4.4

SPP — Standpipe pressure, 4.15.3, 4.15.4

TCOA — Total counts due to oxygen activation

TD — Total depth, 4.15.3, 5.5.11

TVD — True vertical depth, 4.3.2.6, 4.15.3, 4.15.4

WOB — Weight on bit, 3.4.2, 4.17.1, 4.17.2

Elements

Al	Aluminum, 1.3.2, 4.7.1, 4.8.1, 4.10.1, 4.11.1
Am	Americium, 1, 2.1
B	Boron, 4.10.1
Ba	Barium, 4.10.1, 4.10.2, 4.10.5.2, 4.10.6
Be	Beryllium, 1, 2.1
C	Carbon, 4.6.2.3, 4.10.1, 4.11.1
Ca	Calcium, 4.2.2.1, 4.10.1, 4.10.3, 4.10.5.1, 4.10.5.2, 4.10.6, 4.11.1
Cd	Cadmium, 4.6.3, 4.7.1
Cl	Chlorine, 1.1, 4.6.2.4, 4.7.1, 4.7.2, 4.8.3, 4.10.1, 4.10.2, 4.10.3, 4.10.5.2, 4.10.6, 4.11, 4.11.1, 4.11.3, 5.3, 5.4.5, 5.5.10
Cr	Chromium, 4.10.1, 4.10.5.2
Cs	Cesium, 1, 1.1, 2.1, 4.4, 4.9, 4.12.1, 4.12.5
Fe	Iron, 4.7.1, 4.10.1, 4.10.4, 4.10.5.1, 4.10.5.2, 4.10.6, 4.11.1
Gd	Gadolinium, 4.10.1, 4.10.4, 4.10.5.1, 4.10.5.2, 4.10.6, 4.11.1
H	Hydrogen, 4.6.1, 4.6.2, 4.6.2.1, 4.6.2.4, 4.6.3, 4.7.1, 4.7.2, 4.7.3, 4.8.2, 4.9.1, 4.10.1, 4.10.2, 4.10.3, 4.10.5.2, 4.10.6, 4.11.1, 5.4.5, 5.4.5.3, 5.5.10
He	Helium, 1.1, 4.6.1, 4.6.2.3, 4.6.3, 4.7.1, 4.7.4, 4.8.4
K	Potassium, 4.2.1, 4.2.2, 4.2.2.2, 4.2.2.3, 4.2.3, 4.2.5.1, 4.2.5.2, 4.10.1, 4.10.2, 4.10.5.2, 4.10.6, 5.5.5
Mg	Magnesium, 1.3.2, 4.10.1, 4.10.4, 4.10.5.1, 4.10.5.2, 4.10.6, 4.11.1, 5.4.5.2, 5.5.2
N	Nitrogen, 4.2.2.1
Ni	Nickel, 4.10.1, 4.10.5.2
O	Oxygen, 3.2.4, 3.3.3, 4.2.2.1, 4.2.2.3, 4.2.5.1, 4.2.5.2, 4.6.2.3, 4.10.1, 4.11.1
S	Sulfur, 4.10.1, 4.10.2, 4.10.4, 4.10.5.1, 4.10.5.2, 4.10.6, 4.11.1
Si	Silicon, 4.6.2.4, 4.10.1, 4.10.3, 4.10.4, 4.10.5.1, 4.10.5.2, 4.10.6, 4.11.1
Th	Thorium, 4.2.1, 4.2.2.1, 4.2.3, 5.5.5
Ti	Titanium, 4.10.1, 4.10.4, 4.10.5.1, 4.10.5.2, 4.10.6
U	Uranium, 4.2.1, 4.2.3, 4.10.3, 5.5.5
W	Tungsten, 4.2.1

Channel Names

ASAL	Apparent salinity, 4.10.3, 4.10.5.2
BPHI	HI-based best thermal neutron porosity, 1, 4.7.1, 4.7.2, 4.7.3, 4.7.4, 4.7.5.2, 4.7.6, 4.8.1, 4.8.2, 4.8.3, 4.11.4, 4.11.5.1, 5.5.3
DBPHI	HI-based best thermal neutron porosity correction, 4.7.2, 4.7.5.2
DCAV	Azimuthal density caliper, 1, 4.12.4.2, 4.12.5
DEVI	Near-bit inclination, 1
DHAP	Annular pressure, 1, 4.15.5.2
DHAT	Annular temperature, 1, 4.16.3.2
DRBB	Bottom-quadrant density correction, 4.4.2.2
DRHN	Density correction, 4.9.2, 4.9.4, 4.9.5.2
DWxx Wxxx	Elemental spectroscopy, 1
GRMA	Azimuthal gamma ray, 1, 4.2.4, 4.2.5.2
IDDP	IDD tool path, 4.4.2.2, 4.4.5.2
IDDQ	IDD quality image, 4.4.2.2, 4.4.5.2, 4.5.5.2
IDDR	Image-derived density correction, 4.4.2.2, 4.4.5.2
IDPE	Image-derived PEF, 4.4.2.2, 4.5.5.2
IDQR	Quality ratio, 4.4.2.2, 4.5.5.2
IDQT	Quality threshold parameter, 4.4.2.2, 4.4.5.1, 4.5.5.1
IDRO	Image-derived density, 4.4.2.2, 4.4.5.2, 5.4.5.1
IDU	Volumetric PEF, 4.4.2.2, 4.5.5.2
PEB	Bottom PEF, 4.4.2.2, 4.5.4, 4.9.1
PEF/U	Photoelectric factor/volumetric PEF, 1
PESD	Stabilized static pressure, 4.15.4, 4.15.5.1
PESD_TM	Time at which stabilized static pressure occurs, 4.15.4, 4.15.5.1
PMAX	Maximum static pressure, 4.15.4, 4.15.5.1
PMIN	Minimum static pressure, 4.15.4, 4.15.5.1
PxxH AxxH	Propagation resistivity, 1
QIDD	Quality factor over the four sectors centered on the tool path, 4.4.2.2, 4.4.5.2, 4.5.5.2
QRBB	Quality factor over the four sectors centered on the bottom of the hole, 4.4.2.2, 4.4.5.2, 4.5.5.2
RHOB	Gamma-gamma density, 4.4.2.4, 4.4.5.2, 4.9.1, 5.5.3
RHON	NGD, 1, 4.9.1, 4.9.4, 4.9.5.2, 5.5.3
ROBB	Bottom-quadrant density, 4.4.2.2, 4.4.4, 4.4.5.2
SHKxx VIBxx	Three-axis shocks and vibrations, 1
SIFA	Sigma, 1, 4.11.4, 4.11.5.2
TNPH	Thermal neutron porosity, 1, 4.4.2.2, 4.7.2, 4.8, 4.8.1, 4.8.2, 4.8.3, 4.8.4, 4.8.5.2, 4.8.6, 5.5.3
UB	Bottom volumetric PEF, 4.4.2.2, 4.4.4
UCAV	Azimuthal ultrasonic caliper, 1, 4.13.5.2

Length

Multipy number of column A by line B to obtain

A \ B	Centimeters	Feet	Inches	Kilometers	Nautical Miles	Meters	Mils	Miles	Millimeters	Yards
Centimeters	1	30.48	2.540	10^5	1.853×10^5	100	2.540×10^{-3}	1.609×10^5	0.1	91.44
Feet	3.281×10^{-2}	1	8.333×10^{-2}	3281	6080.27	3.281	8.333×10^{-5}	5280	3.281×10^{-3}	3
Inches	0.3937	12	1	3.937×10^4	7.296×10^4	39.37	0.001	6.336×10^4	3.937×10^{-2}	36
Kilometers	10^{-5}	3.048×10^{-4}	2.540×10^{-5}	1	1.853	0.001	2.540×10^{-8}	1.609	10^{-6}	9.144×10^{-4}
Nautical miles	-	1.645×10^{-4}	-	0.5396	1	5.396×10^{-4}	-	0.8684	-	4.934×10^{-4}
Meters	0.01	0.3048	2.540×10^{-2}	1000	1853	1	-	1609	0.001	0.9144
Mils	393.7	1.2×10^4	1000	3.937×10^7	-	3.937×10^4	1	-	39.37	3.6×10^4
Miles	6.214×10^{-6}	1.894×10^{-4}	1.578×10^{-5}	0.6214	1.1516	6.214×10^{-4}	-	1	6.214×10^{-7}	5.682×10^{-4}
Millimeters	10	304.8	25.4	10^5	-	1000	2.540×10^{-2}	-	1	914.4
Yards	1.094×10^{-2}	0.3333	2.778×10^{-2}	1094	2027	1.094	2.778×10^{-5}	1760	1.094×10^{-3}	1

Area

Multipy number of column A by line B to obtain

A \ B	Acres	Circular Mils	Square Centimeters	Square Feet	Square Inches	Square Kilometers	Square Meters	Square Miles	Square Millimeters	Square Yards
Acres	1	-	-	2.296×10^{-5}	-	247.1	2.471×10^{-4}	640	-	2.066×10^{-4}
Circular mils	-	1	1.973×10^5	1.833×10^8	1.273×10^6	-	1.973×10^9	-	1973	-
Square centimeters	-	5.067×10^{-6}	1	929	6.452	10^{10}	10^4	2.590×10^{10}	0.01	8361
Square feet	4.356×10^4	-	1.076×10^{-3}	1	6.944×10^{-3}	1.076×10^7	10.76	2.788×10^7	1.076×10^{-5}	9
Square inches	6,272,640	7.854×10^{-7}	0.1550	144	1	1.550×10^9	1550	4.015×10^9	1.550×10^{-3}	1296
Square kilometers	4.047×10^{-3}	-	$10-10$	9.290×10^{-8}	6.452×10^{-10}	1	10^{-6}	2.590	10^{-12}	8.361×10^{-7}
Square meters	4047	-	0.0001	9.290×10^{-2}	6.452×10^{-4}	10^6	1	2.590×10^6	10^{-6}	0.8361
Square miles	1.562×10^{-3}	-	3.861×10^{-11}	3.587×10^{-8}	-	0.3861	3.861×10^{-7}	1	3.861×10^{-13}	3.228×10^{-7}
Square millimeters	-	5.067×10^{-4}	100	9.290×10^4	645.2	10^{12}	10^6	-	1	8.361×10^5
Square yards	4840	-	1.196×10^{-4}	0.1111	7.716×10^{-4}	1.196×10^6	1.196	3.098×10^6	1.196×10^{-6}	1

Volume

Multipy number of column A by line B to obtain

A \ B	Bushels (Dry)	Cubic Centimeters	Cubic Feet	Cubic Inches	Cubic Meters	Cubic Yards	Gallons (Liquid)	Liters	Pints (Liquid)	Quarts (Liquid)
Bushels (dry)	1	-	0.8036	4.651×10^{-4}	28.38	-	-	2.838×10^{-2}	-	-
Cubic centimeters	3.524×10^4	1	2.832×10^4	16.39	10^6	7.646×10^5	3785	1000	473.2	946.4
Cubic feet	1.2445	3.531×10^{-5}	1	5.787×10^{-4}	35.31	27	0.1337	3.531×10^{-2}	1.671×10^{-2}	3.342×10^{-2}
Cubic inches	2150.4	6.102×10^{-2}	1728	1	6.102×10^4	46,656	231	61.02	28.87	57.75
Cubic meters	3.524×10^{-2}	10^{-6}	2.832×10^{-2}	1.639×10^{-5}	1	0.7646	3.785×10^{-3}	0.001	4.732×10^{-4}	9.464×10^{-4}
Cubic yards	-	1.308×10^{-6}	3.704×10^{-2}	2.143×10^{-5}	1.308	1	4.951×10^{-3}	1.308×10^{-3}	6.189×10^{-4}	1.238×10^{-3}
Gallons (liquid)	-	2.642×10^{-4}	7.481	4.329×10^{-3}	264.2	202	1	0.2642	0.125	0.25
Liters	35.24	0.001	28.32	1.639×10^{-2}	1000	764.6	3.785	1	0.4732	0.9464
Pints (liquid)	-	2.113×10^{-3}	59.84	3.463×10^{-2}	2113	1616	8	2.113	1	2
Quarts (liquid)	-	1.057×10^{-3}	29.92	1.732×10^{-2}	1057	807.9	4	1.057	0.5	1

Mass and Weight

Multipy number of column A by line B to obtain

A \ B	Acres	Circular Mils	Square Centimeters	Square Feet	Square Inches	Square Kilometers	Square Meters	Square Miles	Square Millimeters	Square Yards
Acres	1	-	-	2.296×10^{-5}	-	247.1	2.471×10^{-4}	640	-	2.066×10^{-4}
Circular mils	-	1	1.973×10^5	1.833×10^8	1.273×10^6	-	1.973×10^9	-	1973	-
Square centimeters	-	5.067×10^{-6}	1	929	6.452	10^{10}	10^4	2.590×10^{10}	0.01	8361
Square feet	4.356×10^4	-	1.076×10^{-3}	1	6.944×10^{-3}	1.076×10^7	10.76	2.788×10^7	1.076×10^{-5}	9
Square inches	6,272,640	7.854×10^{-7}	0.1550	144	1	1.550×10^9	1550	4.015×10^9	1.550×10^{-3}	1296
Square kilometers	4.047×10^{-3}	-	10–10	9.290×10^{-8}	6.452×10^{-10}	1	10^{-6}	2.590	10^{-12}	8.361×10^{-7}
Square meters	4047	-	0.0001	9.290×10^{-2}	6.452×10^{-4}	10^6	1	2.590×10^6	10^{-6}	0.8361
Square miles	1.562×10^{-3}	-	3.861×10^{-11}	3.587×10^{-8}	-	0.3861	3.861×10^{-7}	1	3.861×10^{-13}	3.228×10^{-7}
Square millimeters	-	5.067×10^{-4}	100	9.290×10^4	645.2	10^{12}	10^6	-	1	8.361×10^5
Square yards	4840	-	1.196×10^{-4}	0.1111	7.716×10^{-4}	1.196×10^6	1.196	3.098×10^6	1.196×10^{-6}	1

Pressure or Force per Unit Area

Multipy number of column A by line B to obtain

A \ B	Atmospheres[†]	Bayres or dynes per Square Centimeter[‡]	Centimeters of Mercury at 0°C[§]	Inches of Mercury at 0°C[§]	Inches of Water at 4°C	Kilograms per Square Meter[††]	Pounds per Square Foot	Pounds per Square Inch[‡‡]	Tons (short) per Square Foot	Pascals
Atmospheres[†]	1	9.869×10^{-7}	1.316×10^{-2}	3.342×10^{-2}	2.458×10^{-3}	9.678×10^{-5}	4.725×10^{-4}	6.804×10^{-2}	0.9450	9.869×10^{-6}
Bayres or dynes per Square Centimeter[‡]	1.013×10^{6}	1	1.333×10^{4}	3.386×10^{4}	2.491×10^{-3}	98.07	478.8	6.895×10^{4}	9.576×10^{5}	10
Centimeters of mercury at 0°C[§]	76	7.501×10^{-5}	1	2.540	0.1868	7.356×10^{-3}	3.591×10^{-2}	5.171	71.83	7.501×10^{-4}
Inches of mercury at 0°C[§]	29.92	2.953×10^{-5}	0.3937	1	7.355×10^{-2}	2.896×10^{-3}	1.414×10^{-2}	2.036	28.28	2.953×10^{-4}
Inches of water at 4°C	406.8	4.015×10^{-4}	5.354	13.60	1	3.937×10^{-2}	0.1922	27.68	384.5	4.015×10^{-3}
Kilograms per square meter[††]	1.033×10^{4}	1.020×10^{-2}	136.0	345.3	25.40	1	4.882	703.1	9765	0.1020
Pounds per square foot	2117	2.089×10^{-3}	27.85	70.73	5.204	0.2048	1	144	2000	2.089×10^{-2}
Pounds per square inch[‡‡]	14.70	1.450×10^{-5}	0.1934	0.4912	3.613×10^{-2}	1.422×10^{-3}	6.944×10^{-3}	1	13.89	1.450×10^{-4}
Tons (short) per square foot	1.058	1.044×10^{-5}	1.392×10^{-2}	3.536×10^{-2}	2.601×10^{-3}	1.024×10^{-4}	0.0005	0.072	1	1.044×10^{-5}
Pascals	1.013×10^{5}	10^{-1}	1.333×10^{3}	3.386×10^{3}	2.491×10^{-4}	9.807	47.88	6.895×10^{3}	9.576×10^{4}	1

[†] One atmosphere (standard) = 76 cm of mercury at 0°C

[‡] Bar

[§] To convert height h of a column of mercury at t°C to the equivalent height h_0 at 0°C, use $h_0 = h \{1 - [(m - l) t / 1 + mt]\}$, where m = 0.0001818 and l = 18.4×10^{-6} if the scale is engraved on brass; l = 8.5×10^{-6} if on glass. This assumes the scale is correct at 0°C; for other cases (any liquid) see *International Critical Tables*, Vol. 1, 68.

[††] 1 gram per square centimeter = 10 kilograms per square meter

[‡‡] psi = MPa × 145.038

psi/ft = 0.433 × g/cm3 = lbf/ft3/144 = lbf/gal/19.27

Density or Mass per Unit Volume

Multiply number of column A by line B to obtain

A \ B	Grams per Cubic Centimeter	Kilograms per Cubic Meter	Pounds per Cubic Foot	Pounds per Cubic Inch	Pounds per Gallon
Grams per cubic centimeter	1	0.001	1.602×10^{-2}	27.68	0.1198
Kilograms per cubic meter	1000	1	16.02	2.768×10^{4}	119.8
Pounds per cubic foot	62.43	6.243×10^{-2}	1	1728	7.479
Pounds per cubic inch	3.613×10^{-2}	3.613×10^{-5}	5.787×10^{-4}	1	4.329×10^{-3}
Pounds per gallon	8.347	8.3×10^{-3}	13.37×10^{-2}	231	1

Temperature

°F	$1.8°C + 32$
°C	$5.9(°F - 32)$
°R	$°F + 459.69$
K	$°C + 273.16$

About the author

Roger Griffiths is a Schlumberger technical advisor in petrophysics and well placement. He is currently Head of Petrophysics for Drilling & Measurements, having previously been involved in the concept development, engineering, field testing and commercial introduction of the EcoScope service. He holds an honors degree in mechanical engineering from the University of Melbourne, Australia.